Becoming an Agile Software Architect

Strategies, practices, and patterns to help architects design continually evolving solutions

Rajesh R V

Packt>

BIRMINGHAM—MUMBAI

Becoming an Agile Software Architect

Group Product Manager: Aaron Lazar
Publishing Product Manager: Shweta Bairoliya
Senior Editor: Nitee Shetty
Content Development Editor: Tiksha Lad
Technical Editor: Rashmi Subhash Choudhari
Copy Editor: Safis Editing
Project Coordinator: Francy Puthiry
Proofreader: Safis Editing
Indexer: Tejal Daruwale Soni
Production Designer: Shankar Kalbhor

First published: February 2021

Production reference: 1170221

Published by Packt Publishing Ltd.
Livery Place
35 Livery Street
Birmingham
B3 2PB, UK.

ISBN 978-1-80056-384-1

www.packt.com

To my lovely wife, Saritha, and our kids, Nikhil and Aditya; they sacrificed their interests to support my passion.

To my parents, Ramachandran Nair and Vasanthakumari, and extended family members; their enthusiasm fueled my energy.

Contributors

About the author

Rajesh R V is a seasoned IT architect with over 20 years of extensive experience in various technologies.

As the head of architecture at the Emirates Group, Rajesh has helped the company transform, reshape, and re-organize into a high-performing organization by adopting SAFe and Agile architecture practices.

Rajesh has a deep passion for technology and architecture. He also architected the **Open Travel Platform** (**OTP**) that earned the Emirates Group the prestigious 2011 Red Hat Innovation Award.

He previously wrote the best-selling books, *Spring Microservices and*, *Spring 5.0 Microservices*, and reviewed *Service-Oriented Java Business Integration* by Packt.

I want to thank everyone at Packt and the reviewers; their in-depth knowledge of this subject helped improve the quality. Thanks to all who supported and encouraged me on my journey to become an Agile software architect.

About the reviewers

With more than a decade of experience in both fields, **Gaurav Mishra** is an expert in user interface (UI) development and user experience (UX) design. He is comfortable working with any type of technology, and employs a growth mindset.

He has mentored many students around the world, providing workshops and training in UI development, UX design, and Drupal.

With a passion for building products and services from scratch, Gaurav has ultimately played a key role in the success of many organizations. He likes to challenge the status quo to bring out the best from a team and to reshape an organization's culture.

In his free time, Gaurav enjoys all genres of music from Indian classical to club.

Ken Cochrane is an experienced technologist, architect, author, and leader. He has extensive professional experience leading software development teams, developing and implementing scalable web solutions, and delivering high-quality, highly functional web applications that are currently in use by millions of people worldwide. Ken was previously a member of the founding teams of Docker and CashStar. He is now currently the senior director of enterprise architecture for WEX, a global leader in financial technology solutions. Ken coauthored the book *Docker Cookbook - Second Edition,* published by Packt in 2018. He currently resides in Southern Maine with his wife, Emily, and two sons, Zander and Maddox.

I would like to thank my wife, Emily, and my two boys, Zander and Maddox, for giving me the time to work on this book; also, my parents for buying me my first computer, and letting me spend so much of my free time on it.

Table of Contents

Section 2:
Transformation of Architect Roles in Agile

3

Agile Architect – The Linchpin of Success

4

Agile Enterprise Architect – Connecting Strategy to Code

5

Agile Solution Architect – Designing Continuously Evolving Systems

Section 3: Essential Knowledge to Become a Successful Agile Architect

6
Delivering Value with New Ways of Working

7
Technical Agility with Patterns and Techniques

8

DevOps and Continuous Delivery to Accelerate Flow

9

Architecting for Quality with Quality Attributes

10

Lean Documentation through Collaboration

11
Architect as an Enabler in Lean-Agile Governance

Section 4: Personality Traits and Organizational Influence

12
Architecting Organizational Agility

13

Culture and Leadership Traits

Other Books You May Enjoy

Index

Preface

Enterprises worldwide are significantly accelerating their digital transformation programs to achieve the state of business agility. Building a strong foundation to resist change is no longer a viable strategy. In today's increasingly turbulent market conditions, business agility is the holy grail for many enterprises to be more productive and responsive. Organizations embracing business agility place Agile software development at the center of their IT strategy to bring immediate value.

Agile development practices, underpinned by Lean-Agile principles, focus on the speed of delivery to support a high rate of business changes by eliminating impediments to flow. However, a substantial focus on speed challenges many of the traditional architecture principles and practices. This book guides and enlightens you to understand how architecture development practices can be effectively readjusted in Agile software development projects without compromising speed and quality. We will also explore how to position architects appropriately in Agile software projects to deliver the maximum possible value.

This book focuses on two key architect roles – enterprise architect and solution architect – and explains their duties in Agile software development projects with numerous examples.

The book then provides a series of strategies, best practices, and patterns to deliver value without compromising on delivery speed or quality. Later in this book, we will cover the critical role of architects in architecting organizations for agility. At the end of this book, we will cover several personal and interpersonal qualities required for architects' success.

Who this book is for

This book is for architects currently working on Agile development projects or aspiring to work on Agile software delivery, irrespective of the methodology they are using. You will also find this book useful if you're a senior developer or a budding architect looking to understand the Agile architect's role by embracing Agile architecture strategies and a Lean-Agile mindset.

What this book covers

Chapter 1, Looking through the Agile Architect's Lens, provides a framework and navigation tool to explore this book easily.

Chapter 2, Agile Architecture – The Foundation of Agile Delivery, explains the concepts and principles of Agile architecture and compares it with traditional architecture.

Chapter 3, Agile Architects – The Linchpin to Success, uses several metaphors to reinforce Agile architects' roles and responsibilities.

Chapter 4, Agile Enterprise Architect – Connecting Strategy to Code, establishes the duties of modern enterprise architects in Agile software development.

Chapter 5, Agile Solution Architect – Designing Continuously Evolving Systems, focuses on solutions architects' operating methods in Agile software development projects.

Chapter 6, Delivering Value with New Ways of Working, provides techniques for architects to be successful in Agile delivery environments.

Chapter 7, Technical Agility with Patterns and Techniques, introduces several patterns and practices to achieve technical agility.

Chapter 8, DevOps and Continuous Delivery to Accelerate Flow, explores the importance of architects in DevOps and continuous delivery.

Chapter 9, Architecting for Quality with Quality Attributes, discusses various quality models, tools, and approaches to ensure that teams deliver quality products to customers.

Chapter 10, Lean Documentation through Collaboration, talks about alternate approaches to documentation and familiarizes you with the required documentation concepts.

Chapter 11, Architect as an Enabler in Lean-Agile Governance, blows apart the myths around governance in Agile software development and introduces Lean governance principles.

Chapter 12, Architecting Organizational Agility, draws attention to the need to architect organizations around the flow of work.

Chapter 13, Culture and Leadership Traits, discusses the need to change by introducing new personal and interpersonal qualities required for architects.

To get the most out of this book

To understand the concepts covered in this book easily, you need to have prior knowledge of basic Agile development practices. Besides this, software architecture fundamentals are essential as the book builds on the foundational architecture practices.

Download the example code files

You can download the supporting files for this book from GitHub at `https://github.com/PacktPublishing/Becoming-an-Agile-Software-Architect`. In case there's an update to these assets, it will be updated on the existing GitHub repository.

We also have other code bundles from our rich catalog of books and videos available at `https://github.com/PacktPublishing/`. Check them out!

Download the color images

We also provide a PDF file that has color images of the screenshots/diagrams used in this book. You can download it here:

`https://static.packt-cdn.com/downloads/9781800563841_ColorImages.pdf`

Conventions used

There are a number of text conventions used throughout this book.

Bold: Indicates a new term, an important word, or words that you see onscreen. For example, words in menus or dialog boxes appear in the text like this. Here is an example: "A scenario from Snow in the Desert's **Automatic Vehicle Tracking System (AVTS)** is captured as an example in the preceding diagram, mapping **Source**, **Stimulus**, **Artifact**, **Environment**, **Response**, and **Response measure**. A mini-QAS is a simplified version that eliminates **Environment** and **Artifact**."

> Tips or important notes
> Appear like this.

Get in touch

Feedback from our readers is always welcome.

General feedback: If you have questions about any aspect of this book, mention the book title in the subject of your message and email us at customercare@packtpub.com.

Errata: Although we have taken every care to ensure the accuracy of our content, mistakes do happen. If you have found a mistake in this book, we would be grateful if you would report this to us. Please visit www.packtpub.com/support/errata, selecting your book, clicking on the Errata Submission Form link, and entering the details.

Piracy: If you come across any illegal copies of our works in any form on the Internet, we would be grateful if you would provide us with the location address or website name. Please contact us at copyright@packt.com with a link to the material.

If you are interested in becoming an author: If there is a topic that you have expertise in and you are interested in either writing or contributing to a book, please visit authors.packtpub.com.

Reviews

Please leave a review. Once you have read and used this book, why not leave a review on the site that you purchased it from? Potential readers can then see and use your unbiased opinion to make purchase decisions, we at Packt can understand what you think about our products, and our authors can see your feedback on their book. Thank you!

For more information about Packt, please visit packt.com.

Section 1: Understanding Architecture in the Agile World

In this section, you will gain clarity on various aspects of Agile architecture, how it differs from traditional architecture practices, and how the Agile architecture helps organizations to deliver a continuous sustainable flow of work.

This section contains the following chapters:

1
Looking through the Agile Architect's Lens

"Less is more."

– Ludwig Mies van der Rohe (Architected Barcelona Pavilion, Expo 1929)

Agile software development is an effective practice that helps organizations continuously deliver the maximum possible business value with organizational agility. The 14th Annual State of Agile Report 2020 shows that an astounding 95% of the 1,121 respondents adopted Agile software development practices. The sharp increase in Agile adoption indicates why organizations cannot shy away from embracing Agile software development practices in their business. Most of these organizations adopt Agile software delivery practices for accelerating software delivery by successfully managing the changing priorities of the business. Architecting systems using Agile values, principles, and practices are incredibly crucial to improve productivity and deliver quality products.

This book introduces the Agile Architect's Lens—a comprehensive representation of the knowledge of key segments and leading practices for Agile architects to successfully operate in a Lean-Agile organization. The Agile Architect's Lens has twelve focal points, as shown in the diagram. Each focal point that can you see in the following figure captures tried and tested approaches, patterns, and guidelines for architects to design continually evolving solutions:

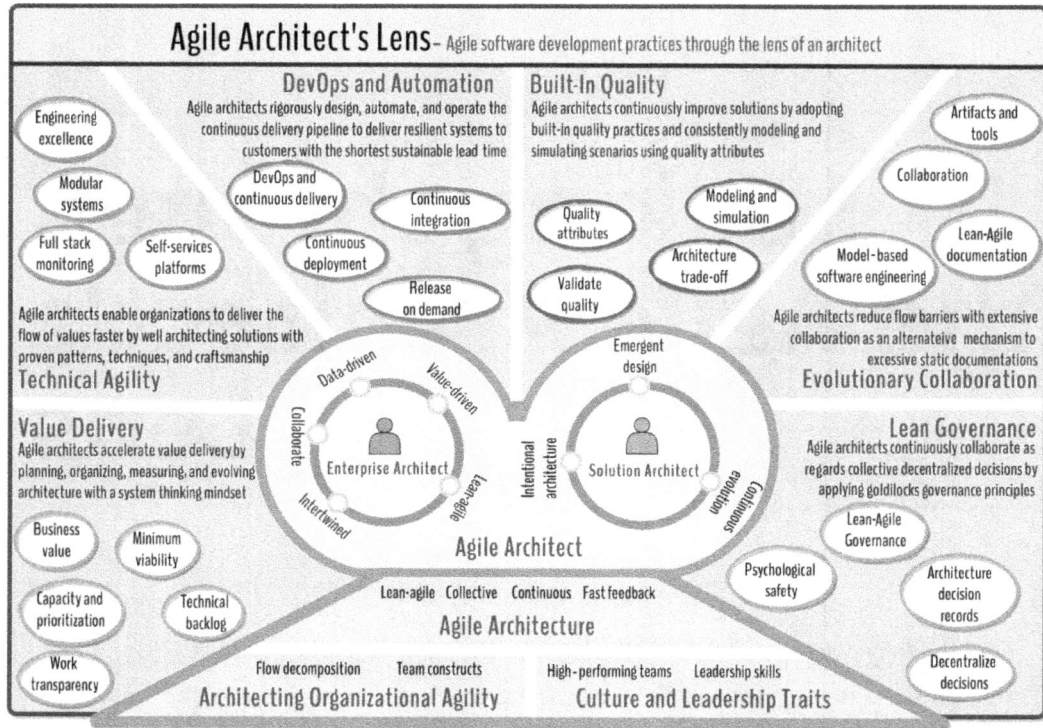

Figure 1.1 – The Architect's Lens

The Agile Architect's Lens act as a single pane of glass, guiding light, and topic navigator for readers to easily explore this book.

Agile architecture is a set of practices, principles, and guidelines for developing architecture as an intrinsic element of Agile software development. Using Agile architecture concepts, Agile teams can confidently drive sustainable, fit-for-purpose solutions to meet customer needs. We thus begin by focusing on the Agile Architecture focal point of the Agile Architect's Lens.

Agile architects need special skills to operate efficiently in Agile organizations. The collaborative nature of Agile software development needs a new set of metaphors to understand expectations from Agile architects. A well-architected solution enables organizations to deliver value with the shortest sustainable lead time, which significantly enhances business operations. Our next logical focus will therefore be on the Agile Architect focal point, followed by covering two specific architect roles crucial to Agile development projects – the Enterprise Architect and Solution Architect focal points of the Agile Architect's Lens.

Agile architects are responsible for bridging strategies and providing a shared vision to all stakeholders in the continuous delivery flow. Agile architects have to acquire technical knowledge in specific areas to address commonly occurring challenges efficiently. Hence, the knowledge kitty has to have the right blend of process excellence and technical excellence. We will therefore shift our attention next to covering the Value Delivery, Patterns and Techniques, DevOps and Automation, Built-in Quality, Evolutionary Collaboration, and Safety Nets focal points of the Agile Architect's Lens.

In addition to the knowledge kitty, Agile architects need to undergo a radical transformation in behavior and mindset to achieve operational excellence. Servant leadership is one of the foremost characteristics that Agile architects need to have, but it is just a starting position. In our journey ahead, we expand on the Culture and Leadership Traits focal point of the Agile Architect's Lens.

Business agility is one of the key motivations for many organizations to opt for digital transformation. The lean practices of Agile software delivery can boost organizational agility. However, this needs considerable effort to organize systems and people around values to be nimble enough to respond quickly to changes in business strategies. Architecting organizations around values is an essential aspect of organizational agility. We cover this in our final focal point of the Agile Architect's Lens, Architecting Organizational Agility.

In summary, digital transformation is a continuous activity for many organizations. Agile software delivery is the foundation on which the organizational agility essential to support successful transformations can be built. Architecture, and especially architects, play an extremely critical role not just in Agile software delivery but also in architecting organizations for agility. While technical excellence is paramount for architects, working beyond traditional boundaries with an intrinsically motivated team of multi-disciplinary individuals requires a significant shift in mentality.

Let's start our exciting journey by exploring our first focal point, Agile Architecture, in our next chapter.

2
Agile Architecture – The Foundation of Agile Delivery

"Simplicity is the ultimate sophistication."

– Leonardo da Vinci

In this era of innovative and disruptive digital transformation, Agile software development is no longer a marketing buzzword. Agile software development methodologies have enabled many organizations at scale (at the required size) to significantly accelerate their ambitious journey to accomplish and sustain organizational agility. Agile architecture and the new Agile ways of developing architecture and design are incredibly important in delivering an uninterrupted, continuous flow of values to meet the needs and wants of customers. However, architecture and architects need a radical reshaping to be relevant in the world of Agile software development.

This chapter takes you through the evolution of software development and analyzes the lessons learned, which help in reinforcing the vital need of architecture in Agile software development initiatives.

In this chapter, we're going to cover the following main topics:

- The journey leading to Agile software development
- Agile development and traditional architecture – an oxymoron?
- Agile and architecture – the battle between speed and sustainability
- Comparing different scaling Agile frameworks
- Measuring Agile architecture
- Lessons learned from Snow in the Desert

This chapter focuses on the **Agile Architecture** focal point of the **Agile Architect Lens**, as shown in the following diagram:

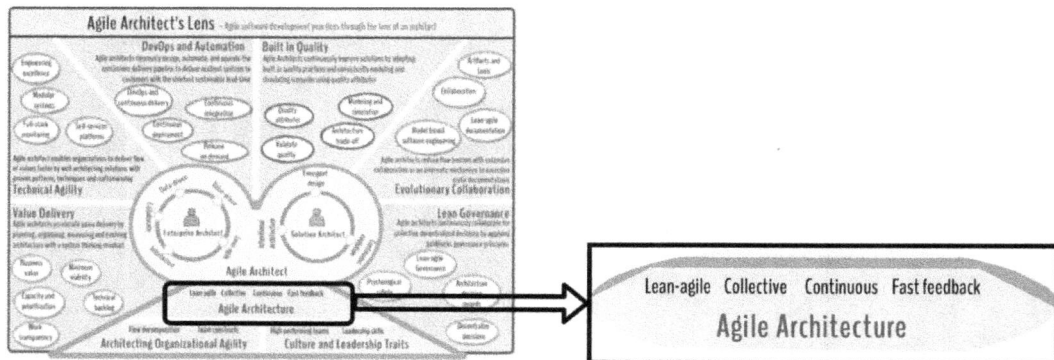

Figure 2.1 – The Agile Architect Lens – focal point

Technical requirements

Additional materials such as posters and Agile architecture assessment parameters referenced in this chapter are available to download from `https://github.com/PacktPublishing/Becoming-an-Agile-Software-Architect/tree/master/Chapter2`.

The journey leading to Agile software development

Interestingly, unlike many other natural evolutions, software development has gone through many forms, shapes, and, notably, cycles over the years. One of the significant parameters in software development evolution has been the risk of failure. The likelihood of risk dramatically reduced as a result of this evolution.

Software development is a young discipline, with only just under seven decades of legacy. Grady Booch, Chief Scientist at IBM, in his speech *History of Software Engineering*, at the Association for Computing Machinery, observed that the first mention of software was by John Turkey back in 1952.

The Agile method of software development has become increasingly popular in the last decade, but the traces of Agile development are deeply rooted in history. The paper *Iterative and Incremental Development: A Brief History* published by IEEE highlighted the existence of Agile development even before that. The article stated that, back in the 1960s, NASA's Project Mercury ran with an iterative approach of very short, half-day cycles and used test-driven development practices. The more interesting observations came from the *Software Engineering* report at a conference sponsored by the NATO science committee held back in 1968. Kinslow, a computer systems consultant, articulated very well in this paper that the software design process is an iterative one. Ross, another attendee of the conference from MIT, also highlighted the deadly symptom of traditional software development as first specifying what you are going to do and then doing it. Both Kinslow and Ross were pointing toward what is today called the Agile software development approach.

Most of the software development during the 1960s followed the waterfall approach, and the earlier statements came from frustration at the failures and challenges of waterfall-style software development. Sequentially staged software development was first proposed by Dr. Winston W. Royce back in the 1970s, which appeared in an IEEE publication, *Managing Development of Large Software Systems*. Later, the term "waterfall" was introduced by Bell and Thayer in the second International Conference on Software Engineering referring to Royce's approach. Royce, in his paper, explained a step-by-step approach, as shown in the following diagram, as a better way of software development:

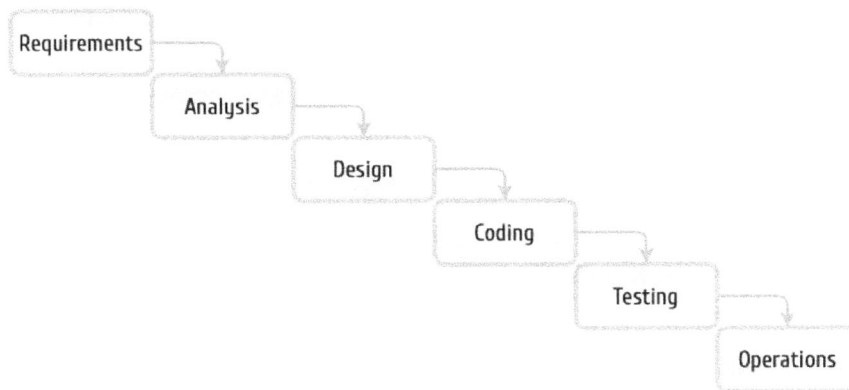

Figure 2.2 – Sequential approach to software development

The most interesting part is that Royce called out the risk of this sequential step-by-step approach in the same paper, highlighting that the testing phase only occurs at the end of the development cycle.

As Royce expected and clearly articulated, the biggest challenge in the waterfall model of software development is the risk associated with late validation. As shown in the following diagram, the risk is much higher in the waterfall approach since we confirm the design too early in the development cycle, with almost no means for immediate feedback. When we start receiving feedback, it is too late as the cost of rework is going to be unimaginably huge:

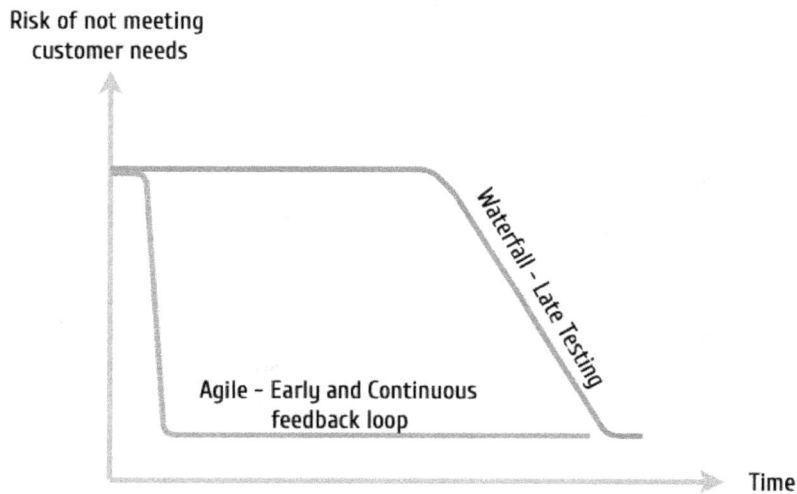

Figure 2.3 – Risk of not meeting customer expectations

The most attractive aspect of iterative software development is managing potential risks and, therefore, better management of the cost of delays and reworks.

Even though there were many traces of Agile software development before the 1990s, Agile software development practices eventually became popular as an aftereffect of the failures of many software development projects in the 1990s. As a result of this hue and cry, an alternate set of lightweight methods emerged, such as Scrum, **Extreme Programming (XP)**, **Feature Driven Design (FDD)**, and a few more. Later, in 2001, a group of Agile software development practitioners joined together and developed a manifesto for Agile software development, documenting a set of values and principles.

What is Agile software development?

The most important document for Agile software development is the manifesto for Agile software development. The manifesto is positioned as uncovering better ways of developing software, but it does not prescribe any approach or process for Agile software development.

Agile software development is a software development practice based on a set of values and principles defined in the manifesto for Agile software development. Manifesto statements are pretty simple, abstract, and general, for the right reasons. But the marketing buzz around it led to many interpretations and complicated explanations.

One of the most radically simple models to explain Agile is *The Heart of Agile*, proposed by Alistair Cockburn, one of the signatories of the manifesto for Agile software development. He summarized Agile with four imperatives: Collaborate, Deliver, Reflect, and Improve, as shown on the left-hand side of the following diagram:

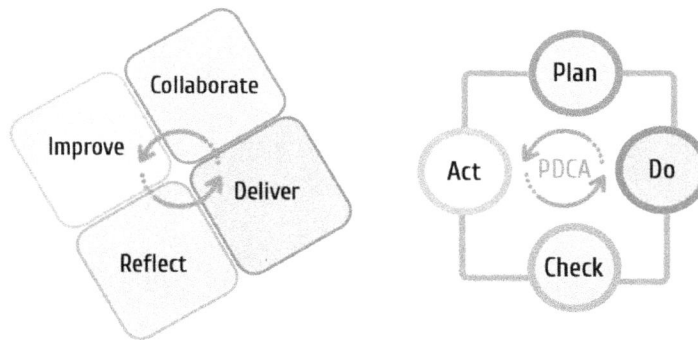

Figure 2.4 – The Heart of Agile and PDCA cycles

The meaning of collaborate, deliver, reflect, and improve is as follows:

- **Collaborate**: Collaboration in this context is a mix of trust, culture, motivation, and the act of collaboration between people.

- **Deliver**: Deliver means delivering business value with a continuous, uninterrupted flow, that essentially helps to earn more revenue and learn things such as market potential, among others.

- **Reflect**: Reflect is about collecting and examining subjective data from people and processes, and objective information from the ecosystem of the business and its customers.

- **Improve**: Improve is about solving problems based on past learnings and experimenting, to explore further and acquire new knowledge.

The Heart of Agile is pretty much similar to the **Plan-Do-Check-Act** (**PDCA**) cycle applied in the context of Agile, as shown on the right-hand side of the preceding diagram. PDCA was first introduced by Walter Shewhart in his book *Statistical Method from the Viewpoint of Quality Control* back in 1939. It is also called the Shewhart cycle and the Deming cycle. PDCA is a continual process improvement model for organizations to plan actions, execute those actions, study deviations from the plan, and act on what has been learned from the study.

In reality, Agile software development means two key things:

- Decompose problems in to small bite-size chunks so that PDCA can be applied repeatedly.

- Continually optimize the PDCA cycle time, without compromising purpose and quality.

We see these two key Agile philosophies in the following diagram:

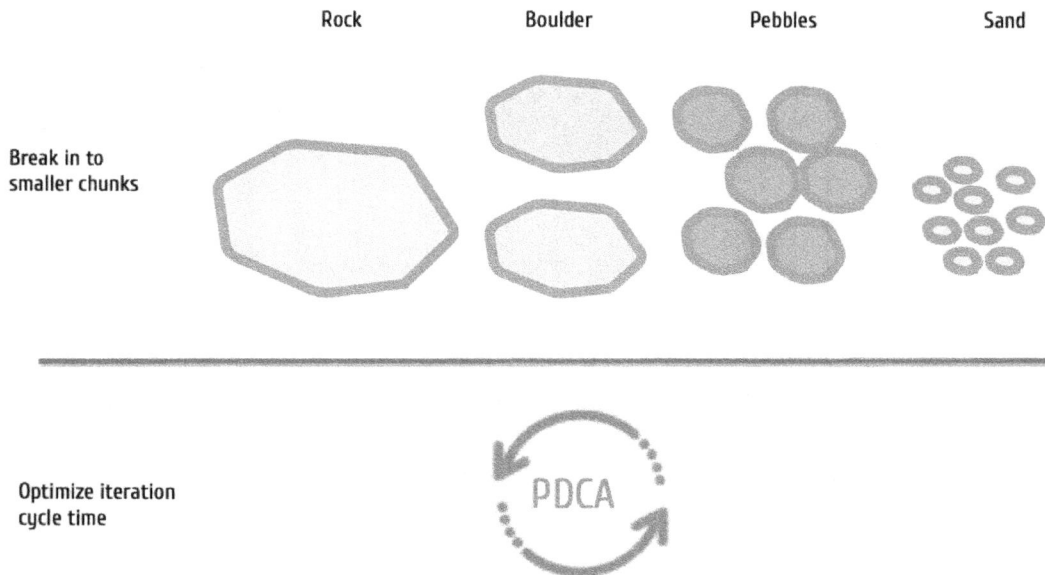

Figure 2.5 – Two key concepts of Agile: small batch size and cycle time

The PDCA cycle time can be improved dramatically by continuously monitoring and optimizing activities across different steps and, more importantly, diligently applying Lean principles.

Is Lean different from Agile?

We always use the words *Lean* and *Agile* almost interchangeably. But what is Lean? How is it related to Agile?

Back in the 1990s, when many large software projects started failing, many industry gurus looked at the reasons for failure and opportunities to improve. Up until then, software engineering was always compared with the construction industry as they both used engineering practices. However, there is a substantial difference between these two. In the construction industry, there is always a detailed plan. Once the architect signs off the plan, it goes for approval from the authorities before engineers start construction. Unfortunately, we had replicated this in software engineering with a stage-gate process with audits and approvals.

Finally, when things fell apart, the hunt for best practices extended to adjacent industries. The manufacturing industry was doing quite well with Lean manufacturing principles. The **Toyota Production System** (**TPS**) was leading this space, focusing considerably on enumerating and eradicating wastage, shortening lead times, relentless improvements, moving work to the fixed capacity workforce, using just-in-time production, and delaying critical decisions.

The following diagram shows key parts of the Lean manufacturing process, with a focus on minimizing lead time from order to cash:

Figure 2.6 – Concepts of Lean manufacturing

As you will have noticed, Agile and Lean have many characteristics in common. The reason is, many Agile software development frameworks got inspired by the Lean thinking of manufacturing and adopted many of those practices into software development even before the manifesto for Agile software development was constructed in 2001.

Agile software development has the best of both worlds: all Lean principles are also reflected in Agile software development practices. Effectively, many of those principles help to reduce the PDCA cycle time in Agile software development, which brings enormous benefits to software development, businesses, and customers.

Five benefits of Agile software development

There are many benefits when adopting Agile software development at scale. Some of the benefits, such as a set of motivated individuals and an empowered team, are considered to be indirect benefits, essentially contributing to one of the direct benefits.

The top five benefits of Agile-based software development are as follows:

- **Reduced risk**: The iterative development, faster release cycles, visible and transparent progression, and short feedback loops increase predictability, reduce risk, and eliminate unnecessary cost.

- **Better productivity**: Agile methodologies focus on continuous sustainable delivery of business values to maximize customer satisfaction by removing forces that impose delays in the software delivery flow. It also helps the organization to organize teams systematically without burdening them with management overheads.

- **Improved quality**: The people-centric approach, with a high degree of collaboration together with fast learning cycles of inspect, feedback, and adapt mechanisms improve the quality and maintainability of products.

- **Shorter cycle time**: The quick turnaround nature of Agile development helps businesses to respond quickly ahead of the competition. The *test and learn* approach in Agile assists faster and cheaper innovation cycles.

- **Business agility**: The ability to smoothly manage frequent priority changes in business as a mechanism by design, dramatically improves business agility to respond rapidly to changes in trading conditions and market dynamics.

As we see here, Agile brings many benefits, and Agile architecture inherits principles and practices from Agile software delivery to develop architecture as an integrated software development activity.

Agile software development is most important for organizations to successfully swim through the increasingly difficult environment. We will explore Agile software development and how architecture fits into the flow of work in the next section.

Agile development and traditional architecture – an oxymoron?

The perception of paradox came out of contrasting approaches adopted by Agile software development and the waterfall model. While Agile software developments are heavily rooted in iterative development, traditional developments follow rigid sequential steps.

To comprehend the cause of this perception, it is imperative to define what we mean by *software architecture* and its purpose.

Architecture as a continuum

Architecture is often misunderstood and misused in the software industry. There are many architecture definitions referenced in the document *What is Your Definition of Architecture*, published by the Software Engineering Institute at Carnegie Mellon University.

The very first definition of architecture came from Fred Brooks, writing about computer architecture in his book, *Planning a Computer System: Project Stretch 1962*, inspired by structural engineering:

> *"Computer architecture, like other architecture, is the art of determining the needs of the user of a structure and then designing to meet those needs as effectively as possible within economic and technological constraints."*

A more concise definition of software architecture is from *IEEE-1471-2000*:

> *"Architecture is the fundamental organization of a system embodied in its components, their relationships to each other, and to the environment, and the principles guiding its design and evolution."*

The IEEE definition is the most logical and straightforward definition of architecture and has been followed widely ever since it was defined.

The **Rational Unified Process** (**RUP**) provides a more elaborate definition of architecture as follows:

> *A set of significant decisions about the organization of a software system, the selection of the structural elements and their interfaces by which the system is composed together with their behavior as specified in the collaboration among those elements, the composition of these elements into progressively larger subsystems, the architectural style that guides this organization, these elements, and their interfaces, their collaborations, and their composition. Software architecture is not only concerned with structure and behavior, but also with usage, functionality, performance, resilience, reuse, comprehensibility, economic and technological constraints and tradeoffs, and aesthetics.*
>
> *– RUP (Kruchten, 1998)*

Martin Fowler, in the IEEE journal *Who Needs an Architect?*, defined architecture as important stuff that is expensive to change, whereas Grady Booch called it *significant decisions, where the cost of change measures significance.*

There is no debate on these definitions of architecture itself in the context of Agile software development methodologies. However, any architecture definition that treats architecture as "the foundation", "hard to change", "things that you need to know upfront", or "get it right early" are up for debate and no longer stack up in Agile development.

The purpose of architecture is to shape a solution that meets near-term customer needs, and address any technical risks that could hinder operations. A simple definition of software architecture in the context of Agile development is: *architecture is a series of irreversible decisions for the development of a sustainable technology solution.* The keywords in this definition are the following:

- **Irreversible**: Irreversible decisions are significant decisions that are hard to change without incurring considerable cost. For example, you decided to build an order management system on an off-the-shelf ERP. It is almost impossible to reverse that decision halfway through without an expensive change.

- **Sustainable**: Ability to maintain the state of the solution beyond immediate needs with an optimal run cost.

- **Technology solution**: A working software component or a hardware component or, in most scenarios, a mix of both, consisting of one or more products or services.

- **Decisions**: Decisions are related to defining the right solution for a given purpose, and the organization of the solution, context, elements, and relationships between various elements.

The preceding architecture definition from RUP touched on behavior. Behavior could be anything from the messaging style of an interface, such as synchronous or asynchronous, to the flow of data represented with a dynamic interaction diagram. This often leads to another question: whether the behavior is a question of architecture or design.

The distinction between architecture and design is not always clear. Simon Brown, who introduced *C4 models*, observed that the "*significant decisions are architecture, and everything else is design.*" Grady Booch's viewpoint is that "*all architecture is design, but not all design is architecture.*" The definition from Tom Hollander, Microsoft evangelist, is more appropriate in the Agile software development context. Hollander described architecture as "*a continuum where coarse-grained is architecture and fine-grained is design.*"

Architecture can be represented as a continuum, as shown in the following diagram:

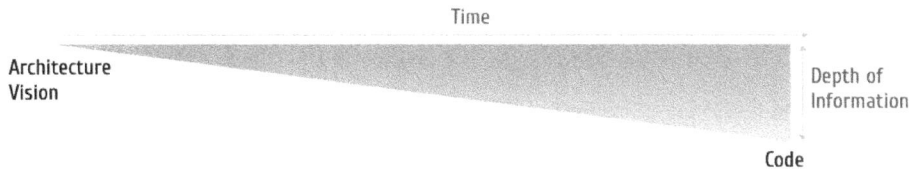

Figure 2.7 – Architecture represented as a continuum

As shown, there is no real cut-off point between architecture and design or high-level and low-level. The number of people who touch architecture also increases as we go from left to right in the continuum. However, in the traditional waterfall, both architecture and design activities are performed as part of the upfront design stage.

Traditional development methods frontload architecture efforts

Traditional architecture is an upfront activity that is deterministic and finite. In traditional projects, architects had the time and luxury of clearly knowing the requirements at the very early phase of the project to define the architecture upfront. This approach commits technologies and design decisions too early, which are unvalidated until close to going live. Often this is done without understanding the purpose and the business drivers, and, more importantly, draws unnecessary attention to many imaginary requirements. Neal Ford from ThoughtWorks observed that "*spending too much time upfront won't guarantee that you will know everything that you need to know about the system to future proof architecture.*"

The following diagram shows the timelines for architecture decisions in a typical waterfall project as well as in Agile software development:

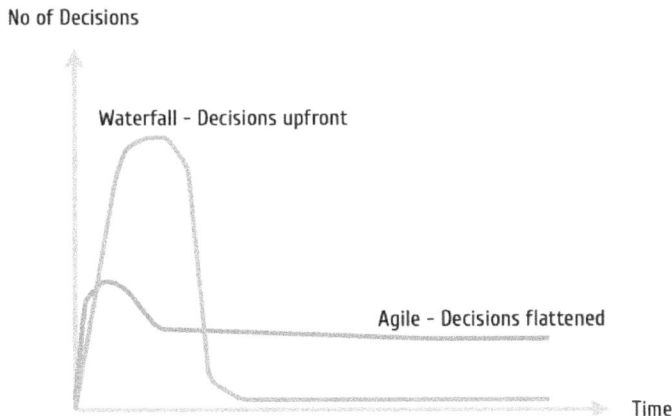

Figure 2.8 – Architecture decision timelines in waterfall and Agile

As shown in the diagram, the Agile architecture approach flattens the architecture efforts over time compared to the upfront decisions in traditional development.

Big Upfront Design (BUFD) is treated as one of the anti-patterns when developing architecture in Agile software development projects. In iterative development, upfront architecture is unfeasible as the requirements are not known until we are close to the end of the iteration. Architecturally significant requirements may surface even at a later stage, which in effect impact the resiliency of the architecture and cause significant rework.

Overengineering is a symptom of upfront designs. In the early 2000s, the abstract database factory was a common pattern, as many architects believed that they needed to cater to multiple databases such as MySQL, Oracle, and so on, but in reality, the need does not exist. Overengineering makes the system significantly complex and expensive to develop, operate, and evolve.

Before Agile, architecture was treated as a solid foundation, and therefore the notion of architecture decisions had to be taken upfront in the project life cycle. As a result, architects used to spend a lot of time creating abstractions. The notion of an architecture foundation came from building physical architecture where foundations are critical and have to be strong. Alistair Cockburn observed that "*some technologies exist today to replace the foundation of buildings without even moving the tenants out.*"

In my own experience from one of the largest successful home-grown projects, back in early 2000, we spent a good initial 6 months for requirements analysis. During this period, as architects, we spent time and effort in building an architecture foundation and a development framework. We had to do that as EJB was underperforming, and Spring was just in its nascent stage. We committed all decisions upfront, including the selection of technologies. It was successful in the end, mainly because of the spirited mindset of the team. Two years down the line, when we deployed in production, the technologies were already obsolete, which severely impacted the longevity of the solution.

One of the key lessons from this project resonates with Dietzler's law, which we see in the following diagram:

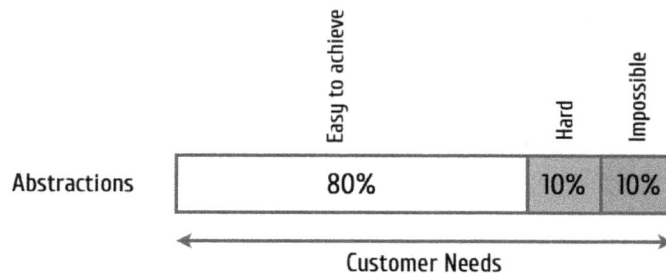

Figure 2.9 – Dietzler's law applied to abstraction

Creating an abstraction upfront will make the developer's life easier for 80% of the customer requirements, 10% is hard to achieve, and the last unanticipated 10% is impossible to achieve. However, customers always want 100%. The last 10% will break the abstraction and contribute to architecture complexities.

Developing architecture in Agile projects is more demanding than in traditional waterfall projects and therefore needs special attention and treatment. In applying the lessons learned from the failed story of Kodak, where it failed to embrace innovations, teams must fearlessly adopt Agile architecture.

We have learned that traditional architecture is not suitable for Agile software development. We will explore the importance of architecture in Agile and the reasons why it is termed the *foundation* of Agile software development in the following sections.

Agile architecture – architecting with Agile practices

Agile architecture is a combination of doing architecture using Agile methodologies, as well as building resilient architecture to meet evolving customer needs. Agile architecture treats architecture as a continuous flow of decisions delivering sustainable solutions that maximize business value. It is a disciplined way of iteratively developing and managing architecture in line with the values, principles, and practices of Agile software development.

The art of Agile architecture is described as follows:

- Anticipating, discovering, and stacking architecture decisions with an overarching intent
- Systematically solving problems by considering options
- Careful prioritization to minimize the cost of delay
- Continually evolving and improving solutions
- Transparently evaluating and addressing possible technical risks
- Maintaining velocity from demand to delivery

Agile architecture avoids large upfront design activities and provides guardrails for feature teams to avoid offshoot from the right directions without slowing down delivery flow.

Balancing speed and sustainability

Agile means quick and easy, whereas good architecture provides sustainability; Agile architecture balances speed and sustainability. There is a notion that Agile is all about fast and cheap, and everything else is secondary. Often development teams want to go just too fast. But developing sustainable solutions needs checks and balances. Fred Brooks made an interesting observation in his book *The Mythical Man-Month*:

"Good cooking takes time; some tasks cannot be hurried without spoiling the result."

Kevlin Henney, the author of the book *97 Things Every Programmer Should Know*, shared the analogy of a speeding car. Driving a car at 150 km an hour is always great, but only if in the right direction. Otherwise, it creates more damage than benefits as backtracking will consume time and money. Kevlin Henney also shared another great metaphor of the four forces that keep aircraft in the air and compared them with Agile software development. A modified version of this metaphor is shown in the following diagram:

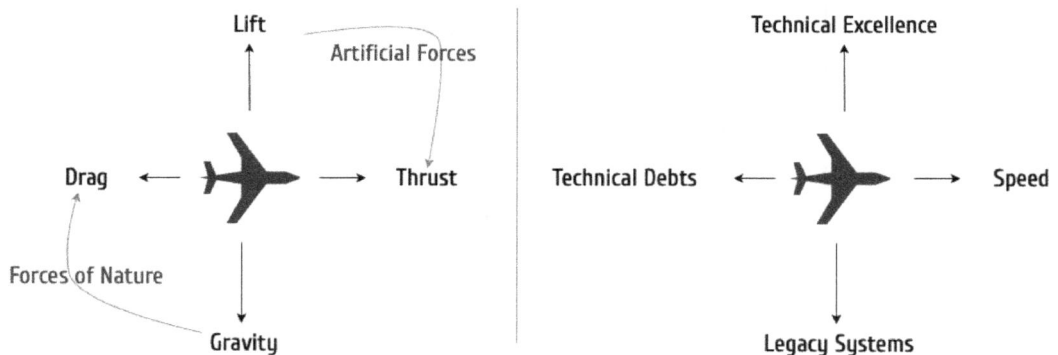

Figure 2.10 – Speed and sustainability – aircraft metaphor

Traveling faster without balancing sustainability will create more tech debts, which eventually drag the project back. This is where sustainability, and hence architecture, is immensely important. The architecture keeps the required checks and balances to ensure delivery is heading in the right direction by continually managing technical debts and preparing the architecture for the next set of iterations.

In Agile development, BUFD is not acceptable, but at the same time, having no upfront architecture is also not the right thing to do. Simon Brown, in his book *Software Architecture for Developers*, quoted Dave Thomas's observation that "*the big upfront design is dumb but no design upfront is even dumber.*"

Developing a little architecture upfront, but how much?

While big upfront architecture is an anti-pattern for Agile architecture, it is inevitable to have some elements of architecture upfront. In earlier days, an example of such an upfront decision was the technology stack and programming language. However, with the evolution of microservices, this is no longer an issue.

How much architecture upfront depends on the nature, size, and complexity of the project. In larger and critical projects, not doing enough architecture upfront leads to higher exposure to risk and may lead to failure. Philippe Kruchten observed this in the IEEE paper *Software architecture and Agile software development: a clash of two cultures?* as ignoring the architecture focus for certain classes of systems will lead projects to hit a wall and collapse.

In addition to nature, size, and complexity, there are several other forces proposed by Michael Waterman, James Noble, and George Allan in the IEEE paper *How Much Up-Front? A Grounded Theory of Agile Architecture*. The paper emphasizes the importance of considering technical and environmental factors such as the organizational context and domain-related parameters, as well as social factors such as the experience and skills of the architects.

When there are multiple solution options, the **Value, Cost, Risk (VRC)** model can calculate the upfront design score. The one that is highest is the solution with the right upfront design. The formula is shown in the following diagram:

$$\text{Upfront Design} = \frac{\text{Maximum Possible Business Value}}{\text{Cost of Development} + \text{Risk of Rewrite or Discard}}$$

Maximum Possible Business Value = relative scale of 1 to 5, 1 for small and 5 for high
Cost of Development = relative scale of 1 to 5, 1 for low cost and 5 for high cost
Risk of Rewrite = relative scale of 1 to 5, 1 for low risk and 5 for high risk

Figure 2.11 – VRC model for calculating the amount of upfront architecture

A good metaphor to explain how much upfront design to have is building an airport runway. Usually, massive investments are directly proportional to revenue generation. In the case of airports, the expansion will by and large depend on the growth in passenger volumes. This metaphor is shown in the following diagram:

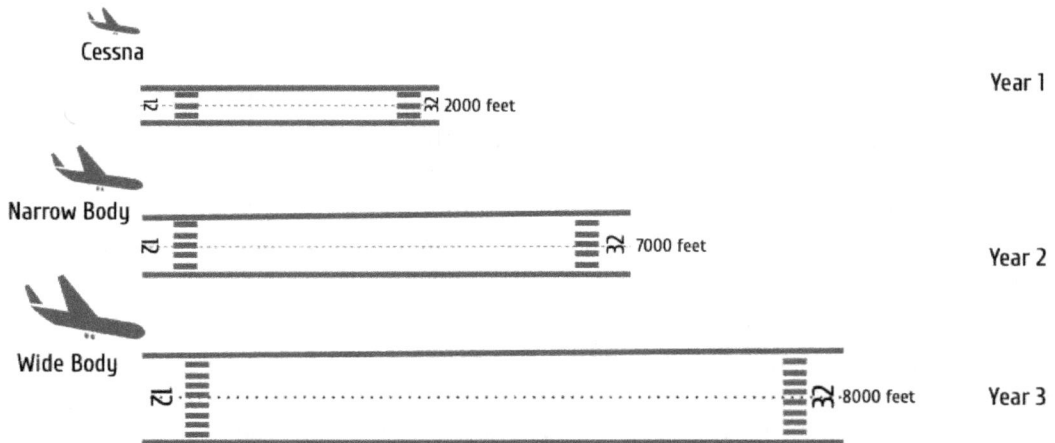

Cessna

2000 feet

Year 1

Narrow Body

7000 feet

Year 2

Wide Body

·8000 feet

Year 3

Figure 2.12 – Extending the runway based on immediate needs

Let's say the initial plan for the first year is to only handle Cessna-type aircraft, which only need a runway length of 2,000 feet. There is no point in building an 8,000-foot runway upfront as it is not going to add any business value. In the second year, the airport wants to handle narrow-body aircraft. This means we need to widen and lengthen the runway to 7,000 feet. We do this incrementally while operations are in progress. In the third year, the airport wants to accept wide-body aircraft. This means we will have to further widen and extend the runway to 8,000 feet. Again, runway upgrades will be done incrementally while aircraft movements are in progress. In this metaphor, we need to have an architecture intent to build an 8,000-foot runway, which can accept wide-body aircraft over 3 years, but we build incrementally based on what the airport needs at each point, which balances the present and the future.

Simon Brown summarized the elements of upfront design quite nicely in his article *Contextualizing just enough up front design*, published at codingthearchitecture. com. He observed that upfront architecture brings clarity on the big picture, how significant elements fit together, and identified the key risks to be mitigated, such as quality attributes, environment, and so on.

A disciplined approach, following a set of principles for developing Agile architecture, helps to bring agility without leading to heavy rework. We will discuss this in the next section.

Agile architecture principles

Agile architecture revolves around five core principles. Culture is one of the most important factors in Agile architecture and is considered as embedded in every principle.

The following diagram captures the five key Agile architecture principles:

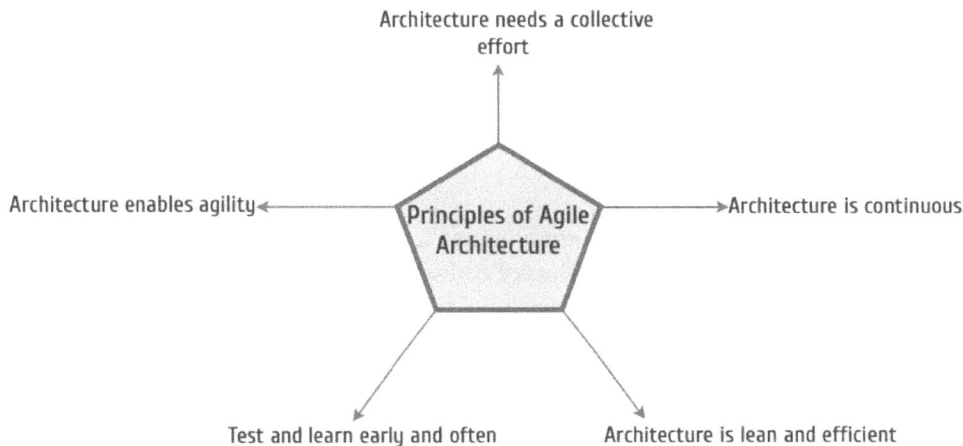

Figure 2.13 – Agile architecture principles

These five principles act as guiding forces for any successful transformation to Agile architecture. As we know, Agile software development practice does not prescribe anything, so the implementations can take different shapes.

Architecture needs a collective effort

Architecture development needs collective effort from all team members. One of the principles from the manifesto for Agile software development is "*the best architectures, requirements, and design emerge from self-organizing teams.*" This also resonates with the *Heart of Agile*, where *Collaborate* is one of the keywords. Agile architecture moves away from prescriptive to collaborative architecture development with no more throwing architecture over the fence to developers.

Three key aspects around this principle are as follows:

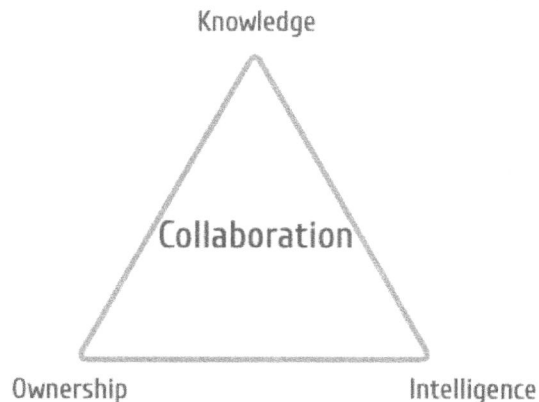

Figure 2.14 – Agile collaboration triangle

Let's understand each of these principles:

- **Collective intelligence**: In Agile architecture, architecture is collaboratively developed by all team members. The collective intelligence of all the team members is far better than an individual's knowledge and intelligence.

- **Collective ownership**: Collaborative development brings collective ownership. The nature of human beings is they take better care of things that they own. The team strives for success for those decisions taken collectively by them.

- **Collective knowledge**: Collaboratively built solutions eliminate the need for extensive knowledge sharing, documentation, and communication. This also improves the resiliency of the team in case they lose an individual's service.

This principle increases the cohesiveness of the team, which results in better quality, reduced maintenance cost, faster delivery cycles, and reduced risk of losing knowledge.

Architecture is continuous

If architecture is a continuum, then architecture development is a continuous activity. One of the principles of the manifesto for Agile software development emphasizes *"customer satisfaction through early and continuous delivery of valuable software."*

A continuous architecture needs an intent called **intentional architecture** and an incremental evolution called **emergent design**, as shown in the following diagram:

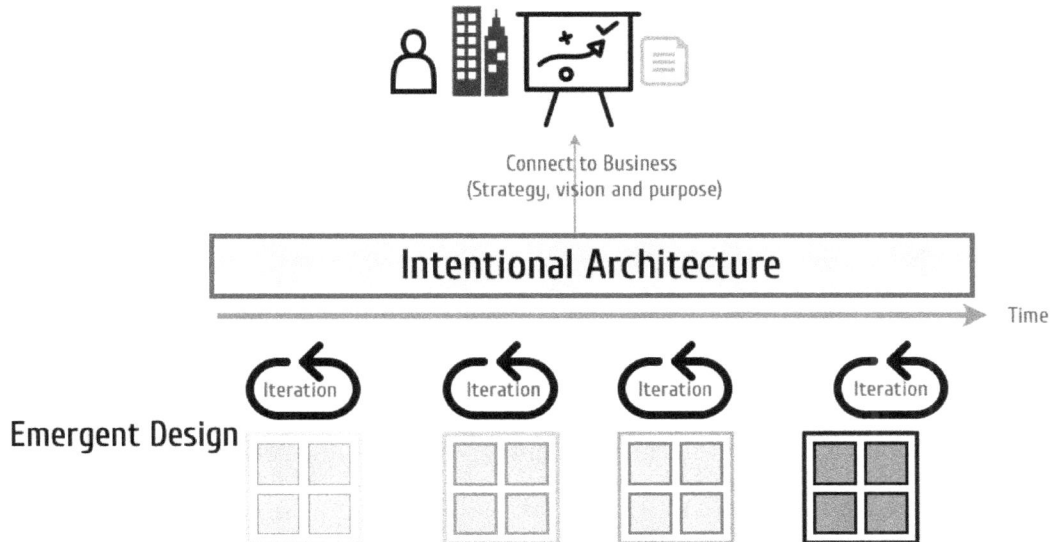

Figure 2.15 – Intentional architecture and emergent design

The concepts of intentional architecture and emergent design are as follows:

- **Intentional architecture**: Intentional architecture defines the big picture and shared understanding of the purpose, vision, contextual guardrails, and guidelines that the development teams must know to protect, align, and balance the customer's and organization's interests. It also provides economic conditions for trade-offs and sets shared goals when multiple teams are working on a solution.

- **Emergent design**: With intentional architecture in place, development teams evolve the design of the solution continuously over several iterations to meet the immediate needs of the customer. The emergent design is a natural outcome of the response to new customer demands, as well as addressing findings coming from constant refactoring efforts.

The *bridge girder erection* machine is a great real-world metaphor for intentional architecture and emergent design. The working principle for the bridge girders is that pillars are already developed, and girders are built as the machine moves forward. The pillars represent intentional architecture, and girders represent the emergent design.

Architecture is Lean and efficient

Architecture development must not perform non-value-added activities that potentially dent the continuous delivery flow. The parent principle for this is *"at regular intervals, the team reflects on how to become more effective, then tunes and adjusts its behavior accordingly."* Relentlessly looking for the eradication of potential wastage is one of the key concepts adopted from Lean manufacturing. The team must systematically determine and drive continuous improvement on the end-to-end cycle time.

By applying this principle, the team can identify and utilize untapped capacity by shifting focus from non-value-adds to the flow of values. The architecture wastage comes predominantly from the following sources:

- **Avoid hand-offs**: One source of wastage in traditional architecture development is from hand-offs. The architect creates architecture documentation and throws it over the wall to the team, as opposed to collaborative development. Clarity will deteriorate as the information flows from architects to developers and results in additional back-and-forth conversations. The collaborative solution design approach in Agile architecture powers the continuous flow of work without hand-offs.

- **Keep it simple**: Traditional architecture assumes many imaginary future requirements without understanding the full context and business vision. The Abstract Factory example discussed earlier is a classic scenario of overengineered complex architecture. Cost and lead time increase as the complexity of architecture goes up. Agile architecture is always a function of business value and is based on near-term customer needs.

- **Alignment over governance**: Agile teams need faster decision cycles and are usually autonomous and empowered to make appropriate decisions. Often, traditional governance boards meet infrequently, causing teams to wait longer for decisions, which interrupts the flow. The basis for Agile architecture is alignment, utilizing constant feedback over fixed bureaucratic governance.

- **Communication over documentation**: Creating massive documentation upfront is one of the challenges in traditional architecture development. This not only wastes time and effort to build documentation but also contributes to the cost of maintenance. Agile architecture promotes free-flowing communication, collaborative development, and the use of visual aids.

Every team is different in terms of skills, culture, technologies, and the nature of their work. Self-organized teams find themselves an ideal way to collaborate for better results. They use inspection and feedback to facilitate continuous process improvements.

Test and learn architecture early and often

Architecture has to be continuously refactored and evolved using the frequent test, learn, and refactor cycles to maintain quality. The team explores every possible scenario to improve steps in the delivery by bringing efficiency as well as enhancing the quality. One of the development principles in the manifesto for Agile software is *working software is the primary measure of progress* and is the motivation for this principle.

Key aspects related to this principle are the following:

- **Test architecture early and often**: One of the significant risks in the waterfall model is the late validation of the software. In Agile architecture, teams use every opportunity to test architecture using automated test scripts as part of the continuous delivery pipelines starting from the very first iteration, possibly in production alongside business features. Use spikes to prove the design when the team is not confident.

- **Learn and refactor constantly**: Using fast, integrated learning, identify potential shortcomings of the design and refactor as early as possible as part of the day job to develop long-lasting fixes. The team has to constantly solve problems in production, with triage connecting to backlog items, which leads to refinement of solutions by refactoring, redesigning, and stepping up monitoring.

Feedback-driven learning is a great mechanism to evolve architecture continuously and to be in a better position to respond to changes quickly.

Architecture enables agility

Architecture and design have to be adaptable to absorb requirements coming even at a later stage of the project. One of the principles in the manifesto for Agile software development is *"welcome changing requirements even late in the development"* and is linked to this architecture principle. To achieve this, the architecture has to be evolvable.

Evolutionary architecture is the term coined by Neal Ford and the team at ThoughtWorks to define evolvable architecture. They described evolutionary architecture as an architecture that supports incremental guided changes as a first-class citizen.

Key aspects of this principle are the following:

- **Automate**: Agile architecture promotes automation as an integral part of the solution. This includes **Continuous Integration** (**CI**), **Continuous Delivery** (**CD**), and releases on demand. Use extensive, automated telemetry to understand architecture behaviors and quality attributes in production.

- **High on technical excellence**: Growth mindset, continuous investment in learning, and passion for breakthrough innovations are essential team behaviors to enable constant improvement in technology, techniques, practices, and patterns. Faster time to delivery by adopting proven patterns as code, using data for trade-off, and building cost and risk-aware architecture are positive signs of good Agile architecture practices.

- **Reduce blast radius**: Using techniques such as microservices to break architecture into smaller chunks that are modular and loosely coupled. This will help to reduce the blast radius in case of issues or changes such as technology upgrades.

These five principles are agnostic to any Agile framework, and guide architecture and design in Agile software development projects. In the next section, we will examine how some of the Agile scaling frameworks approach Agile architecture.

Comparing different enterprise Agile frameworks

The *14th Annual State of Agile Report 2020* (https://stateofAgile.com) indicates that Scaled Agile Framework has been adopted far more than other Agile scaling frameworks. For this book, we will consider comparing **Scaled Agile Framework (SAFe)**, **Disciplined Agile (DA)**, and **Large-Scale Scrum (LeSS)**.

Scaled Agile Framework (SAFe)

SAFe (www.scaledAgileframework.com) is built on four values: alignment, built-in quality, transparency, and program execution. Program execution is particularly important in SAFe and is accomplished with a team construct called **Agile Release Trains (ART)**. There are four levels of team constructs in SAFe: portfolio, solution, program or essential, and team.

SAFe defines Agile architecture as a set of values, practices, and collaborations for architecting a sustainable system. The focus areas of Agile architecture in SAFe are a continuous flow of values, collaboration, intentional architecture, emergent design, and design simplicity.

Three key SAFe principles that are related to Agile architecture are as follows:

- Assume variability; preserve options.

- Build incrementally with fast, integrated learning cycles.

- Decentralized decision-making.

Agile architecture evolves over a period of time to support the needs of current customers and avoid delays related to phase-gate processes and big upfront designs. SAFe uses the term, the **Architectural Runway** as a key focus area in Agile architecture. The Architectural Runway is built to evolve architecture with a series of enablers just in time to support upcoming business features.

Agile architecture in SAFe also emphasizes DevOps, continuous integration, continuous deployment, and release on demand. It also proposes to organize teams around business values with value streams. SAFe also talks about the alignment of architecture with the enterprise architecture technology strategy and roadmap.

Disciplined Agile (DA)

DA (www.pmi.org/disciplined-Agile) is a process- and goal-driven framework with a strong philosophy of *"not one size fits all."* It further provides guidelines for different categories of Agile projects. The framework proposes three stages: inception, construction, and transition. DA offers a lot for architecture in the inception phase, as well as in the transition phase.

In the inception phase, a lightweight, just-enough-upfront architecture will be developed, consisting of an architecture vision, architecture models, and non-functional requirements. One of the first activities in the construction phase is proving the architecture. DA's belief is that envisioning architecture in the inception phase will help in getting the initial technical direction. DA observes that a good architecture strategy is necessary for the success of the project.

DA is a firm supporter of intentional architecture and emergent design. It states architecture is not a one-time activity; it exists throughout the life cycle of an initiative and evolves through a series of decisions. DA puts a lot of emphasis on appropriate modeling of architecture and design. The architecture modeling requirement is based on the size of the project. DA calls for team collaboration.

During the transition phase, DA emphasizes a continuous delivery life cycle consisting of continuous integration, continuous deployment, and DevOps.

Large-Scale Scrum (LeSS)

LeSS (https://less.works) uses a simple approach for scaling scrum teams based on basic scrum, Lean, and Agile development philosophies with minimal processes. LeSS defines fewer roles compared to other scaling frameworks to avoid additional overheads. Similar to SAFe, LeSS also puts emphasis on system thinking as a fundamental mental model required to implement long-lasting fixes.

LeSS focuses more on emergent design, collaboratively implemented by the team. The inspirational metaphor compares architecture and design with a desire line in a snow park. Desire lines in a snow park are organically created based on how people naturally choose to walk. They also depend on volumes of people. Where to build them, how to build them, how wide, and so on are not planned upfront. LeSS observes that designs are created based on demand rather than speculatively pushed. LeSS gives adequate focus to technical excellence, with strong references to continuous integration, continuous deployment, test-driven development, and test automation.

Three interesting observations from LeSS on architecture are the following:

- The sum of all the source code is the true design blueprint or software architecture.
- The real software architecture evolves (for better or worse) every day of the product, as people do programming.
- The real living architecture needs to be grown every day through acts of programming by master programmers.

LeSS took a team-centered approach for developing architecture and reiterated that the architecture design is an important activity and must happen within the team.

Just like Agile practices, Agile architecture needs to be measured frequently to determine areas of improvement. In the next section, we will explore a simple measurement framework for measuring Agile architecture practice.

Measuring Agile architecture maturity

Measuring and improving is as important as adopting Agile architecture. It provides a simple mechanism for teams to reflect, determine areas of improvement, and take the right steps toward the desired target without top-down enforcement. These assessments are done collectively by the team, and help to quantify the maturity of architecture practices and efficiently drive continuous improvements. The measurement system itself must be Agile and Lean to be efficient and effective. These assessments ideally should be done periodically, such as the end of every third iteration.

There is no global ideal target state for these assessments. Instead, self-organized teams determine what their desired state is not required as a team and work toward achieving their goals. The Agile architecture measurement framework shown in the following diagram is based on the five principles of Agile architecture on a five-point scale:

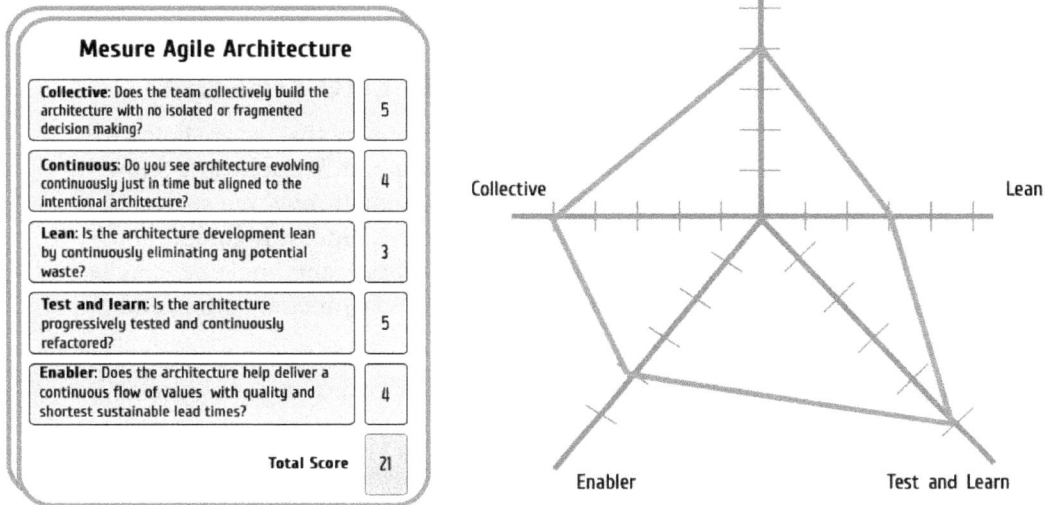

Mesure Agile Architecture

Collective: Does the team collectively build the architecture with no isolated or fragmented decision making?	5
Continuous: Do you see architecture evolving continuously just in time but aligned to the intentional architecture?	4
Lean: Is the architecture development lean by continuously eliminating any potential waste?	3
Test and learn: Is the architecture progressively tested and continuously refactored?	5
Enabler: Does the architecture help deliver a continuous flow of values with quality and shortest sustainable lead times?	4
Total Score	21

Figure 2.16 – Measuring Agile architecture

For each parameter, as shown in the diagram, the team collectively decides which behaviors score one and which behaviors score five. The spider chart on the right shows the visual view of the outcome. Once the overall score is calculated by summing individual scores, using the maturity model shown in the following diagram reflects the current state of the Agile architecture:

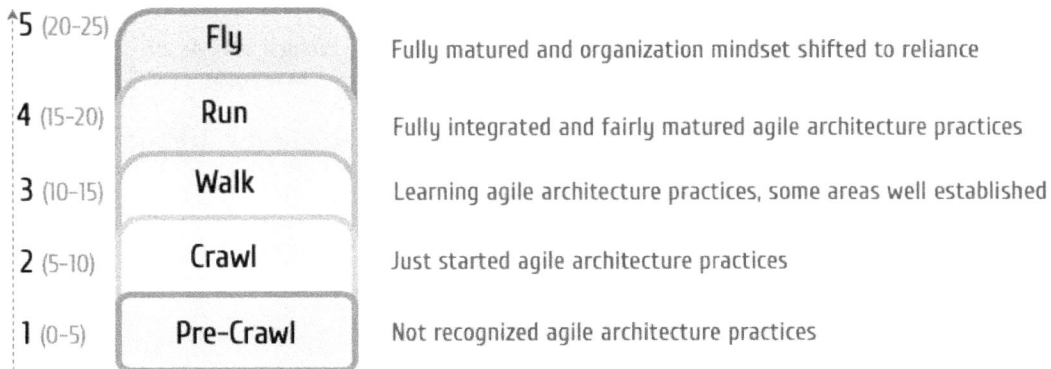

5 (20-25)	Fly	Fully matured and organization mindset shifted to reliance
4 (15-20)	Run	Fully integrated and fairly matured agile architecture practices
3 (10-15)	Walk	Learning agile architecture practices, some areas well established
2 (5-10)	Crawl	Just started agile architecture practices
1 (0-5)	Pre-Crawl	Not recognized agile architecture practices

Figure 2.17 – Agile architecture maturity model

The five maturity levels shown in the preceding diagram are adopted from the *AgilityHealth radar* (https://agilityhealthradar.com).

A number of cues provided for each question to help collective assessment is available in the Git repository for download.

Lessons learned from Snow in the Desert

Snow in the Desert is a fictitious global tour and travel company that we are going to have as a point of reference to get this contextualized viewpoint. We are using Snow in the Desert as a case study for reinforcement of our learning throughout this book. The nature of Agile software development is not prescriptive. Since it is guided, and principle- and practice-driven, there are many ways to embrace and implement Agile engineering concepts. The organization's context mostly determines adoption, and therefore it is hard to have a generic opinionated view.

In this chapter, we are introducing Snow in the Desert and the way Agile architecture is adopted to accomplish organization agility.

The group has four business verticals, as shown in the following diagram:

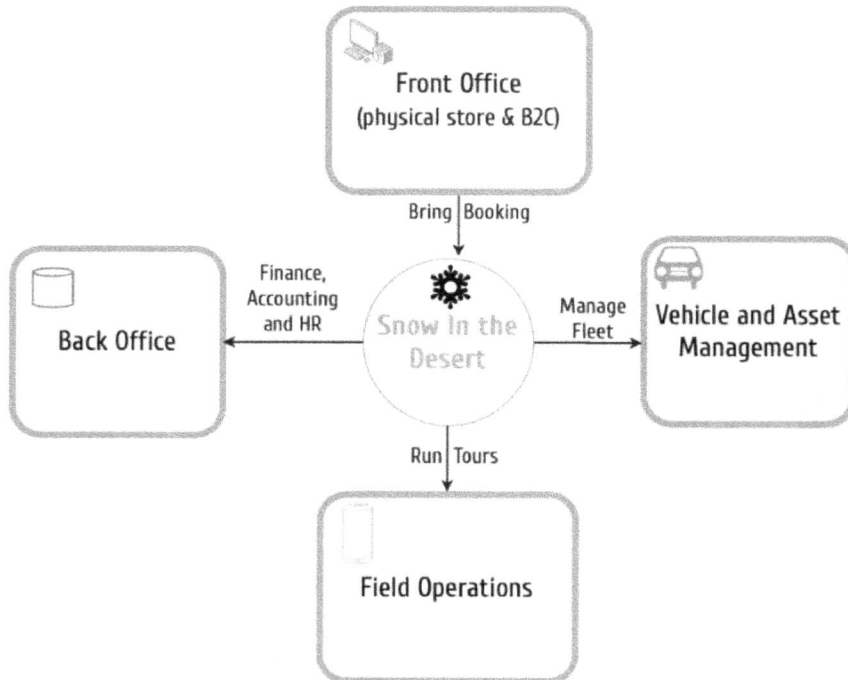

Figure 2.18 – Snow in the Desert business units

Snow in the Desert is a company with a long legacy, having footprints across the globe with close to 2 billion USD revenue per year and 6,000 employees. As shown in the preceding diagram, there are four distinct lines of business. The front office consists of brick and mortar shops as well as online sales. The field operations handle the day-to-day operations, conducting tours. The majority of the field operations are outsourced to tour partners. The vehicle and asset management line of business manages operating fleets such as tour buses. The back-office line of business consists of finance, HR, and other supporting services.

Snow in the Desert has its own central IT department with more than 500 employees. Over 100 systems power the business with a mix of home-grown as well as vendor-managed systems.

With the recent economic crisis, Snow in the Desert conducted a review of the entire business. The report from the management consultant alluded to a fat organization and complexity in business processes as key points of failure. The board of directors kicked off a transformation program to bring back agility. As part of this, the IT department was asked to streamline its processes in line with industry trends. The CTO's vision was to build an IT organization that was nimble, right-sized, and right-shaped to be in a position to respond swiftly to changes in the business strategies.

Traditionally, Snow in the Desert was a heavy waterfall shop. They have been using silos of Agile development for some time. The executive decided to go for a big bang organizational transformation with Lean and Agile product delivery at the center, by choosing SAFe. In the new design, Full SAFe was adopted with a portfolio layer as well as ARTs. The new design resulted in a single portfolio with four ARTs. The following diagram shows those four ARTs, the number of teams, team sizes, and systems associated with each ART:

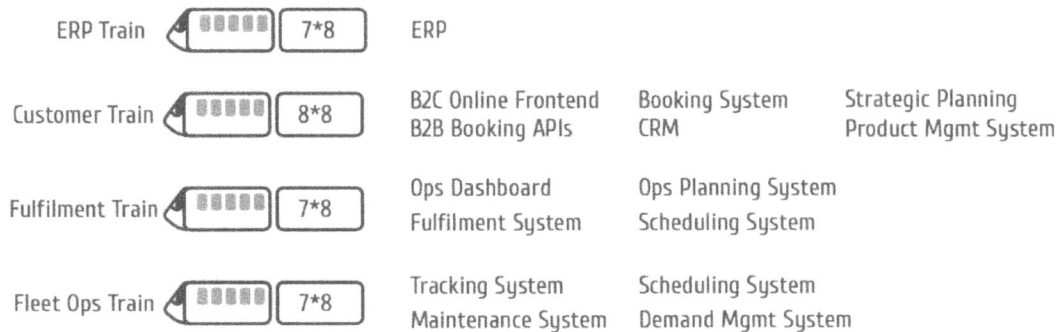

ERP Train	7*8	ERP		
Customer Train	8*8	B2C Online Frontend B2B Booking APIs	Booking System CRM	Strategic Planning Product Mgmt System
Fulfilment Train	7*8	Ops Dashboard Fulfilment System	Ops Planning System Scheduling System	
Fleet Ops Train	7*8	Tracking System Maintenance System	Scheduling System Demand Mgmt System	

Figure 2.19 – ART design at Snow in the Desert

We will discuss the full structure in the subsequent chapters, and more details on design decisions in *Chapter 12, Architecting Organizational Agility*.

Circumventing key challenges in adopting Agile architecture

The most challenging part of adopting Agile architecture and Agile in general is embracing a new culture. Most of the challenges encountered at Snow in the Desert in their transformative Agile journey were to do with the legacy culture and mindset.

The architecture department was disbanded to align the architecture with the ART and the team constructs. However, adoption was precluded by many teething issues. The team and leadership jointly employed mechanisms to circumvent those challenges, as we see here:

- **Missing the art of giving away control**: One of the symptoms noted was architects continued to work in silos, not ready to give away control, empowerment, and autonomy to the team. This created an unhealthy tension between the architects and the team. Some teams with smart people crafted solutions in silos without engaging with the team, specifically the architects. On the contrary, some developers with a legacy mindset were expecting architects to provide finished solutions before they could start to code as they were not confident enough to make their own decisions. To address both problems, a formal solution design ceremony was organized in the middle of every iteration for refining solutions for the next iteration, an idea taken from LeSS. We will discuss this further in later chapters.

- **Falling into the "urgency of now" trap**: One of the other patterns observed during the initial stages of adoption was the lack of forward planning. Teams were not planning ahead sufficiently, irrespective of **Program Increment** (**PI**) planning events, and almost discovered work when they were ready to code. As a result, the architecture decisions were taken haphazardly within the iteration. In most cases, this resulted in tactical and accidental architecture. Even though the initial velocity went high, maintenance issues started creeping in as time elapsed, proved costly, and faced difficulties in responding quickly due to the high impact of change. This issue was addressed by formulating a conscious strategy to build a solution roadmap and architecture runway together with the business and other stakeholders. The team set themselves a goal for maintaining the health of the architecture runway and it was diagnosed at the end of every PI consistently. A proper definition of ready, along with the solution design workshop, contributed to improving this situation. We will discuss this more in *Chapter 6, Delivering Value with New Ways of Working*.

- **Dealing with commercial off-the-shelf (COTS) products**: Snow in the Desert has many COTS products deployed in their live system environment and are interested in more and more COTS products to support their operations in the non-differentiated areas. Evaluation for selection, product rollouts, and incremental development was considerably challenged as many of these vendors were not on Agile practices. Introducing an MVP first approach for everything, including COTS, was one of the solutions adopted to extract value faster and to reduce risk on contractual commitments. Many lightweight Lean-Agile processes and ceremonies were introduced to improve partnerships with vendors. This is an area mostly not so clear in many of the scaling Agile frameworks.

- **Working with legacy systems**: One of the biggest challenges was also to work with legacy systems that were not responsive to many of the technological advancements such as automation, microservices, and so on. Considerable investment was required for continuous technical improvements on these live systems. At many times, the business value of such large-scale improvements outweighed the cost of implementation. A top-down program to improve the technical condition with value-driven incremental changes was one of the solutions to this problem. We will discuss this more in *Chapter 5, Agile Solution Architect – Designing Continuously Evolving Systems*.

- **Balancing business and technical backlog items**: In the initial period of adoption, there has been a considerable struggle between mixing functional backlog items and technical backlog items when drawing down for iterations or PIs. Business functionalities have been given precedence over technical features in prioritization exercises as they directly impact business and, therefore, better value for money. Product owners always want to pack functional backlog ahead of technical backlog, resulting in the aging of technical backlog items, in turn piling up tech debts. This is effectively addressed by implementing top-down capacity allocation models with dedicated buckets for different backlog items. We will discuss this more in *Chapter 6, Delivering Value with New Ways of Working*.

These symptoms and many others may appear in any organization's journey toward adopting Agile architecture. Documenting detailed process steps, a **Responsibility Assignment Matrix** (**RACI**), and role boundaries are the worst approaches for fixing Agile architecture adoption challenges, which prevent teams from self-organizing. The golden mantra is to identify change agents within the team and make them champions in the transformation journey. Change agents take further steps to repeatedly communicate, reinforce, and demonstrate a Lean-Agile mindset. Keep expanding the circle of influence to include more and more people as the journey takes shape.

Agile architecture process flow and poster

The high-level architecture flow of work is documented and is available for download from the Git repository using the following link: e.

We will cover this flow of work in *Chapter 3*, *Agile Architects – The Linchpin to Success*, and *Chapter 4*, *Agile Enterprise Architect – Connecting Strategy to Code*.

The Agile architecture poster is also available for download from the Git repository using the following link: `https://github.com/PacktPublishing/Becoming-an-Agile-Software-Architect/tree/master/Chapter2`.

Both the process flow and Agile architecture posters are prepared based on the SAFe approach to adopting Agile architecture.

Summary

To respond rapidly to changes in the business, IT organizations are expediting enterprise Agile framework adoptions at scale. Traditional architecture development methods don't add value in the fast-moving iterative Agile software development, and cannot survive any longer. Therefore, Agile architecture practices are important. Instead of big upfront designs, iterative and just-enough architecture using intentional architecture and emergent design is the way forward.

Agile architecture development is a collaborative effort. It must support the continuous flow of values with a Lean-Agile mindset to actively eradicate any non-value-added features and activities. The team must adapt and practice continuous refactoring with repeated and fast learning cycles. Architecture should enable agility in product delivery by vigorously seeking opportunities to adopt proven patterns and techniques such as DevOps, continuous delivery, microservices, and so on. Agile architecture practice must be monitored and measured all the time to determine improvement steps. Successful adoption of Agile architecture can produce fit-for-purpose, sustainable, and quality solutions faster, at optimal cost.

As we have learned in this chapter, Agile architecture is a team effort. This perspective creates many questions about the role of architects in Agile. The next chapter will focus on the role of architects in Agile software development and addresses some of the challenges.

Further reading

- **History of Software Engineering with Grady Booch**: `https://www.youtube.com/watch?v=QUz10Z1AfLc`

- **Iterative and Incremental Development, A Brief History**: `https://www.craiglarman.com/wiki/downloads/misc/history-of-iterative-larman-and-basili-ieee-computer.pdf`

- **Software Engineering, NATO Science Committee Report**: `https://www.scrummanager.net/files/nato1968e.pdf`

- **Managing The Development of Large Software Systems**: `http://www-scf.usc.edu/~csci201/lectures/Lecture11/royce1970.pdf`

- **14th The Annual State of Agile Report**: `https://explore.digital.ai/state-of-Agile/14th-annual-state-of-Agile-report`

- **Heart of Agile**: `https://heartofAgile.com`

- **IEEE1471 and Systems Engineering**: `http://www.mit.edu/~richh/writings/ieee1471-and-SysEng-(draft).pdf`

- **The Role of an Architect in an Agile Team**: `https://channel9.msdn.com/Events/TechEd/Australia/2010/ARC204`

- **What is your definition of software architecture?**: `https://resources.sei.cmu.edu/asset_files/FactSheet/2010_010_001_513810.pdf`

- **Software Architecture and Agile Development**: `https://resources.sei.cmu.edu/asset_files/Presentation/2010_017_001_23424.pdf`

- **Contextualizing Just Enough Upfront Design**: `http://www.codingthearchitecture.com/2012/01/05/contextualising_just_enough_up_front_design.html`

- **How much up-front? A grounded theory of Agile Architecture**: `http://citeseerx.ist.psu.edu/viewdoc/download?doi=10.1.1.702.4489&rep=rep1&type=pdf`

Section 2: Transformation of Architect Roles in Agile

This section will help you understand different architect roles, how they are different in Agile, and how enterprise and solution architects can add value by positioning themselves appropriately in terms of Agile development.

This section contains the following chapters:

- *Chapter 3, Agile Architects – The Linchpin to Success*
- *Chapter 4, Agile Enterprise Architect – Connecting Strategy to Code*
- *Chapter 5, Agile Solution Architect – Designing Continuously Evolving Systems*

3
Agile Architect – The Linchpin of Success

"Architects, absorbed as they are in contemporary problems of design, devote little time to questioning the assumptions underlying their work."

- Sotirios Kotoulas (Supervised Wuskwatim Hydroelectric dam in Northern Canada)

In the last chapter, we learned about collaboration to achieve collective ownership being one of the key Agile architecture principles for developing evolving solutions. Collective ownership ensures all team members equally own decisions and share responsibilities to drive toward achieving the architecture vision.

While the team self-organizes in a way that delivers maximum value with agility, critical architecture decisions need to be taken carefully to avoid long-lasting impacts on the quality and sustainability of solutions. Therefore, it is crucial to have one member of the team hold the baton while the rest of the team swarms around and contributes actively to make correct, pragmatic decisions with consensus. The person who has the baton is typically an architect who can think outside the box with full awareness of the business context and purpose. Architects operating in Agile software delivery have to unlearn the traditional operating models and consistently make use of a new set of skills and knowledge with a lean-Agile mindset.

This chapter will focus on challenges in the traditional architect role when operating in Agile software delivery. A number of metaphors are used to explain Agile architects' roles, responsibilities, and ways of working. A critical point to examine is how different Agile scaling frameworks treat architects differently.

In this chapter, we're going to cover the following main topics:

- Understanding the challenging environment of architects
- Why self-organizing teams propel the no architect movement
- Architects are vital in Agile – the question is who plays that role?
- Explaining an Agile architect's behavior and duties with metaphors
- How different scaling frameworks position architects
- Learnings from Snow in the Desert

This chapter focuses on the *Agile Architect* focal point of the *Agile Architect Lens*:

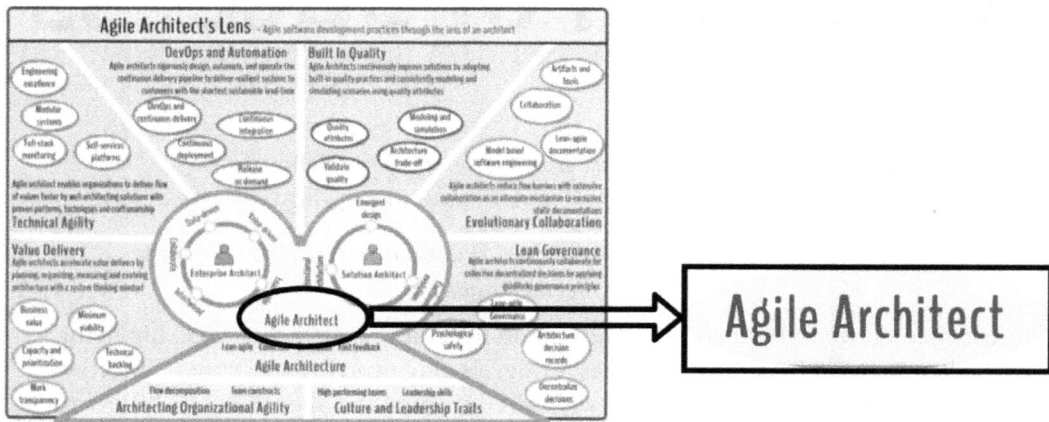

Figure 3.1 – Agile Architecture Lens – focal point

Technical requirements

Additional materials related to this chapter are available at the following link for download:

```
https://github.com/PacktPublishing/Becoming-an-Agile-Software-Architect/tree/master/Chapter3
```

Understanding the challenging environment of architects

Today's lean and dynamic software development environments pose many challenges for architects. Agile software development has amplified these challenges since many traditional practices do not go very well with Agile delivery. Confusing architect roles, commanding and controlling ways of working, bureaucratic governance, and tons of disconnected documentation are causes for concern. An in-depth analysis of these challenges follows in the next section and will help us learn about the sources of pain to rediscover and reposition architects to deliver better value to an organization.

A myriad of architect roles have impacted clarity

The term *architect* opens a Pandora's box of confusion. The architect role is still one of the most misunderstood roles in the software development world. Unlike some other roles, such as product manager or Scrum master, there is no unified role definition for architects in Agile software development. While many internet giants are getting rid of architect titles from their organizations, architect designations across the software industry remain endless.

Used to represent any type of architect, *software architect* is the general umbrella term used in the software industry. The *Wikipedia* definition of a software architect is as follows:

> *"A software architect is a software expert who makes high-level design choices and dictates technical standards, including software coding standards, tools, and platforms. The leading expert is referred to as the chief architect."*

As you can imagine, some of the keywords in this definition, such as *makes*, which refers to decisions, *high-level*, and *dictates* are antipatterns in an Agile culture. IASA Global, an association of IT architects, defines an IT architect as a technology strategist for a business. IASA uses five pillars to represent an IT architect in its **Information Technology Architecture Body of Knowledge (ITABoK)**.

These five pillars are illustrated in the following diagram:

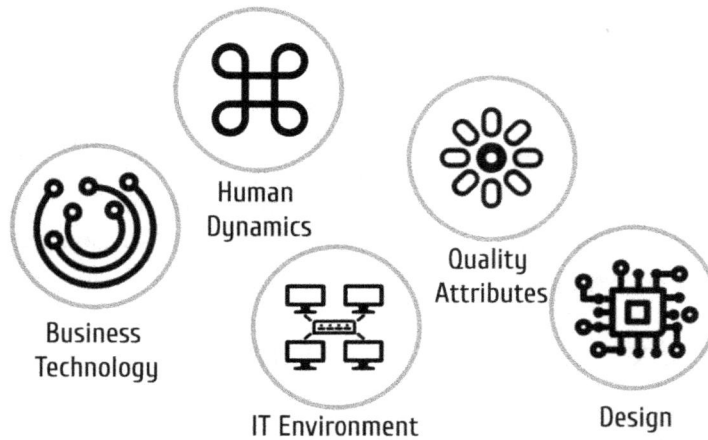

Figure 3.2 – Five pillars of an IT architect's responsibilities

An IT architect is a person who operates on the five pillars shown in the preceding diagram, which are further explained here:

- **Business Technology strategy**: Defines how technology aligns with the business goals and understanding the operations of the business

- **Human dynamics**: Emphasizes leadership and communication qualities such as managing and influencing people and their interactions

- **Quality attributes**: Focuses on addressing the quality of solutions that are measured and monitored

- **Design**: These are attributed to key elements are decisions including justifications, rationale, trade-offs, architecture styles, patterns, and so on, developed in response to business needs

- **IT environment**: Deals with how operations are carried out, building new capabilities and making decisions such as buy versus build; also concerns the technology in use, and innovations

IASA further breaks down the IT architect role into three concrete roles – *Enterprise Architect*, *Solution Architect*, and *Technical Architect*. SFIA, the global skills and competency framework, defines architect roles as *Enterprise Architect*, *Solution Architect*, and *Business Architect*.

In reality, there are numerous architect job titles used for different purposes by different organizations. This range varies from narrowly scoped architect titles such as NFR architect, lead API architect, and so on to broad level titles such as pre-sales architect, IT architect, and so on. Architect titles vary between types of organizations. Business IT companies use more standard job titles closer to IASA and SFIA definitions, whereas service companies use architect titles more as a medium for marketing. Technology companies use architect titles closer to the technologies they operate in, such as cloud solution architect. Product companies often use specific titles matching their products, such as Snowflake architect.

Countless architect titles with varying responsibilities have contributed adversely to the source of confusion as there is no single frame of reference.

Architects slow down delivery

In traditional software development, often, architects are far away from the ground, yet make architecture decisions. As a result, architects' low perceived value tempted many organizations to remove architects from their projects.

The following diagram illustrates a few potential challenges from the traditional software development world:

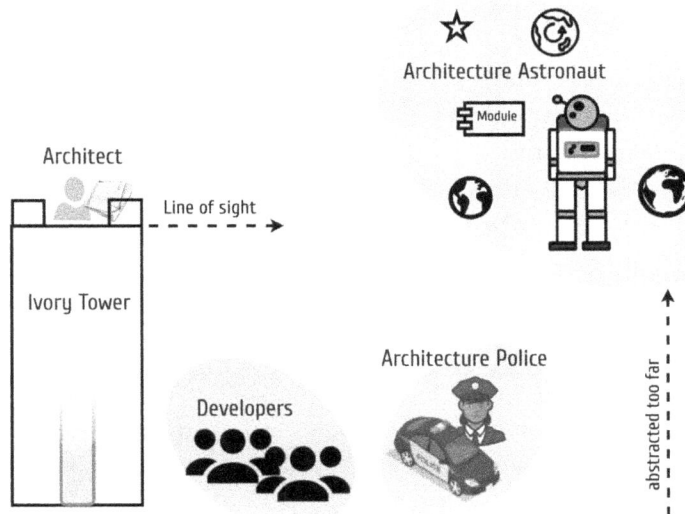

Figure 3.3 – Symptoms that slow down team velocity

The three key symptoms of architects slowing down delivery are described in the following sections.

Ivory tower architects can't see the ground

The term **ivory tower** refers to privileged people disconnected from the facts and practicalities of the real world. Ivory tower architects work in a centrally managed group, away from the team, without having the project's context but still making decisions based on what they think. Ivory tower architects are dogmatic in their approach, often using obsolete standards and a bunch of unproven, non-consumable architecture documentation to command and control developers.

This ivory tower approach is strikingly different from the practices and principles of Agile architecture, such as decentralized decision making, collective ownership, and simplicity. Ivory tower architects wait for phase gates to introspect, and then inspect alignment, substantially slowing down the continuous flow.

Architecture astronauts can't write code or design programs

In a similar line as ivory tower architects, Joel Spolsky called architects displaying particular behavior **architecture astronauts**. He mentioned in his article *Don't Let Architecture Astronauts Scare You* observing that *architects abstracting themselves far away run out of oxygen*. Spolsky further explained this behavior as *architects who have not written code for ages, wholly disconnected from reality, yet produce nice, high-level pictures that do not convey anything useful*.

In Agile architecture practice, decisions are based on collective intelligence with no handoffs. It is a general belief that traditional architects, over time, get outdated and disconnected from technologies, and hence they no longer understand the nitty-gritty of technologies. When these architects define solutions, they only scratch the surface and therefore carry higher degrees of risk.

Architecture policing only adds roadblocks

This analogy of architecture policing is also in line with previous scenarios. These types of architects focus their time and energy on stage-gated governance using intensive bureaucratic processes. They focus on zero tolerance and blind enforcement of compliance using standards and policies over context, purpose, value, and business impact.

Agile architecture promotes alignment over governance using rapid feedback cycles. Since policing types of architects focus on governance, they hardly spend time on business strategy alignment and communicating vision and purpose to the team. Doing so further weakens the team's ability to deliver business solutions with a sense of purpose.

The perception developed from these symptoms raised questions about the architect's role in self-organizing teams. We will elaborate on this in the next section.

Self-organizing teams propel the no architect movement

One of the manifesto principles for Agile software development states that *the best architectures and designs emerge from self-organizing teams*. Members of the team have to share their knowledge and efforts to make the right architecture decisions.

Many organizations and teams interpreted this manifesto principle differently. As a result, two movements emerged in the early days of Agile software development. One tried to eliminate architects from Agile projects, whereas the other movement promoted architects and portrayed engineers just as developers.

The proponents of eliminating architects justified their position by pointing to the Scrum methodology, where there is no explicit architect role defined. Scrum defines only three roles – product manager, Scrum master, and development team. This movement also pointed out the affordability of architects in teams. Agile teams typically consist of seven to eight members, and therefore the argument was that a dedicated architect in every team impacts velocity.

As a result, many Agile projects ran without architects. When there is no architect to hold teams technically together with a shared vision, they lack a holistic and connected view of the solution.

The birth of the accidental architect

With no architect in place, developers code without thinking about architecture and design. Developers make decisions as they code. As a result, architecture tends to be treated as a post-delivery documentation activity of what has already been implemented.

Grady Booch coined the term **accidental architecture** back in 2006 in the IEEE journal *The Accidental Architecture*. He observed that accidental architecture appears in development projects that do not start with architecture. Booch echoed Philippe Kruchten's observation that *the architecture evolves through a long and rapid succession of suboptimal decisions taken often in the dark*. Booch concluded with a statement that *accidental architecture are not evil things but are necessary for the organic growth of systems. When there is no focused attention and skills to see the intention and business context of the architecture, these suboptimal decisions can lead to catastrophic disasters*. Following this line of thought, Brian Foster and Neal Ford coined the term **accidental architect**, at the *O'Reilly Software Architecture Conference*, as someone who takes architecture decisions without a formal architect title.

The *Winchester Mystery House* in San Jose, California, is a classic metaphor of an accidental architect in the real world. It is an architectural wonder and is known to be one of the world's most haunted houses. It was built by Sarah Winchester in 1886 as advised by a medium after she lost her daughter and husband. The advice from the medium was to build the house to home all the spirits that haunted her. The mansion was built with no architecture, no vision, and no blueprint. The mystery house has 160 rooms, 2,000 doors, 10,000 windows, 47 stairways, 13 bathrooms, and 6 kitchens with one of the most abnormal constructions. Staircases span across multiple levels before ending without exit, doors face walls, and pathways lead to dead ends.

New generation developers easily get inspired by new technologies. Working with new technologies keeps them motivated. Often, developers adopt these shiny new technologies without looking holistically using systems thinking. Even though some of these technologies can bring immediate benefits, projects may suffer from long-term cost, quality, and maintenance challenges. Developers lose focus under pressure, and they shift their passion as they find new attractive technologies.

As shown in the following diagram, at a certain point, complexity goes beyond what developers can manage:

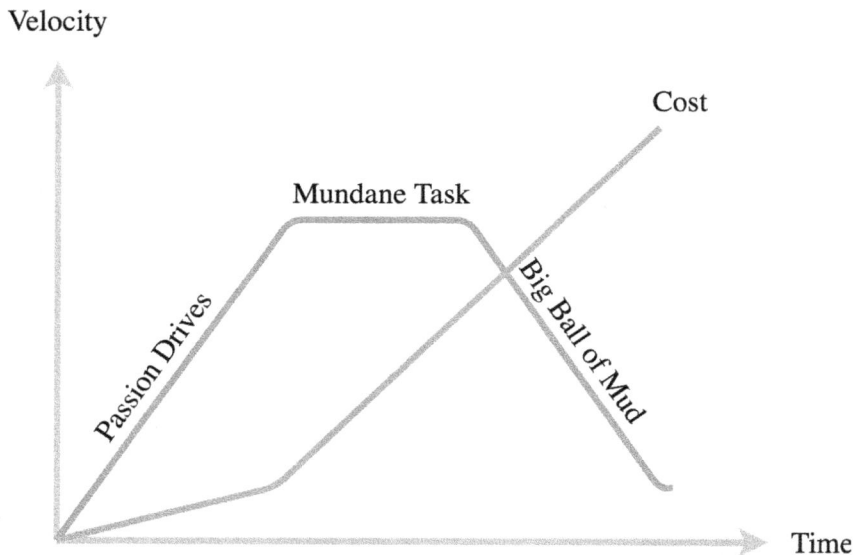

Figure 3.4 – Impact on velocity with no architect

As shown, over a period, simple maintenance may turn into days and weeks, and deployments become nightmares. This maintenance challenge resonates with Brian Foote and Joseph Yoder, documented as the *Big Ball of Mud* pattern. They observed that, *despite the best intentions and efforts of developers, a lot of software turns into a Big Ball of Mud.*

Neal Ford, Rebecca Parsons, and Patrick Kua mentioned in their book *Building Evolutionary Architectures: Support Constant Change*, quoting Rich Hickey, the creator of Clojure, that *developers understand benefits but fail to consider trade-offs*. Neal Ford and the team added to that, stating *architects must understand both benefits and trade-offs to build better engineering practices*. Yegor Bugayenko, founder and CEO of Zerocracy, mentioned in his book *Code Head volume 1* that *teams can effectively resolve technical conflicts if there is an explicit architect role in the team*.

These statements echo that the architect's role in a team is an absolute necessity, not a nice-to-have. We will establish this further in the next section.

Architects are vital in Agile – the question is who plays that role?

The survey published by *6point6.com* in 2017 revealed the importance of architects in Agile projects. As per the survey, 68% of CIOs agreed teams require more architects when Agile development is used at scale. CIOs reinforced that strategy, architecture, and oversight play an important role in the success of enterprise-scale Agile delivery, and therefore architects are essential.

Raymond Slot's Ph.D. thesis *A method for valuing architecture-based business transformation and measuring the value of solutions architecture* underlined the importance of solution architects.

Slot observed the following when architects are assigned to projects:

- A reduction of 19% in project budget overrun and a 40% reduction in time overruns.
- Due to better predictability, budget overruns reduced from 38% to 13%.
- Customer satisfaction increased from 0.5 to 1 on a scale of 1 to 5.
- Success increased by 10% with an improvement in overall technical fit.

Slot's thesis demonstrated substantial positive effects on projects when solution architects are in place. Architecture is often one of the most critical risks in project delivery. Therefore, special attention from architects is immensely important to avoid delay, rework, and failures.

Organizing Agile architects – none, volunteered, designated, or dedicated

Collective ownership drives better solutions as it encourages ideas from everyone. But this is possible only by collaboratively making decisions with members of the team. Successful collaboration only occurs within a cohesive boundary. Therefore, Agile architects must stay within those cohesive team boundaries.

Depending on the context, architects may be differently organized in different teams, as illustrated in the following diagram:

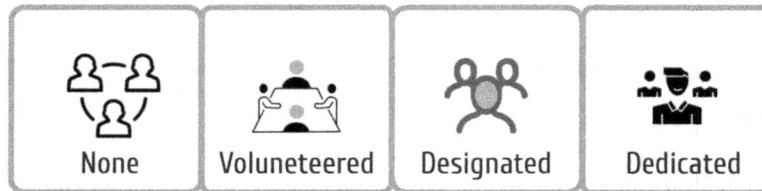

Figure 3.5 – Different ways of organizing architects in teams

In the **None** scenario, there is no architect in a team. As a result, the team focuses on delivering code based on assigned stories with almost no clear thinking about architecture and design. Decisions are taken on the fly by individuals in isolation based on their knowledge and perception. Depending on the individual's capabilities, a high risk of failure may surface in terms of quality and sustainability. Such solutions may lead to accidental architecture and the Big Ball of Mud pattern.

In the **Volunteered** scenario, there is no specific architect in the team. Rather, anyone in the team may play the lead role in architecture and design decisions. Since there is no one person, role switching occurs quite frequently. This approach is very efficient for less complicated and smaller-sized projects such as startup incubation, MVPs, and so on. The success of this model is attributed to the team's maturity and capability.

A **Designated** architect is a part-timer, mostly a senior developer who plays an architect role in the team in addition to regular day-to-day development activities. The designated architect offers technology leadership and also engages everyone in the team cohesively to deliver high-quality products. The success of this model depends on the capability of the developer in addition to complexity and scale.

A **Dedicated** architect is a seasoned architect grounded full-time in the team. Dedicated architects are technically sound, often exceptionally good at programming. Dedicated architects split their capacity by mostly spending time on driving architecture collaboratively with the team. At the same time, a small amount of capacity will be reserved for sharing some development pain, such as code optimization, debugging, automation, pattern development, and so on. This approach is a good fit for projects with higher complexity and scale where architects need more brain space to think beyond the work in progress and constraints of today and the required coordination between teams.

Different roles played by Agile architects

This book focuses on two critical roles of an Agile architect – enterprise architect and solution architect. In addition to these two recognized roles, architects may play other roles such as part-time developers, product owners, and external stakeholders such as specialist architects.

Various roles played by Agile architects are depicted in the following diagram:

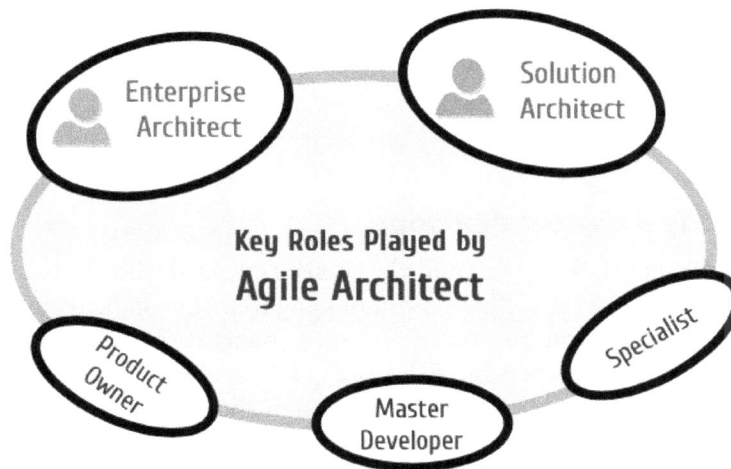

Figure 3.6 – Different roles played by Agile architects

The different roles are explained in the following sections.

Agile architect as an enterprise architect

Enterprise architects in Agile software delivery connect strategy to code by closely participating in development activities. Enterprise architects are responsible for helping teams shape up the solution vision by balancing business strategy with IT strategy. Enterprise architects act as active advisors to the business and IT leadership on strategy and investments. They also ensure the system landscape is healthy and evolves continually by appropriately modernizing, rationalizing, and decommissioning systems.

Compared to the traditional world in Agile software delivery, enterprise architects have different roles, responsibilities, and ways of working. We will further explore the enterprise architect role in *Chapter 4, Agile Enterprise Architect – Connecting Strategy to Code.*

Agile architect as a solution architect

Solution architect is one of the most critical roles in Agile software development, primarily responsible for supporting the incremental evolution of sustainable solutions with minimal redesign costs. It is the responsibility of solution architects to ensure the quality of the software delivered to customers is of the highest standard. Solution architects foster technical skills and also act as the glue within and across teams.

We will discuss more about solution architects in *Chapter 5, Agile Solution Architect – Designing Continuously Evolving Systems.*

Agile architect as a master developer

In Agile software delivery, architects should always keep engaged with the team throughout the development cycle. This continuous engagement will help architects to understand the challenges on the ground and adjust architecture and design accordingly. Architects are not regular developers but can write quality code when needed. They frequently help developers to debug and optimize design and code. Architects need to keep a portion of their capacity to look after current and near-term architecture needs, and therefore architects limit code-level activities.

Agile architect as a product owner

The **product owner** role is specialized, and it needs attention and is important. However, in some cases, architects need to play the product owner role. Architects playing the product owner role are mainly required at the beginning of an initiative, such as the sprint 0 or inception phase, where teams spend time on ideating architecture, design, and technology for the solution. It is a common scenario for architects to play the product owner role until the solution is shaped up and ready for development.

In some cases, if the team is developing technical products, platforms, or technical components, solution architects act as the technical product owner for those teams.

Agile architect as a Scrum stakeholder

Specialist architects such as security architects, business architects, information architects, and so on may stay outside of the team as day-to-day engagement with the team is not necessarily required. Moreover, it is not economically viable to have specialists in every team when implementing Agile development at scale. In such cases, specialist architects act as a consultant to the team as a Scrum stakeholder. Special planning is required to engage these architects as they must not delay development teams' velocity.

Essential skills for Agile architects to be relevant

The ultimate responsibility of Agile architects is to do the right thing in a way that delivers significant value to the business and its customers.

To accomplish this goal, Agile architects need to have four critical skills, which are captured in the following diagram:

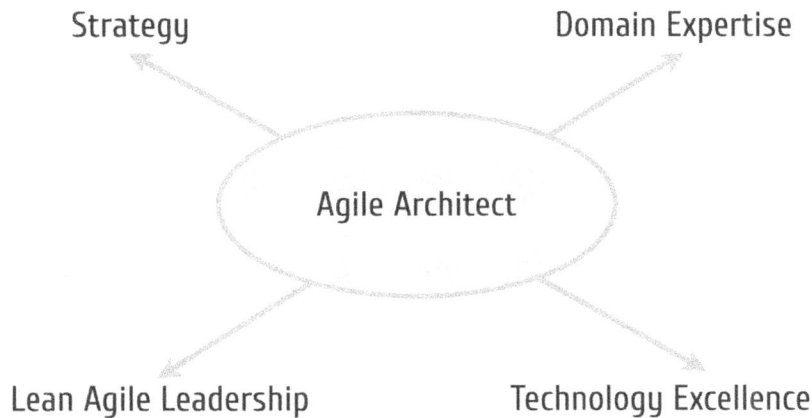

Figure 3.7 – Different skills of an Agile architect

As depicted in the preceding diagram, four skills that define architects' success are the ability to work at the strategic and organizational leadership level, expertise in the business domain, lean-Agile leadership, and technical excellence. The following list explains this better:

- **Domain Expertise**: Agile architects with deep domain expertise can have meaningful conversations with a business to clearly understand its purpose, vision, and objectives. It helps Agile architects to align and blend business strategies with IT strategies with a sense of pragmatism. It also helps Agile architects to understand the business value and business impact of architecturally significant requirements and quality attributes.

- **Technical Excellence**: Agile architects have to be at the forefront of technologies. Technical craftsmanship helps Agile architects in bringing out realizable innovative ideas to solve burning business issues. In addition to that, technical excellence helps to build cost-effective, sustainable technical solutions that fit the purpose of business. It also allows architects to earn respect among the developer community. We will discuss more of this in later chapters.

- **Lean Agile Leadership**: Most organizations see architects as trusted advisors and leaders at the highest level of the organizational decision process. An architect is someone who can speak the business language, IT leadership language, and the language of the developer community. We will discuss this in *Chapter 13, Culture and Leadership Traits*.

- **Strategy**: Agile architects should have broader knowledge than core technical concepts. Architects should handle business strategy and IT strategy, its implications, and investment decisions with a holistic, big picture view.

Most traditional skills are no longer valid and have to be unlearned. Architects need to ruthlessly prioritize activities to support the continuous flow of values to deliver long-lasting solutions. To enable the flow of work, architects need to continually focus on many areas such as delivering business value over architecture diagrams, technical agility over rigid designs, balancing sustainability with quality attributes, communication over documentation, and alignment over centralized decision making. We will cover these topics in later chapters.

So far, we have learned the importance of architects in Agile software delivery. Now we are in an excellent position to explore the duties and behavioral changes of Agile architects.

Explaining the Agile architect's behavior and duties with metaphors

In many organizations, such as *Spotify*, culture is one of the biggest differentiators. The key change in culture is the shift from ownership to selfless collaboration and sharing. The collaborative mindset is deeply connected to the new ways of working of architects in Agile software development. Culture, behavior, and workstyles determine the success of an individual as well as the outcomes delivered.

The changing behavior of the Agile architect

Agile architects continuously collaborate with all stakeholders, with long-term success in mind, to support forward engineering plans, decision making, and alignment to an architecture vision. A collaborative culture and mindset foster better communication, knowledge sharing, trust, and respect among team members and architects.

Almost all Agile software development methodologies strongly advocate self-organized teams with creative ways of collaborating over institutionalized boundaries between many roles. Since no two teams are identical, even within the boundaries of a single organization, the need for collaboration and guardrails around self-organizing must evolve from within teams. Unified teams confidently deliver faster results with higher levels of quality. John Kotter, professor of leadership, in his book *Leading Change*, called this a *guiding coalition*. He observed, *a guiding coalition operating as an effective team can process information far more quickly*.

The following diagram shows a spectrum of different approaches architects employ to define solutions:

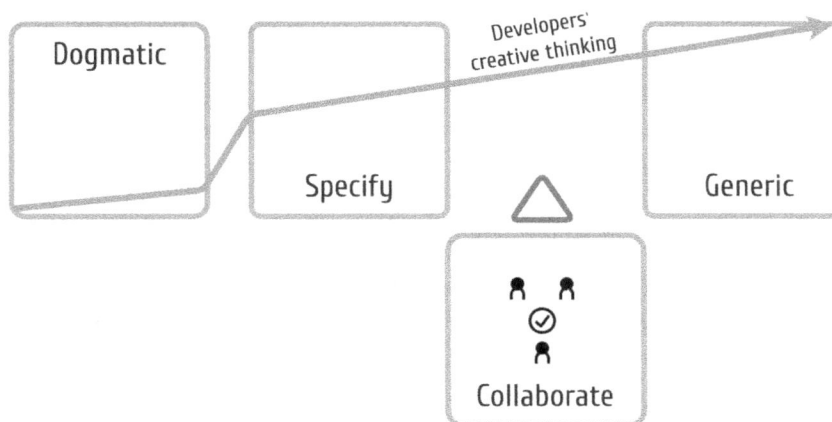

Figure 3.8 – Different approaches to a solution definition

Architects are **Dogmatic** in the traditional world, creating designs with no room for developers to apply their creative thinking, almost like a construction architect. Then comes the **Specify** approach, as shown in the preceding diagram, where designs are delivered as a set of specifications, including API contracts, design patterns, a technology stack, architecture blueprints, and so on. While this offers more freedom for developers than the dogmatic model, this still limits developers' creativity as they always need to stick to the given specification. Since, in most cases, architects only share specifications, not the underlying purpose, developers lose opportunities to apply their intellectual abilities to explore better alternate solutions. The other extreme of **Dogmatic** is that architects remain too **Generic** by only setting standards and guidelines, giving developers full freedom to innovate. Staying generic may lead to accidental architecture, as we discussed earlier.

Agile architecture needs another innovative model for architecture development, which is missing from this spectrum. In a highly collaborative approach, teams can appropriately adjust solutions through direct communications, mutual understandings, and continuous feedback. Frequent collaboration balances architecture decisions by valuing engineers' voices and applying the expertise of architects.

To ensure higher cohesion and collaboration between architects and developers, the architect must be part of the team as just another member, not an elite or superior person. Architects need to demonstrate egoless collaboration with a serving and enabling mindset. *The Mythical Man-Month*, authored by Frederick Brooks back in 1975, has some great suggestions on how architects must change their behaviors to achieve the best outcomes. Brooks reminds us that *the developer has the creative responsibility for implementation. Hence the architect should only suggest solutions, not enforce decisions.* Fred goes on to ask architects to *be prepared to accept counter approaches, which are equally good and give credit for such improvement suggestions.*

In summary, architects must learn the art of giving away control by strictly moving away from the traditional command and control leadership style. We will discuss more on this topic in *Chapter 13, Culture and Leadership Traits*. A new set of metaphors, which we'll learn in the next few sections, are helpful to understand the collaborative behavior of Agile architects.

Gardener – continuously nurture and grow

Agile architecture is a continuous activity. The book *The Pragmatic Programmer: From Journeyman to Master*, authored by Andrew Hunt and David Thomas, brought in the gardener metaphor to represent an architect as an alternative to a construction architect.

The authors expressed their view that architecture development is like gardening, which is a continuous process and is organic. The garden needs to have an initial plan – the intent or master plan. The gardener starts with that plan, but some parts of the plan may not work, which triggers replanting and rearrangement. Some plants may need to be taken out as they are no longer relevant or attractive. Overgrown plants need to be cut or pruned. Weeds need to be removed and fertilizers added to those plants that need special care. Small amounts of daily care and nurturing are required to maintain the garden in good shape and condition.

The gardener has the ingrained responsibility of taking care of plants by working in the garden every day as a caretaker. Like gardeners, architects are responsible for nurturing and maintaining the health of systems through continuous refinement.

Interior designer – listens to customers and glues ideas together

Interior designer is another good metaphor for understanding the new behavior of Agile architects. An interior designer always demonstrates collective and collaborative behavior between the designer and the customer.

The worst behavior an interior designer can exhibit is a stubborn character by forcing their ideas on the customer. A good interior designer is a people-focused person, an excellent listener, and a good communicator. Interior designers always listen to the customer's needs and their purpose. They then assess the environment to check the viability and further enhance the customer's ideas by applying their subject matter expertise before sharing what is possible. The interior designer continuously adjusts solution thinking as the customer shares new ideas. These adjustments are typically made over a series of collaborative and open conversations. When something is technically impossible, the interior designer will openly express such challenges to avoid reputation damage.

Kathy Kuo, an interior designer, observed that in most cases, interior designers work with customers and listen to their ideas. Sometimes customers may have their opinions based on what they have seen somewhere. The interior designer's job is to keep an open and positive mindset to get a good grasp of those ideas and play back to customers so they feel comfortable by stitching good ones together with a flair of design and confidence. Most customers, probably, will straight away accept those solutions because it came from them.

The interior designer's approach completely resonates with what architects have to do in Agile software developments. In many cases, smart developers may have bright ideas. Architects just need to connect dots, look at holistic alignment, make minor adjustments through feedback, and, most importantly, encourage them with a sense of confidence.

Elevator – champions in communicating top to bottom

The elevator architect metaphor presents the holistically connected nature of Agile architects. The elevator metaphor comes from the book *The Software Architect Elevator*, authored by Gregor Hohpe. Gregor compared the enterprise with a large, multi-floor building where the topmost floor is the business – the penthouse, and the bottom floor is where the developers sit – the engine room. There are many floors in between the business in the penthouse and the developers in the engine room. These floors are filled with many management staff. The challenge with such an organization is the disconnect between business and developers as messages are wrongly interpreted and sometimes even corrupted as they pass through different floors.

Gregor then positions architects as the connecting mechanism, the elevator that rides between the top floor and the bottom floor. The elevator architect runs up and down and passes messages to different floors without losing their essence.

Architects are well-positioned to do this job as they can very well translate business strategy to IT strategy and code the strategy together with developers. Potentially, architects are the only ones who speak both languages very well – strategy and code.

Sous chef — a leader on the floor but hands-on too

The sous chef metaphor represents architects as the best-in-class programmers who can code and offer technical leadership.

In large restaurants, the sous chef is the second in command in the kitchen. The executive chef is responsible for the overall restaurant but is usually disconnected from the kitchen floor and focuses more on management. The sous chef is hands-on and manages almost all operations in the kitchen. These include handling recipes, directing food presentations, keeping the kitchen in order, training new chefs, and, more importantly, ensuring the food that goes to the customers is of the best quality standards. They are also responsible for ensuring the kitchen equipment is in proper working condition. The sous chef may even fill in for the executive chef occasionally.

Like a sous chef, an Agile architect is an expert in the field, leads technical activities, is conscious of the quality of outcomes, is hands-on, and can code. The architect, like the sous chef, has a full view of the system landscape, current and near-term evolution, and often has attention spans across multiple teams.

The challenging duties of the Agile architect

Donald Rumsfeld, former US Secretary of Defense, coined the term *known unknowns and unknown unknowns*. Known unknowns are risks related to things that you know, and you can plan for, such as the risk of using new technology for a purpose. The purpose might change, or the technology may not perform as intended. An unknown unknown is something that you don't even know because it never existed before, such as the COVID-19 pandemic that has changed the world. In Agile software development, an architect has to deal with not only the known knowns but also the known unknowns and unknown unknowns.

The following diagram captures various scenarios that Rumsfeld pointed out in his known unknown model:

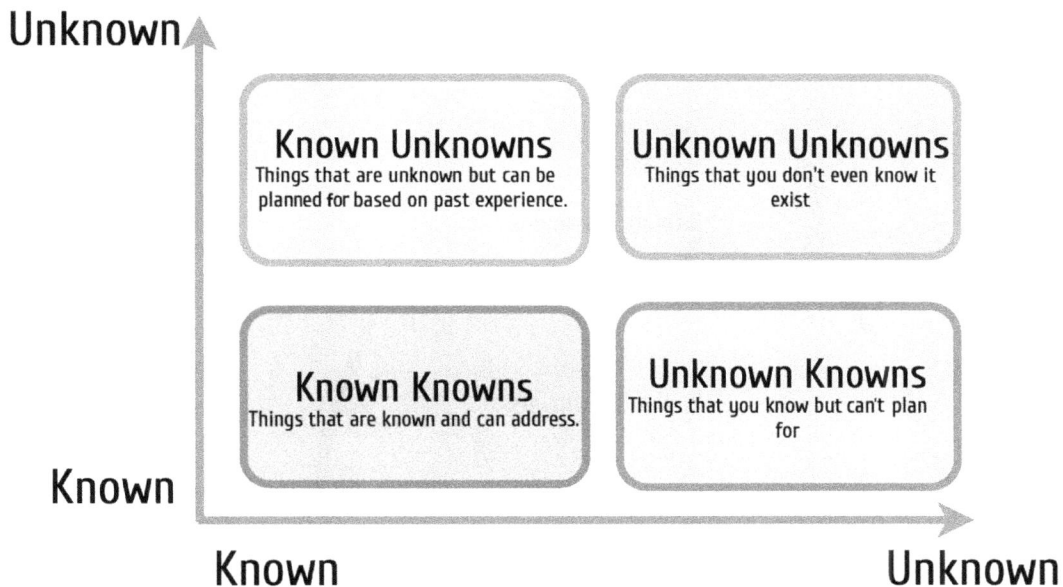

Figure 3.9 – The known unknown model

Unknown Knowns are due to a lack of skills, capabilities, or foresight. **Known Knowns** are handled adequately using the available data and facts. **Known Unknowns** are things that are thought about but not implemented today based on anticipation, such as planning for incremental scaling. **Unknown Unknowns** are really hard to visualize and are generally handled by building evolvable architectures. Evolvability may not always guarantee that the system can handle unknown unknowns, but it can reduce the surface area of impact.

In Agile development, known unknowns and unknown unknowns make the architect's life more tedious and challenging. A number of metaphors are used in this chapter to explain different perspectives on the demanding duties of Agile architects.

The iceberg — unknowns are more than known

Architects in Agile software development need vision, technical excellence, and a flexible mindset to think beyond what is known today. Architects need to be trained and prepared to respond swiftly to demand as they discover new knowledge.

The following diagram is an illustration of an architect's challenges with the help of an iceberg:

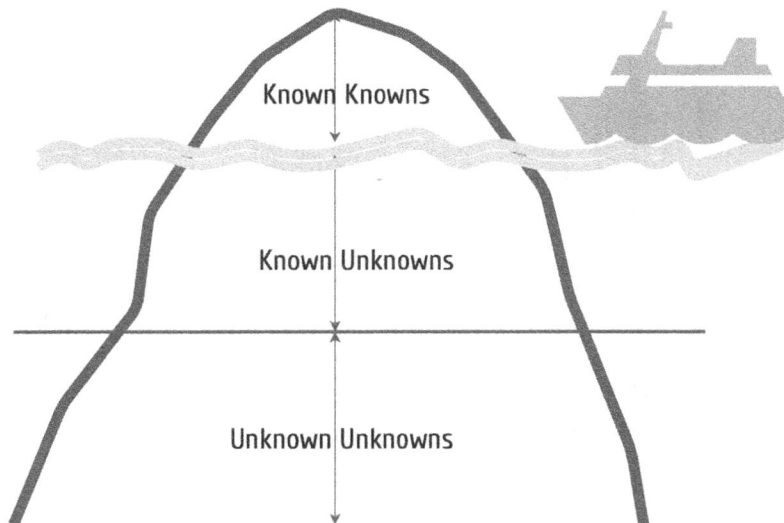

Figure 3.10 – Iceberg model for visualizing unknowns

It is similar to the iceberg shown in the diagram, where only 20% of the iceberg is visible and the rest you may have to assume. Measurement of the iceberg is only possible based on the size and shape of the visible part of the iceberg. You may be able to visualize a smaller portion of an iceberg underwater, but a larger portion deep under the sea cannot be imagined. The sizing of the iceberg is extremely complicated as it varies with the formation.

Agile architecture promotes architecture evolution in terms of value delivery proportionate to changes in customer needs. You may have only around 20% of known requirements and 80% of assumptions at the beginning. The assumptions will be eliminated as clarity is gained on customer needs, as shown in the following graph:

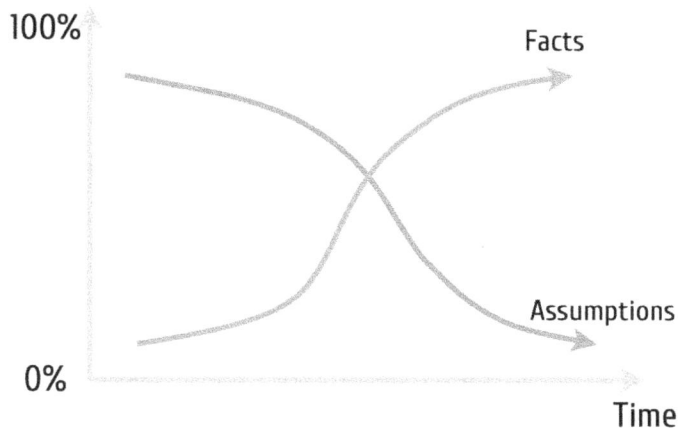

Figure 3.11 – An Agile architect's clear assumptions over a period of time

This infinite scope in Agile solution development, mostly dealing with unknown requirements, poses enormous challenges for Agile architects. Incredibly in-depth knowledge in the domain, mastery of technical agility, and extensive market intelligence is exceptionally critical to foresee the unknowns. Besides, developing a genuinely evolvable architecture that reduces the blast radius is a vital aspect to consider. Lastly, using techniques such as hypothesis-based solution development and delaying decisions to the last responsible moment are useful techniques in Agile software development.

The horse and buggy – always the horse ahead of the buggy

Architects in Agile software development have an inherent responsibility not to submerge in the urgency of now. Architects always need to look ahead and spearhead in planning essential changes in architecture and design before they hit and potentially slow down the team's velocity.

In the horse and buggy scenario, the horse always runs ahead of the cart, pulling it forward in the right direction. The horse runs a few steps ahead of the cart, still maintaining a rhythm between the horse and the buggy. Since the horses and carriage are well-connected, one cannot move faster than the other.

The following diagram captures this scenario with a mapping to Agile teams and various roles:

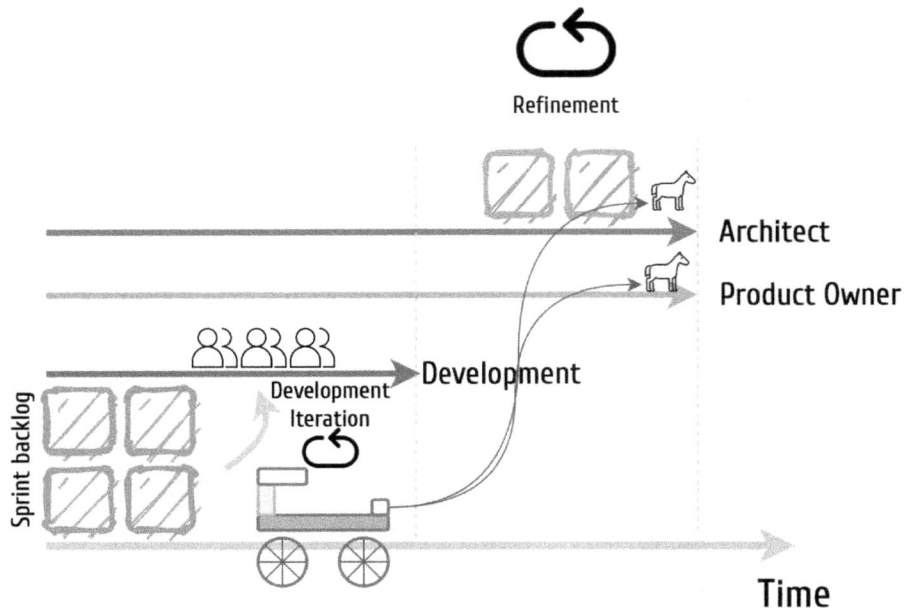

Figure 3.12 – The horse and buggy model

In Agile development, the **Product Owner** and architecture are analogies for the two running horses. These two roles run a little ahead and spend their energy on shaping up the product, based on the vision and roadmap. The product owner and architect work hand in hand, ensuring the product is well rounded with functional and non-functional requirements.

The camera – decide at the last possible moment

The third useful metaphor to understand an architect's challenges is a camera's exposure control settings, typically seen in SLR cameras.

The following exposure triangle captures the three key exposure settings:

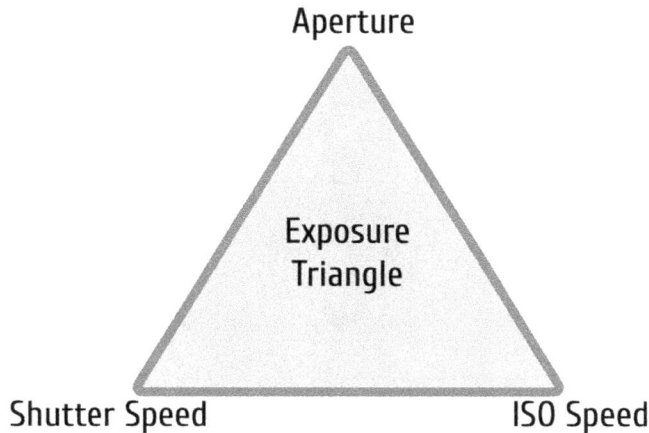

Figure 3.13 – Exposure triangle

A photograph's quality is determined by the exposure triangle – **Aperture**, **Shutter Speed**, and **ISO Speed**, as shown in the diagram. These three parameters together adjust natural light to optimal exposure levels for a better photograph. A professional photographer sets these values just in time before clicking. Since natural light levels change continuously, setting these values much in advance may result in the wrong exposure levels. Taking a picture is even more challenging when you are taking a flying subject, such as an aircraft. You will get the best images when you set exposure levels at the last possible moment. Delaying for too long may not give you sufficient time to adjust the exposure levels perfectly.

It is key for architects to determine the last responsible moment for architecture decisions to be made in Agile software development. Making decisions too early may result in the wrong decisions due to the non-availability of information and, therefore, the high cost of a rework. Making decisions too late may result in delay costs. An architect needs to balance the cost of a rework and delay costs to identify the **Last Responsible Moment** (**LRM**).

Air traffic control – continuously prioritize inflow

An airport with a single runway operation cannot allow more than one aircraft to land or take off within a short span of time. **Air Traffic Control** (**ATC**) uses a *holding pattern* to control the flow of incoming aircraft.

The following diagram captures a typical aircraft holding pattern:

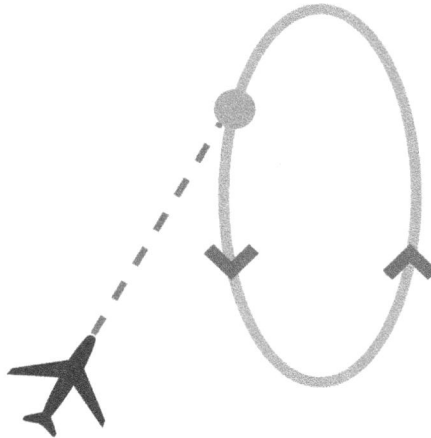

Figure 3.14 – Aircraft holding pattern

An aircraft holding pattern is an approach used for sequencing landing operations. Even if an aircraft is in an airport's vicinity, ATC may not give immediate landing permission, primarily due to air traffic congestion. Airplanes waiting for permission to land may go into a holding pattern, a vertical stack in which every aircraft is vertically separated. The first aircraft entered into the holding pattern will be at the bottom of the stack, and new ones will go to the top of the stack. Generally, airplanes from the bottom of the stack will be allowed to land first. While this is usually the standard pattern, aircraft that need to land urgently will be allowed to bypass other holding airplanes. ATC is authorized to prioritize aircraft based on a number of other parameters as well.

In Agile software development, the architect, together with the product owner, continuously prioritize backlog items for development. Structured methods and practices are used to prioritize both functional and technical backlog items.

The periscope – see things that cannot be seen otherwise

As per the *Oxford Advanced Learners Dictionary*, a periscope is *a special apparatus with a tube and a set of mirrors used typically in a submerged submarine or behind a high obstacle to see things that are otherwise out of sight.*

This scenario is illustrated with the help of the following diagram:

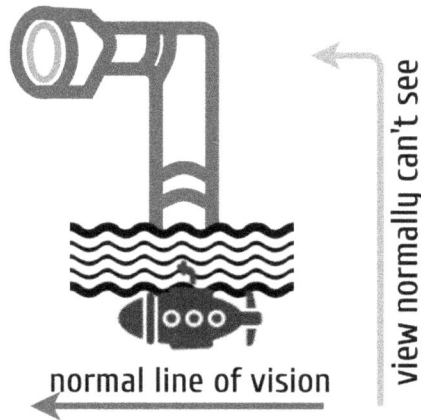

Figure 3.15 – Periscope to see things you usually can't see

Often, architects see things through a different lens than developers. While developers view things from a coding and delivery perspective, focusing on solving their immediate challenges, architects use a different mindset, hovering around business purpose, value, trade-offs, and quality with a long-term and near-term view.

Architects use a system thinking approach to understand the holistic view of the system landscape instead of fragmented solution deployments. System thinking allows architects to occasionally make small adjustments to certain components with the notion that system architecture will achieve eventual integrity over time.

With this clear understanding of how architects' behaviors and duties are changing in Agile software development, let's review how different Agile scaling frameworks handle architect roles.

How different frameworks handle architect roles

There is no architect role in Agile software development methodologies such as Scrum. These frameworks fundamentally reiterate the principle of the Agile Manifesto where the best architectures emerge from self-organizing teams. However, when product development scales across multiple teams, the architect role becomes essential for technical leadership and alignment. Therefore, most of the Agile scaling frameworks explicitly reference Agile architect roles.

Scaled Agile Framework

Scaled Agile Framework (SAFe) treats architecture as one of the major disciplines. There are three architect roles defined in SAFe – enterprise architect, solution architect, and system architect. However, architects are positioned outside teams, which is a significant deviation from the principles of the manifesto for Agile software development.

Enterprise architects operate on the portfolio, responsible for shaping up the technology strategy by aligning strategic themes from the business, assisting with guardrails for lean budgeting, and working with the solution and system architects to produce fit-for-purpose and economically sensible business solutions. Enterprise architects across portfolios frequently connect to share knowledge about portfolio backlogs to gain values from cross-portfolio roadmaps. Enterprise architects are supported by other disciplines of architecture such as business, information, and security architects and may form a team of architects to manage larger portfolios. Enterprise architects keep a constant cohesive link with solution architects and other ART members.

In SAFe, solution architects are placed in large solution trains, responsible for solution alignment and coordination between various ARTs. The troika of the solution train is responsible for the satisfactory delivery of large solutions. The solution management is responsible for the content, the solution architect is responsible for the structure and behavior of the solution, and the solution train engineer is responsible for managing processes and time.

The troika formation repeats at the program or essential level, where solution architects are replaced with system architects. Solution architects and system architects play very similar roles but have different scopes. Large ARTs with many heterogeneous systems and complex technology landscapes may have many system architects.

Disciplined Agile

Disciplined Agile (DA) has a strong emphasis on architects. In line with other Agile development frameworks, DA recommends having a designated architect role be played by a senior developer or one of the most technically experienced developers in the team.

For smaller teams, everyone in the team is responsible for the architecture. In order to address a conflict of opinions and manage consensus among team members, DA suggests a role called **architecture owner**. The architecture owner facilitates agreements and owns architecture decisions. Generally, the designated architect, also termed an Agile solution architect, acts as the architecture owner. The architecture owner does not create architecture in isolation. Instead, the architecture owner facilitates and collaborates with the team to jointly reach architecture decisions and perform architecture modeling. When the team is not aligned, the architecture owner will have the final authority to make architecture decisions. DA also emphasizes the importance of having a good domain understanding for better decision making.

When scaling DA across teams of teams, DA suggests having a virtual coordinating body for architects across teams called architecture ownership or a leadership team, where members are architecture owners of individual teams. This team of architects is generally led by a chief architecture owner. DA also considers enterprise architects as a key participant in the inception, construction, and maintenance of the process goals.

Large-Scale Scrum

Large-Scale Scrum (**LeSS**) does not explicitly position the architect role, encouraging teams to take all decisions. There may be an architect programmer or master programmer person in the team, typically a senior member who may lead design activities.

Here are two interesting observations from LeSS about an architect:

- A software architect who is not in touch with the evolving source code of the product is out of touch with reality.
- Every programmer is some kind of architect—whether they want to be or not. Every act of programming is some kind of architectural act—good or bad, small or large, intended or not.

LeSS observes that architects staying at a high level, like architecture astronauts or PowerPoint architects, is not healthy for good sustainable designs. LeSS focuses more on design than architecture and promotes a no-architect culture described as follows:

> *"Agile architecture comes from the behavior of Agile architecting—hands-on master-programmer architects, a culture of excellence in code, an emphasis on pair-programming coaching for high-quality code/design, Agile modeling design workshops, test-driven development and refactoring, and other hands-on-the-code behaviors."*

We have learned how different frameworks handle the architect role. Before we close this chapter, let's explore some of the learnings from Snow in the Desert.

Learnings from Snow in the Desert

Snow in the Desert started their Agile transformation journey with a textbook adoption of SAFe. At Snow in the Desert, behavior and culture were the most significant barriers to Agile delivery. As per SAFe, the system architect was positioned at the ART leadership level alongside product management and the release train engineer. Although architects are the best programmers, the behavior of individuals and memories of past organization developed friction between Agile architects and the rest of the development teams. This wall of distinction diminished collaboration, causing teams to operate with no shared responsibilities and no alignment with the solution vision.

To overcome this challenge, Snow in the Desert course-corrected their team constructs. The following diagram shows the next evolution of Snow in the Desert:

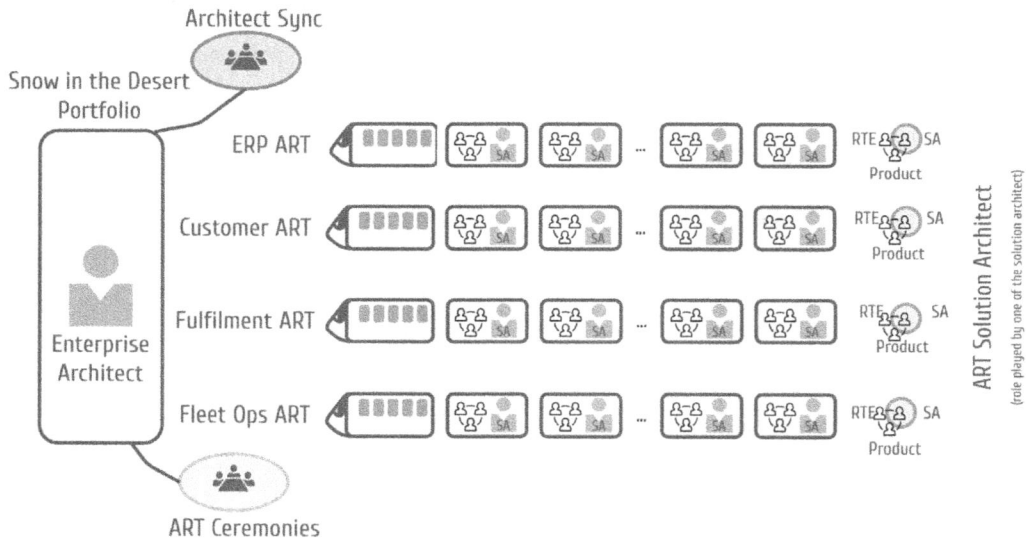

Figure 3.16 – Architect assignments at Snow in the Desert

As shown in the diagram, the system architects were moved from ART leadership to teams. With this change, Snow in the Desert aligned with almost every other Agile development framework by embedding architects in every team. These dedicated architects, mostly master developers, shared some part of the development responsibilities as a member of the team. However, they still reserved a portion of the capacity for architecture activities. These architects provided technical leadership within the team and connected with their peer network across teams for solution alignment and knowledge sharing.

The architect role at the ART leadership level was filled by one of the most capable system architects from one of the teams in that ART, turning it into a virtual role. To reduce confusion and limit the number of organizational roles, at Snow in the Desert, these system architects were called solution architects. Remember, there is no solution train at Snow in the Desert, hence there is no other solution architect role.

Since all four business units are funded and governed together, there is only one SAFe portfolio at Snow in the Desert. The enterprise architect at the portfolio level covers the whole portfolio. The enterprise architect is well connected with all the ARTs by attending ART ceremonies and architect sync meetings. The enterprise architect uses these forums to enlighten and align teams on strategic directions, synchronize architecture runway strategies across ARTs, and collect feedback from teams.

Summary

The playing field for architects has been challenged severely due to the mismatch between traditional ways of architecting and Agile architecture. Elements of the ivory tower, astronauts, and policing behaviors compounded with perceptions and economic considerations added fuel to the no-architect movement that surfaced with Agile software development. As we learned, it is vitally important to have dedicated or designated architects in Agile teams.

While collaboration is an extreme necessity, Agile architects must be substantially and equally strong in technology, domain, leadership, and strategy. The conventional metaphors of a construction architect to a gardener, elevator, sous chef, and interior designer need a radical transformation. The ability to know the unknowns, be prepared to adapt, pave the way for developers, prioritize the technical backlog, timing architecture decisions, and have a solution mindset with system thinking are paramount for Agile architects.

Enterprise architect and solution architect are the pertinent architect roles in Agile software development. In the next chapter, we will cover an enterprise architect's roles and responsibilities in Agile software development.

Further reading

- **Software Architect**: https://en.wikipedia.org/wiki/Software_architect
- **IASA Five Pillars of Architect**: https://itabok.iasaglobal.org/itabok/capability-descriptions/

- **SFIA Skills Framework**: https://sfia-online.org/en/sfia-7/all-skills-a-z

- **Don't Let Architecture Astronauts Scare You by Joel Spolsky**: https://www.joelonsoftware.com/2001/04/21/dont-let-architecture-astronauts-scare-you/

- **Accidental Architecture by Grady Booch**: http://www.inf.ed.ac.uk/teaching/courses/seoc/2006_2007/resources/Arc_Accidental.pdf

- **Becoming an accidental architect by Brian Foster and Neal Ford**: https://www.oreilly.com/radar/becoming-an-accidental-architect/

- **Importance of Architects Survey**: https://6point6.co.uk/insights/an-agile-agenda/

- **Slot, Raymond. (2010). A method for valuing architecture-based business transformation and measuring the value of solutions architecture**: https://www.researchgate.net/publication/254874295_A_method_for_valuing_architecture-based_business_transformation_and_measuring_the_value_of_solutions_architecture

- **Becoming a Better Interior Designer**: https://www.kathykuohome.com/blog/how-to-handle-clients-from-tricky-questions-to-stubborn-personalities/

- **Architect Role in Agile Frameworks**: https://vmmatthes44.in.tum.de/file/1krmqwaddo51/Sebis-Public-Website/-/Investigating-the-Role-of-Architects-in-Scaling-Agile-Frameworks/Investigating%20the%20Role%20of%20Architects%20in%20Scaling%20Agile%20Frameworks.pdf

4

Agile Enterprise Architect – Connecting Strategy to Code

Go and see for yourself to thoroughly understand the situation
(Genchi Genbutsu)

– Toyota Principle 12, The Toyota Way

In the previous chapter, we discussed the importance of an Agile architect by exploring their duties and challenges with a series of metaphors. While Agile architects play many different roles, enterprise architects and solution architects are the most important roles in enterprise-scale Agile software delivery.

The Fourth Industrial Revolution underpins technology as an enabler for businesses to withstand the storm of volatile environments and market instabilities. Rigorous strategic planning and accelerated adoption of lean practices help organizations to achieve the desired state of agility. Enterprise architects perform a significant role in fundamentally enabling business agility through rapid cycles of technology innovations. However, traditional **Enterprise Architecture** (**EA**) methods are profoundly inadequate for digitally transforming businesses driving shorter and flexible demand response cycles. A new set of principles anchored around lean-Agile practices, data-driven decisions, value-driven delivery, and evolutionary collaboration intertwined with the organization's Agile software delivery processes are key to an enterprise architect's success. Enterprise architects are fully connected and work harmoniously with different stakeholders in transforming business ideas into substantial and quantifiable business values. They impartially balance business and IT strategies and relentlessly maintain a well-architected portfolio. Agile EA is still a maturing practice and will surely be the next giant leap in Agile software development.

This chapter will examine how to embrace agility in EA practice with a strong focus on delivering measurable outcomes. We will also introduce and discuss a few radically differentiating principles and duties of Agile enterprise architects. An EA repository is one of the critical pieces of the puzzle in delivering high-value data and insight-driven decisions. Before we close this chapter with the learnings from Snow in the Desert, we will also explore how Agile scaling frameworks treat enterprise architects.

In this chapter, we're going to cover the following main topics:

- The need for lean-Agile enterprise architects
- Why Agile enterprise architects are key to success for business agility
- The EA repository – the gold dust problem
- Measuring value and the Agile maturity model
- How different scaling frameworks position enterprise architects
- Learnings from Snow in the Desert

This chapter focuses on the *Enterprise Architect* focal point of the Agile Architect Lens:

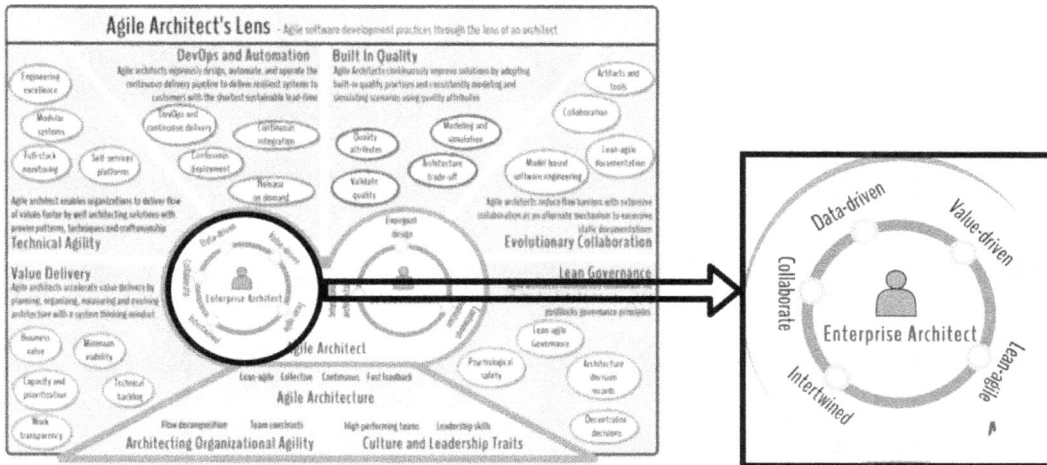

Figure 4.1 – Enterprise Architect – Focal point

Technical requirements

Additional materials related to this chapter are available at the following link for downloading:

```
https://github.com/PacktPublishing/Becoming-an-Agile-Software-
Architect/tree/master/Chapter4
```

The need for change in lean-Agile EA

Enterprise architects significantly contribute to architecting the future state of a business by leveraging the power of innovative modern technologies. IT needs a broader end-to-end vision, strategic alignment, and improved business engagement to enable business agility. Besides, IT needs to adopt lean-Agile software delivery processes and practices to respond rapidly and flexibly to customer demands to continuously keep the business competitive. Agile enterprise architects perform an invaluable role in organizations embracing agility by balancing business and IT strategies by acting as a conduit for creating high cohesion between business and software development teams. The alignment between the business and IT is similar to the connecting bridge in the middle of the *Petronas Towers*.

The following diagram is an illustration of the bridge as well as the balancing act of Agile enterprise architects:

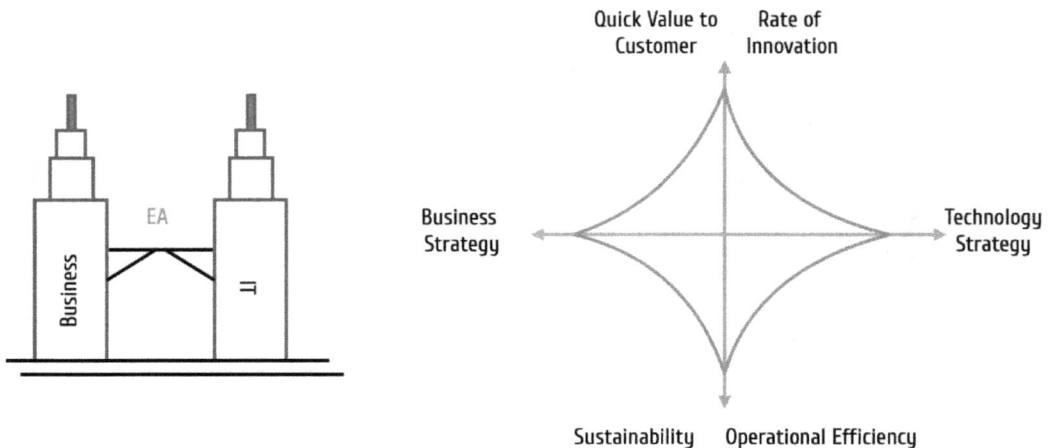

Figure 4.2 – The balancing act of enterprise architects

The challenge with EA in Agile software development is not with its core concepts and values. The implementation approach proposed by traditional EA frameworks in principle hugely differs from Agile software development practices and ways of working. This wide gap created antagonism between EA and Agile software development. The following sections explore a few elements of traditional EA approaches that are not coherent with Agile software development philosophies.

EA frameworks focus on describing enterprises

Brian Burke, research vice president at Gartner, observed that *focusing on a standard EA framework doesn't work*. Many EA frameworks carry a legacy mindset of software developments with long linear lead times disconnected from the continuous flow of values. Yet, they continue to market their services and tools.

Most of the EA frameworks were born before Agile software development practices gained popularity. As a result, most frameworks still use the languages, vocabularies, and practices of traditional enterprises. The evolution of EA is well documented in *The History of Enterprise Architecture: An Evidence-Based Review* by Svyatoslav Kotusev. Moreover, almost all EA frameworks define EA as a *description of the enterprise*, which implies massive documentation.

The starting position for many of the EA frameworks is to methodically document a top-down view of the enterprise, such as organization and business capabilities, and then drill down into more concrete architecture representations. This architecture representation repeats for *as-is*, *to-be*, and one or more transitional steps, as shown in the following diagram:

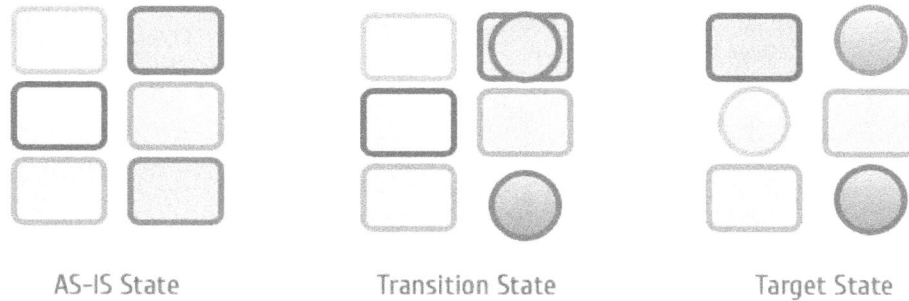

| AS-IS State | Transition State | Target State |

Figure 4.3 – Typical EA roadmap

Top-down documentation consumes a substantial amount of time to complete comprehensive architecture descriptions across the layers of an enterprise with no tangible outcome or business value. Since there is no visible value, enterprise stakeholders disengage from contributing to the EA program, eventually stalling the initiative halfway through. Also, by the time the architecture documentation reaches a reasonable stage for value creation, the data collected may become obsolete as Agile delivery cycles are significantly fast.

Even though, in reality, no two organizations are the same, many enterprises blindly adopt EA frameworks. James McGovern, research director at Gartner, observed *EA teams who use cut-and-paste methods tend to struggle with demonstrating their value because organization DNAs are different*. Using off-the-shelf EA frameworks force-fitted into Agile software development is one of the reasons many enterprises have failed to see the value of EA. Alternately, Gartner introduced the concept of *business outcome-driven EA* to shift the focus from framework adoption to tangible business value delivery.

The city planning metaphor is no longer valid

City planning is the most straightforward and traditionally used metaphor to explain EA. A master plan for a city consists of predefined zones for specific purposes such as schools, parks, residential buildings, and so on.

An illustration of a city plan is captured in the following diagram:

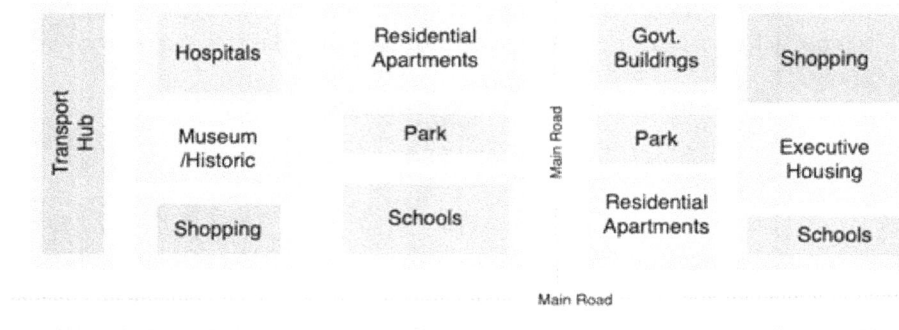

Figure 4.4 – City planning metaphor

As depicted in the diagram, the city is divided into zones for dedicated purposes. Roads and other pathways are designed upfront in a well-connected manner. Once the master plan is approved, buildings will be constructed incrementally according to a given specification that meets the master plan. In this metaphor, the master plan is compared to the EA strategic planning and roadmap, whereas individual building plans within zones are related to solution architecture.

In the previous chapters, we discussed that the construction architect is no longer the best metaphor for Agile architects. Extending that philosophy, the city planning metaphor is just another misleading comparison. City planning is a ground-up activity in most cases with a longer planning horizon. Moreover, demand and incremental development plans are always predictable in traditional software development but that is no longer the case in Agile software development. City planning generally follows a phase-gate approach where the planning team, authorized by government authorities, signs off any plans before construction. These processes are bureaucratic, with almost no exceptions, and are backed by laws.

This metaphor is perhaps well suited to the past, before the Fourth Industrial Revolution, where large enterprises were focused on long-term plans. However, in an increasingly volatile environment where organization strategies are unpredictable and swing around market signals, city planning is no longer a good metaphor for EA.

This does not mean EA in the new world refrains from strategic planning and roadmap activities. These are important value additions to large enterprises. EA planning at the zone level and giving autonomy to the individual zone developers with appropriate guardrails for developing their respective zones is still a viable EA scenario. It is just that the nature of city planning follows rigid plans, and committed detail designs and strict governance make it a less attractive metaphor in Agile software development.

EA operates without purpose

There won't be a single uniform answer to this question – *Why does EA exist in your organization?* EA practice needs a purpose and context. The reason for the existence of EA is solely to meet whatever purpose there may be. The purpose differs from enterprise to enterprise based on the near-term and long-term focus, such as strategies, operating constructs, financial constraints, and so on.

Many organizations shoehorn enterprise architects into the organization structure and process with a standard definition in mind without a clear understanding of the purpose and value EA is expected to deliver. In such cases, the probability of failure is significantly high as enterprise architects cannot show any value without direction.

The diagram captures a number of commonly seen EA purposes across various organizations:

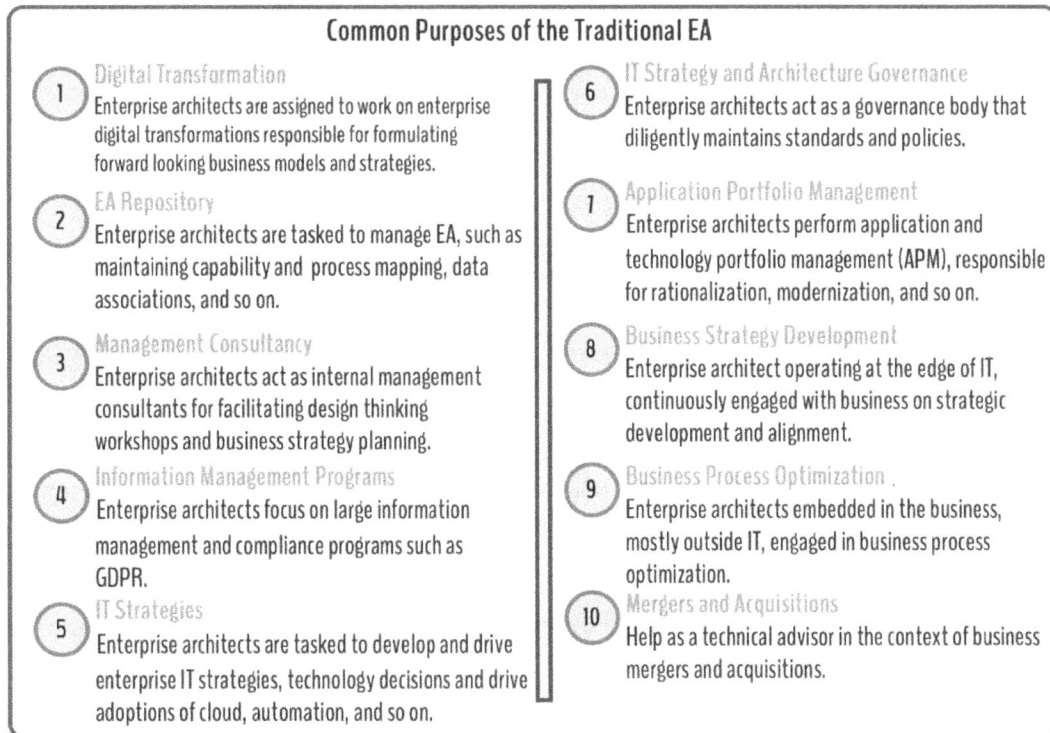

Common Purposes of the Traditional EA

1. Digital Transformation
Enterprise architects are assigned to work on enterprise digital transformations responsible for formulating forward looking business models and strategies.

2. EA Repository
Enterprise architects are tasked to manage EA, such as maintaining capability and process mapping, data associations, and so on.

3. Management Consultancy
Enterprise architects act as internal management consultants for facilitating design thinking workshops and business strategy planning.

4. Information Management Programs
Enterprise architects focus on large information management and compliance programs such as GDPR.

5. IT Strategies
Enterprise architects are tasked to develop and drive enterprise IT strategies, technology decisions and drive adoptions of cloud, automation, and so on.

6. IT Strategy and Architecture Governance
Enterprise architects act as a governance body that diligently maintains standards and policies.

7. Application Portfolio Management
Enterprise architects perform application and technology portfolio management (APM), responsible for rationalization, modernization, and so on.

8. Business Strategy Development
Enterprise architect operating at the edge of IT, continuously engaged with business on strategic development and alignment.

9. Business Process Optimization
Enterprise architects embedded in the business, mostly outside IT, engaged in business process optimization.

10. Mergers and Acquisitions
Help as a technical advisor in the context of business mergers and acquisitions.

Figure 4.5 – Common purposes of the traditional EA

The purposes also determine how enterprise architects are organized to deliver an outcome optimally. There is no one-size-fits-all solution to this problem.

Digging into the current state is an evil

Many enterprises spend time digging into the current state of the enterprise, such as the relationship between capabilities, processes, systems, interfaces, data elements, infrastructure, and so on, mostly with manual efforts. Enterprise architects tend to go down a deep rabbit hole if they focus too much on the current state of affairs.

A more in-depth focus and manual effort to understand the current state of the enterprise by collecting data from experts is a time-consuming activity. In most cases, the result is large-scale non-consumable documentation. These datasets are often sourced from knowledgeable people in the enterprise. Replaying what they already know won't impress those stakeholders.

The Agile Manifesto principle of *working software over comprehensive documentation*, and the LeSS architecture principle, that *the sum of all the source code is the true design blueprint or software architecture*, represent almost opposite sides of the same coin compared to the massive documentation-first approach used in traditional EA practice.

EA disconnected from its intended purpose

In many enterprises, an enterprise architect starts their journey with a purpose. However, due to the culture, construct, and process engraved in the organization's nerve system, EA focuses on one of these three areas:

- Governance
- Abstract strategies
- Operational issues

Rigid policies and standards severely restrict a knowledge worker's creative thinking, and, therefore, the ability to innovate beyond what is known to enterprise architects. Stringent standards and policies are especially destructive if the standards and policies are top-down with infrequent technology adoption cycles. Most of the time, foundation strategies are introduced without assessing the impact on inflight projects and operational systems, which induces tension between teams and enterprise architects. Enterprise architects are totally disconnected from reality and focus on phase gates for architecture assertions that prevent the continuous flow of work and result in delays and distractions in the Agile delivery of software systems.

Too much abstraction and generic architecture building blocks lead to misinterpretations. For example, a *scheduling system* at the highest level of abstraction can be applied to *staff scheduling* or *vehicle scheduling*. Without deep understanding, enterprise architects may argue that the same capability or system could serve both scenarios. These arguments result in a waste of time on discussions and evidence collection to prove the developer's point. Business communication and engagement would become worthless with generic capability-level conversations as they are far from a business's ground-level concerns.

In some organizations, the enterprise architect's focus shifts from strategic planning to day-to-day operations and problem-solving such as reviewing the aftermath of a production issue or project delivery challenges. Focusing on the urgency of now is especially concerning if enterprise architects are sitting well below the software development value chain.

Showing value first is a trap

The symptom of *show me the value* has been repeatedly seen in many organizations where senior leadership is doubtful of what EA can deliver. A lack of understanding of the purpose of EA prompts senior leadership to choose a test-and-learn approach. They probably opt for the least risky minimum investment option, a smaller team with no specialized EA tools, yet let enterprise architects prove their value first.

In these cases, enterprise architects really struggle as they are not fully integrated into the organizational processes and are not positioned as a recognized practice across IT. Lack of sponsorship from senior leadership, inconsistent participation from stakeholders, and the absence of a clear direction eventually lead to failure. Moreover, the success of this largely relies on the capabilities of individuals in the EA team.

Enterprise architects are non-technical

One of the challenges with many EA practices is that enterprise architects are not technically oriented. Even though they mostly come from a technical background, they are not up to date with the technology evolution. They often read headlines about technology innovations but they may not have knowledge that is hands-on or detail-oriented. Such enterprise architects fail to gain trust and respect from the development community.

As a result, the strategies and solutions recommended by the enterprise architects are rejected outright by Agile teams. Instead, they go with their own local solutions, resulting in architecture erosion and, therefore, a significant disconnect between the enterprise strategies and solutions delivered.

In the new world, enterprise architects are supposed to be at the forefront of technologies—identifying industry trends, understanding the pulse, and making wise judgments to adopt technologies to gain a competitive advantage for the business is of utmost importance. Enterprise architects need to have a greater degree of wisdom on these technologies and need to demonstrate the ability to apply them sensibly to capitalize on opportunities in the business.

Agile enterprise architects must altogether avoid these aforementioned anti-patterns and adopt a new set of principles. The next section will go deeper into a recommended set of principles for Agile enterprise architects.

Understanding the principles and duties

In the previous chapter, we used the elevator metaphor to highlight the importance of the architect's role in connecting strategy to code. The up and down movement of the elevator is analogous to the dual operating system of enterprise architects – the **inflow** and **outflow** depicted in the following diagram:

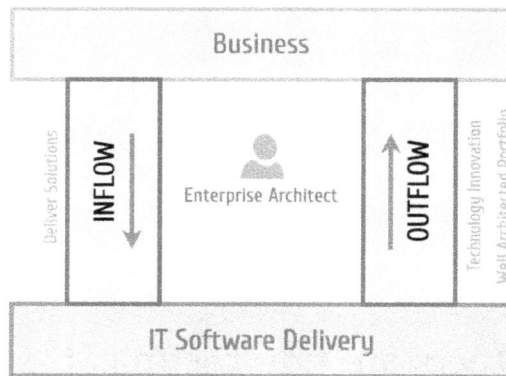

Figure 4.6 – The dual operating system of enterprise architects

As shown in the preceding diagram, Agile enterprise architects consistently need to dedicate time and efforts across both flows of work sufficiently to establish a mutual connection between business and development teams. The *inflow* delivers IT solutions to realize business strategy. The *outflow* fulfills emerging business opportunities with innovative technologies and consistently maintains a well-architected, healthy portfolio.

The metaphor of a large cruise ship versus a speed boat is generally used to represent business agility. Due to their sheer size, it is incredibly tricky for large cruise ships to change their course midway in response to unplanned environmental disruptions. Hence modern enterprises prefer speed boats over cruise ships. If we draw parallels, legacy EA practices resemble cruise ship operations – too heavy and too slow.

The challenge is that large enterprises cannot dismantle the massive structure, processes, and complexities around their legacy operational system landscape. Therefore, they cannot be transformed into speed boats overnight. However, they can incrementally change the machinery, processes, and tools to enable faster maneuvers. Similarly, EA concepts are applicable in Agile enterprises but need a radically different approach and ways of working to succeed.

The principles for Agile enterprise architects to succeed

As mentioned in the manifesto for Agile software development, *the highest priority is to satisfy the customer through early and continuous delivery of valuable software.* To align with this principle, EA practice needs a surgical transformation. To achieve business agility, enterprise architects need to master the art of incremental, tangible value delivery with a razor-sharp focus on eliminating long loops of delivering non-working software artifacts.

The following diagram captures five principles that Agile enterprise architects must adhere to:

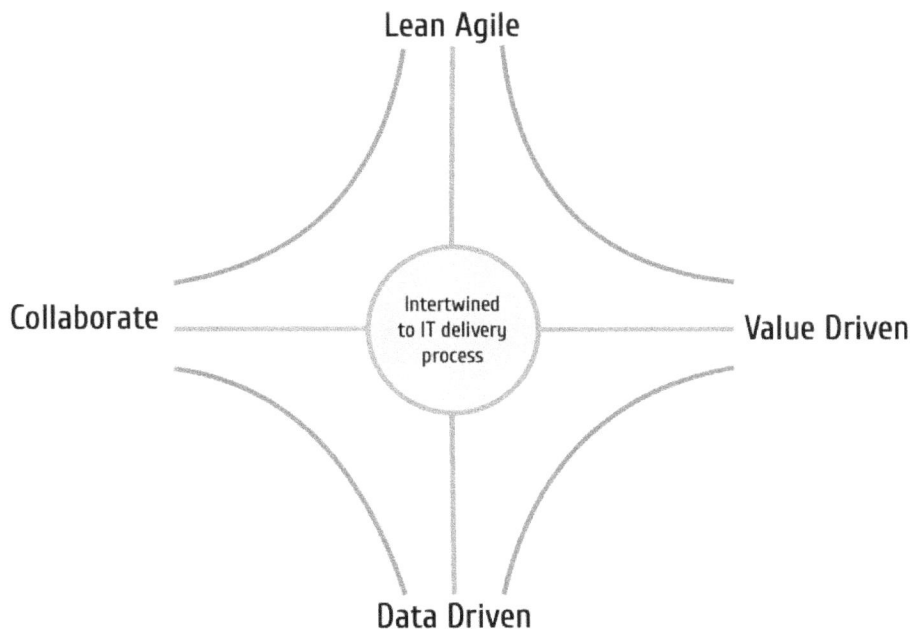

Figure 4.7 – Principles for an enterprise architect's way of working

The principles captured in the diagram are essential for enterprise architects to adapt to the fast-changing paradigm of Agile software development. These five principles are explained in the following sections.

Intertwined processes allow enterprise architects to add value

EA processes and activities have to be well intertwined as part of the integrated IT software delivery process – they must be just like the organization's ways of working. The intertwined processes help enterprise architects to be in the mainstream, fronting the business and supporting development teams. They regularly collaborate and always stay connected with various IT and business stakeholders through established Agile ceremonies.

Enterprise architects need a seat at the table in some of the key enterprise forums and functions, as depicted in the following diagram:

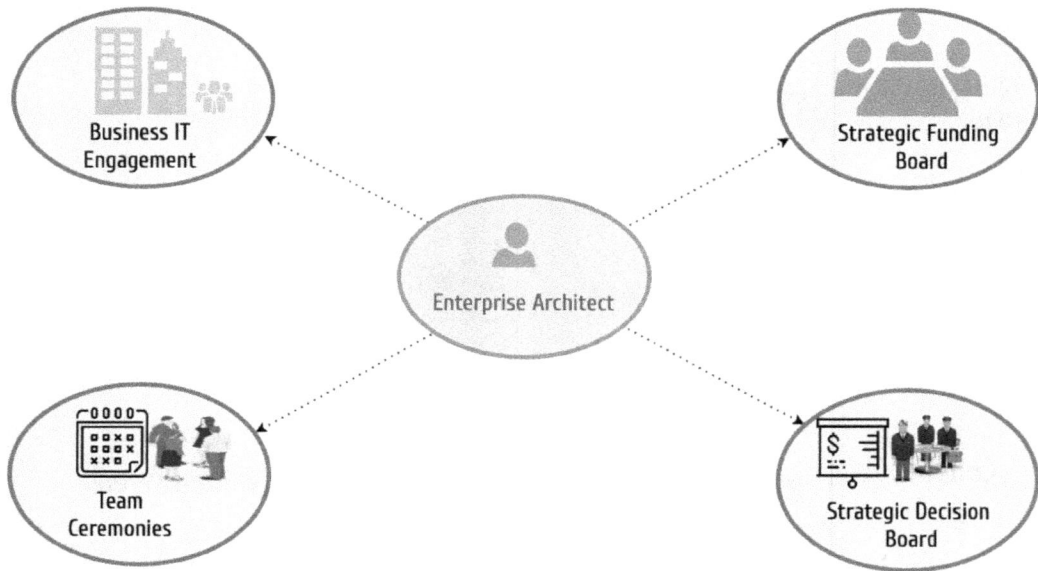

Figure 4.8 – Enterprise architects' key engagements

Enterprise architects need to operate higher up in the value stream, participating in strategic and funding decision boards. At the same time, consistent participation in team ceremonies provides a deep tie to the ground reality of IT DevOps, which is at the bottom of the software delivery value stream. This bottom-line engagement gives a substantial advantage in conversations with business and IT leadership. EA must be perceived as a shared responsibility of all stakeholders linked to the continuous flow of work.

Agile enterprise architects need to think and act lean and Agile

Enterprise architects have to exhibit a lean-Agile mindset to adapt and lead changes in a productive and cost-effective style, ensuring the sustainability and quality of solutions delivered to the business. EA practices must follow lean-Agile values and principles, including Agile architecture guidelines. The following diagram captures a few important points in the context of Agile enterprise architects:

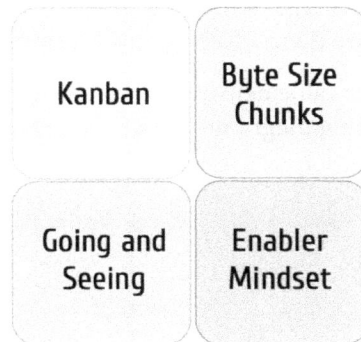

Figure 4.9 – Elements of enterprise architects' lean-Agile operations

Agile enterprise architects always think in terms of **Byte Size Chunks** instead of long-duration low-value deliveries. To deliver innovative products to business or architecture strategies and vision, enterprise architects need to consistently use frequent, smaller iterations of incremental high-value outcomes.

Enterprise architects consciously operate with an **Enabler Mindset**, instead of command and control, to build trust across all stakeholders such as development teams, senior IT stakeholders, and business executives. In a mature EA practice, all stakeholders' mindsets automatically shift to seek help from enterprise architects.

To avoid the perception of ivory tower architects, disconnected from reality, Agile enterprise architects need to oscillate between business engagements and development team ceremonies frequently. Enterprise architects need to **go and see** how developers are coding the solution and visit the business to understand its operations.

Enterprise architects need to adopt Agile ways of working for their internal operations. Transparently visualizing EA work across stakeholders is pivotal in Agile EA. A visual **Kanban**-based EA work management approach helps in moving the work faster to achieve the desired outcome as well as to identify and mitigate risks and impediments as soon as they surface.

Agile enterprise architects focus on delivering value

As in Agile software development practices, an Agile enterprise architect's most important attribute is to deliver incremental value frequently to the business with an intense focus on customer satisfaction. Agile enterprise architects focus on measurable value creation in the form of working software instead of an obsession with documentation. The traditional enterprise architect's measure is based on the data they produce, such as the number of capability maps, process maps, coverage of the system landscape, and so on. In contrast, Agile enterprise architects measure themselves against business KPIs.

There are multiple dimensions for value delivery in the context of Agile enterprise architects, as shown in the following diagram:

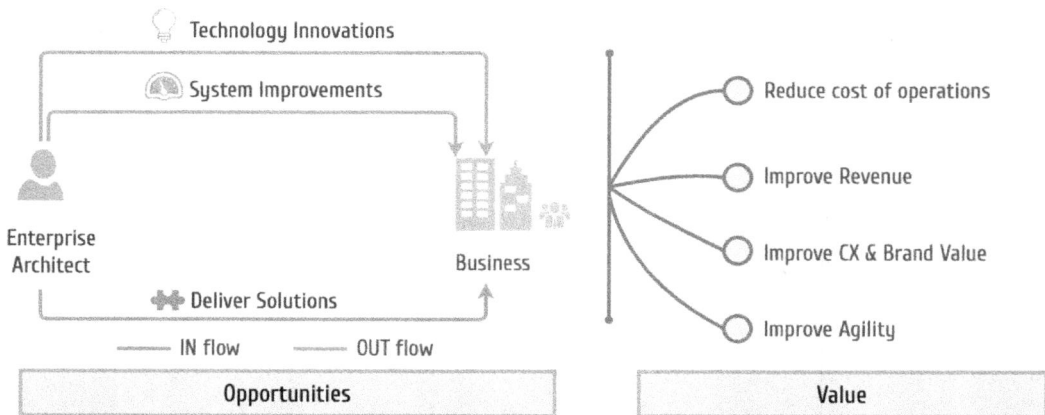

Figure 4.10 – Scenarios of an enterprise architect's value delivery

The left side of the preceding diagram shows opportunities for an enterprise architect to deliver value to the business. The three critical EA flows are as follows:

- **Deliver Solutions**: The most critical value delivery aspect is transforming business ideas into fit-for-purpose and customer-focused solutions with integrated fast learning and improvement cycles. Exploring the best possible, internal or external, solution that maximizes the business benefits with optimal delivery costs is the core responsibility of enterprise architects. They also evaluate and determine the fit of the solution in the existing digital landscape. **Deliver Solutions** is in the *IN flow* – a request coming from the business to the enterprise architect.

- **System Improvements**: Relentless improvement of existing operational systems by collaborating with development and operations teams to support business agility is another critical value delivery flow. **System Improvements** is in the *OUT flow* – the enterprise architect approaching the business with value generation ideas.

- **Technology Innovations**: Proactively selling and deploying technology innovations to solve burning business challenges is the last value delivery flow. Technology innovation is also categorized as the *OUT flow* as the enterprise architect proactively approaches the business.

In the end, enterprise architects measure themselves against a few tangible business values, as shown on the right-hand side of *Figure 4.10*. The business values are measured in terms of reducing the cost of IT as well as business operations, improving revenue through current sources as well as new sources, uplifting the customer experience as well as the brand value, and adopting the business as well as the IT organization's agility.

Enterprise architects make data-driven decisions

Traditional enterprise architects use extremely long loops of massive amounts of volatile data collection, model creation, and insight analysis to arrive at certain decisions. In the end, since much of the documentation is non-consumable, obsolete, and lacks quality, decisions are made based on the enterprise architect's intuition and knowledge on the subject.

Agile enterprise architects invert the decision-making pyramid into a **DIMD** (**Decisions, Insights, Models, Data**) model as shown in the following diagram.

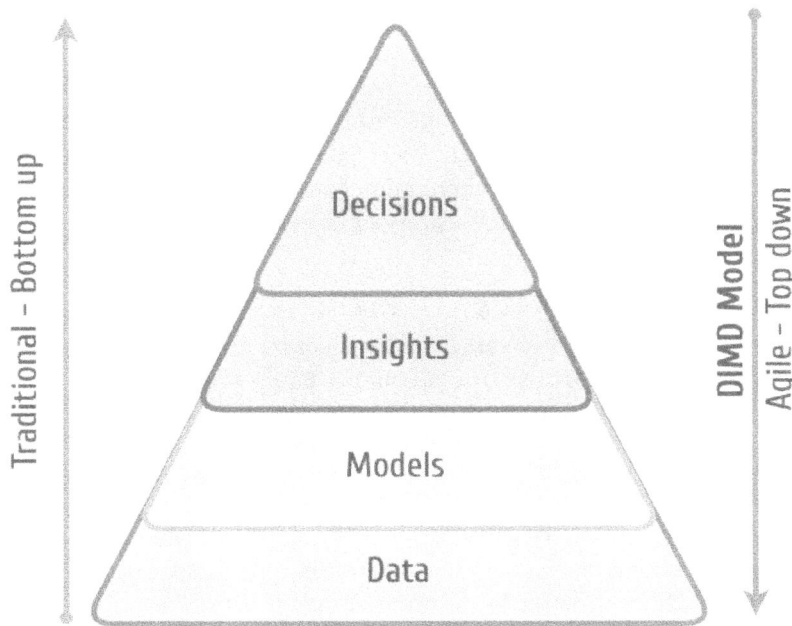

Figure 4.11 – DIMD model for data-driven decision making

Traditional enterprise architects use the bottom-up approach, as shown in the preceding diagram, to produce outcomes. Not many organizations succeed using the traditional approach, merely due to long cycles and the inaccuracy of data.

The following diagram explains the DIMD model with an example:

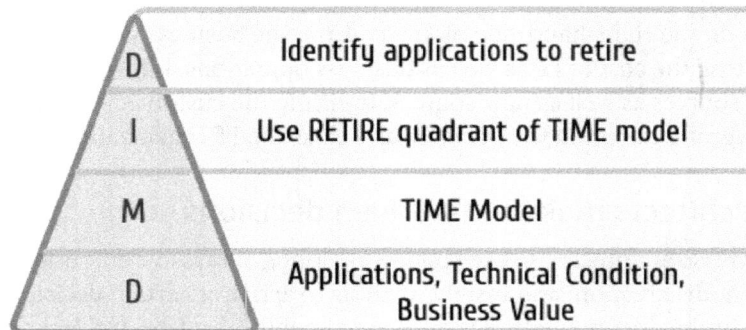

Figure 4.12 – Example to illustrate the DIMD model

In Agile EA, the enterprise architects start with a focused outcome, such as identifying applications to retire, as shown in the diagram. The second step is to establish an approach and parameters for decision making, such as using the **RETIRE** quadrant of the **TIME (Tolerate, Invest, Migrate, Eliminate)** model. Gartner introduced the TIME model to evaluate the application portfolio against business value and technology conditions. TIME is useful to review an application's ability to meet business goals with its current technology state to determine investment potential, then see what model needs to be built. The last step is to see what is the absolute minimum amount of data needed to build a model. In our example, this includes application data, technical conditions, and the business value of those applications. An incremental and MVP mindset with frequent evolutionary enrichments is used until decisions can be made with a certain confidence level.

Data-driven decisions are primarily used for exploring innovation opportunities, potential areas for incremental improvement, strategic planning, and directing strategic investment decisions. Data-driven insights share stories about the enterprise without ambiguity and are an excellent vehicle for communication with both IT and business stakeholders.

Enterprise architects collaborate for better quality outcomes

Enterprise architects sensibly use trust-based relationships with stakeholders to communicate, collaborate, align, and seize opportunities. The following diagram shows the critical points of collaboration for Agile enterprise architects:

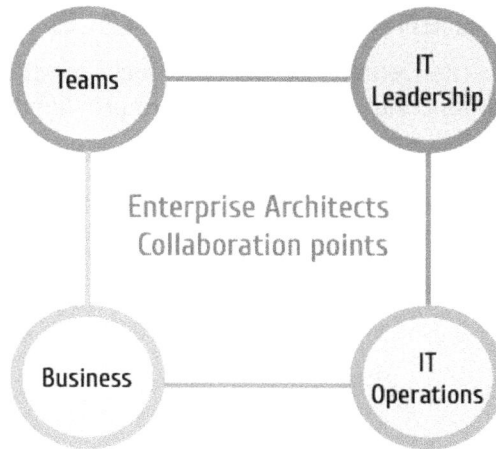

Figure 4.13 – Enterprise architects' collaboration points

The four touchpoints of the enterprise architects' collaboration matrix captured in the diagram are explained here:

- **IT Leadership**: Enterprise architects act as trusted advisors to the IT leadership to articulate and operationalize strategies. Enterprise architects also help in continually improving the IT organization's delivery flow by organizing solutions, systems, and tools around value streams.

- **IT Operations**: Enterprise architects establish a consistent and robust connection with IT operations to understand the operational challenges of live production systems. A good understanding of the operations will help balance investments in new features and technical debts, identify and ring-fence areas for improvement, and discover cost-optimization opportunities and technology adoption decisions.

- **Agile Teams**: Enterprise architects need to have a healthy, trusting relationship with the team. They have to work with a shared understanding of the solution and business objectives. Enterprise architects take the team on their journey, right from the strategy definition to incremental improvements.

- **Business Executives**: Collaboration and cooperation with business executives are fundamental in enabling business agility. Enterprise architects play a crucial role in business strategy formation by sharing and demonstrating the art of possibilities with technologies to evolve and optimize business operations.

Enterprise architects continuously share and align strategies and visions with teams to ensure that the teams deliver solutions with a sense of purpose. In the next section, we will deep dive into the duties of Agile enterprise architects.

The duties of Agile enterprise architects

As discussed in the earlier section, enterprise architects focus on strategic alignment, transforming ideas into business value, and continually improving the health of the solution set, as shown in the following diagram:

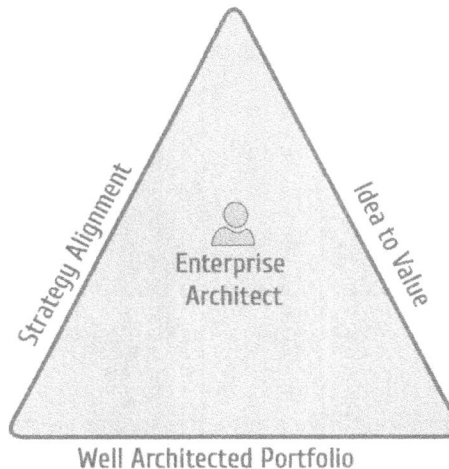

Figure 4.14 – Key duties of an enterprise architect

Setting up well-established portfolio strategies and visions gives development teams a clear view of what the business wants to achieve in the long term. It acts as a reference model for development teams to understand and track how incremental delivery evolves the portfolio toward those long-term objectives. In addition to a vision, enterprise architects maintain and continuously review technology choices, architecture patterns, and development guidelines to keep the technology current. Moving ideas along to produce value with the shortest possible path will enable the business to be more adaptive, resilient, and competitive. While quicker delivery is the most important factor, maintaining a sustainable healthy portfolio helps the business fulfill rapidly changing customer needs more cheaply and more quickly. The following sections explain this in more detail.

Strategic alignment between IT and the business

An enterprise architect balances the interests of both the business and IT teams by playing a neutral advisor, influencer, and negotiator role. The following illustration captures the activities of enterprise architects in strategic alignment:

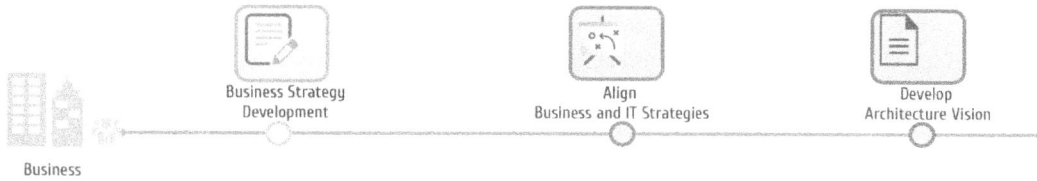

Figure 4.15 – Role of enterprise architects in strategic alignment

While businesses focus on impacts on the business model, strategy, and objectives in a digital transformation, enterprise architects complement system thinking. Enterprise architects collaborate with business stakeholders in strategy development and translate them into a set of executable technology strategies. Similarly, enterprise architects work with IT stakeholders to develop IT technology strategies. IT strategies are generally meant to reduce IT operations' costs and improve the stability, performance, scalability, and sustainability of IT systems. IT strategies, in the long term, help the business to realize benefits. Therefore, it is important to balance between business and IT strategies.

Enterprise architects align and balance business strategies and IT strategies to develop a coherent architecture vision for up to 3 years. Enterprise architects transparently communicate the architecture vision to IT and business stakeholders. Based on the architecture vision, enterprise architects develop funding models such as the *Investment by Horizon* model. *Investment by Horizon* is a sustainable funding model first introduced in the book *Alchemy of Growth*, and in 2000, McKinsey & Company Inc proposed it as the *Three Horizon Model* with recommendations for fund allocations across various types of investments. These funding models provide lean guardrails for strategic funding for all initiatives.

Enabling and collaborating on idea-to-value delivery

Enterprise architects play a critical role in continuous value delivery, including designing the end-to-end flow of work. The following diagram illustrates different aspects of enterprise architects' duties in the idea-to-value flow:

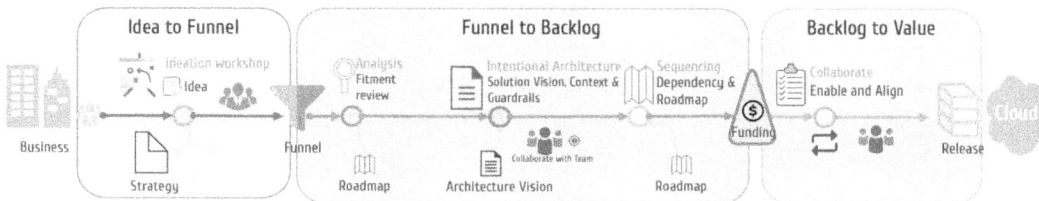

Figure 4.16 – Idea-to-delivery flow

As shown in the diagram, enterprise architects work with the business to discover technology needs that accelerate the meeting of business goals. A typical design workshop with a business leads to a new solution development or the enhancement of existing solutions. In these workshops, enterprise architects use architecture vision, knowledge on innovative technologies, and market insights to ignite the business's thinking to shape up ideas.

These ideas go through a review funnel to secure approvals and land in the development team's product backlog. During this process, enterprise architects analyze the impact of the idea in the context of the overall systems landscape. They also explore the need for additional technical components to deliver high-quality, economically scalable solutions. Enterprise architects also develop international architecture together with the team. Finally, they sequence the initiative by identifying architecture dependencies and risks against an architecture roadmap. Enterprise architects support the business case preparations and influence the strategy funding approvals.

Once the development team picks up backlog items for development, enterprise architects continuously collaborate with the team to establish a shared understanding of the context, purpose, rationale, and business impact. Enterprise architects act as part of the team to deliver quality solutions aligned with the vision and business case.

Continually maintaining a well-architected portfolio

Evolving and growing a healthy portfolio is as important as delivering new and incremental innovations. Enterprise architects proactively devise an architecture roadmap with technology as currency to balance legacy and emerging technologies, protect core systems, simplify the architecture, and leverage new technologies to innovate solutions. The enterprise architect also focuses on rationalization by minimizing the duplication of assets, the consolidation of systems, the retirement of technologies that are no longer delivering value, and minimizing technology licensing and support costs. Enterprise architects act as architecture owners for portfolios by relentlessly improving the health of high-value applications, working closely with development and operational teams.

In summary, enterprise architects focus on three areas – aligning strategies, fast delivery of fit-for-purpose sustainable solutions, and maintaining a well-architected portfolio. In the next section, we will explore and understand the principles to be followed when dealing with the EA repository.

The EA repository – the gold dust problem

Many organizations buy an EA tool as a starting point on their EA journey. Soon they will realize that a new shiny EA tool without data is nothing. The solution is to commission a multi-year project to collect and record data from various parts of the organization into the EA repository. Too much labor to document what people already know without an outcome for a long period is an absolute waste.

For enterprise architects to be highly successful, insights and data-driven decisions are particularly important. However, the availability and completeness of the EA repository is the biggest challenge. Traditionally, enterprise architects spend a significantly large amount of time building an EA repository, which, in the end, still does not guarantee the availability of consistent data for effective decision making. Enterprise architects need to shift the focus from building internal deliverables, such as building an EA repository, to outcomes that deliver business value.

Three antipatterns that Agile enterprise architects must avoid concerning EA repositories are shown in the following diagram:

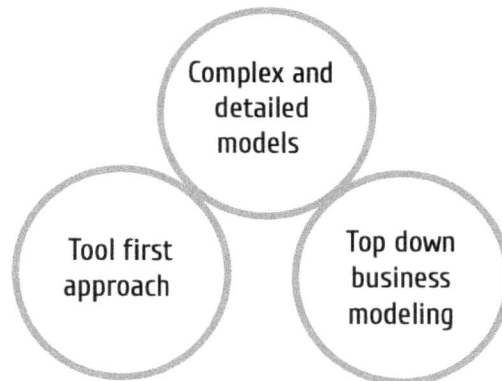

Figure 4.17 – Common pitfalls of building an EA repository

Simple models, insights, and visualizations help to share stories about the enterprise with C-level executives better. The tool-first approach to EA using top-down business modeling is complex, time-consuming, and challenging to realize immediate value. Many of the modern EA tools support network models, and therefore the models can incrementally evolve not just top-down capability to the systems modeling, but in any dimension such as the system-to-investment view.

To harness the power of the EA repository, Agile enterprise architects have to adhere to the following principles:

- **Keep it simple, don't be too scientific**: Follow the DIMD approach, start with simple decision points, and then create models that are not too scientific and perfect. Once insights are visualized, use feedback to improve models. Convert complex dashboard models to insights in terms of risks and cost. For example, a spaghetti integration chart only helps to scare people. Instead, show insights in terms of business risk and the cost of maintaining point-to-point interfaces as a summary.

- **Build it when you need it**: Do not spend time building capabilities or process maps or a system landscape without an outcome or a decision in mind. Use the DIMD approach to build the EA repository iteratively. Start with a decision point, followed by identifying insights, models, and data. Continuously refine the models until enterprise architects are comfortable with the outcome and quality of the decision before moving on to the next decision point.

- **Automate everything possible**: In many cases, the EA repository needs a system landscape, system dependencies, a technology stack, deployment details, running costs, incidents, and so on. Allow these to flow automatically as much as possible from authentic sources of data such as CMDB, ITSM, and ITBM systems.

- **Don't do it, if it cannot be maintained**: Do not capture highly volatile data such as design and code models in the EA repository. The fact is that it may not even be needed for analysis and insights. Do not capture such data in the EA repository as it is too laborious and cannot be maintained consistently.

- **Allow the meta-model to evolve**: Often, EA tools start with a rigid, predefined meta-model. With this approach, teams are forced to capture all data elements specified in the meta-model without exception, resulting in complex data structures that are hard to manage. In Agile, the EA meta-model must also incrementally evolve. The meta-model's starting point must be based on the outcomes and the data elements required to support those outcomes.

While the EA repository is essential, simplicity and iterative development are key to success. The next section explains how to measure the success of enterprise architects.

Measuring enterprise architects

Lean, repeatable, and consistent models are much more effective for measurement in Agile organizations. The Agile enterprise architect's measurement framework shown in the following diagram is based on the principles of enterprise architects on a five-point scale:

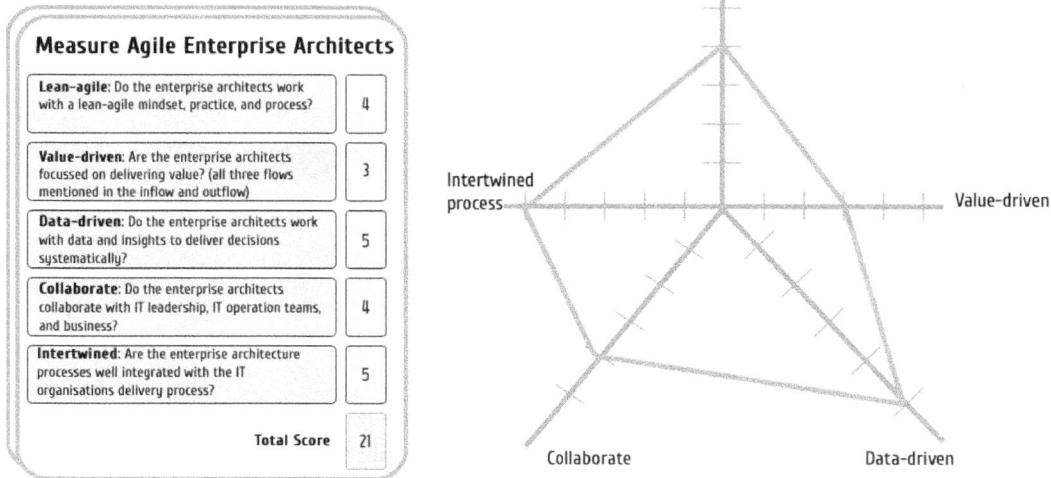

Measure Agile Enterprise Architects

Lean-agile: Do the enterprise architects work with a lean-agile mindset, practice, and process?	4
Value-driven: Are the enterprise architects focussed on delivering value? (all three flows mentioned in the inflow and outflow)	3
Data-driven: Do the enterprise architects work with data and insights to deliver decisions systematically?	5
Collaborate: Do the enterprise architects collaborate with IT leadership, IT operation teams, and business?	4
Intertwined: Are the enterprise architecture processes well integrated with the IT organisations delivery process?	5
Total Score	21

Figure 4.18 – Measurement framework for enterprise architects

For each parameter shown in the diagram, enterprise architects collaborate with other stakeholders to decide on scoring between 1 and 5. The following link captures more cues for marking scores: `https://github.com/PacktPublishing/Becoming-an-Agile-Software-Architect/blob/master/Chapter4/EA-Measure.png`.

The spider chart on the right of the preceding diagram shows a visual view of the outcome. Once the overall score is calculated by summing individual scores, use the maturity model shown in the following diagram to reflect the current maturity state of the enterprise architects:

5 (20-25)	**Fly**	Fully matured and organization mindset shifted to reliance
4 (15-20)	**Run**	Fully integrated and has a reputation for delivering value
3 (10-15)	**Walk**	Agile enterprise architect's services are used in pull mode
2 (5-10)	**Crawl**	Just started realizing value, but in small pots
1 (0-5)	**Pre-Crawl**	Not recognized agile enterprise architecture practices

Figure 4.19 – Agile enterprise architect maturity model

Frequent measurement of enterprise architects, collaboratively with all stakeholders, is advisable. The best way is to use a postcard approach for measuring enterprise architects in review and retro meetings. The next section examines how different enterprise Agile scaling frameworks treat enterprise architects.

The enterprise architect role in Agile frameworks

The enterprise architect role is inevitable when Agile software delivery methods are used at scale in large organizations. The challenges faced without an enterprise architect are perplexed when the system landscape is complex, business departments are distributed, strategic funding is funneled top-down, and investment decisions are taken collectively by IT and business leadership.

Both **Scaled Agile Framework (SAFe)** and **Disciplined Agile (DA)** emphasize enterprise architect roles, whereas **Large-Scale Scrum (LeSS)** has little mention of enterprise architects.

The enterprise architect role in Scaled Agile Framework

SAFe defines the enterprise architect role at the portfolio level of SAFe, closely aligned with Business, Epic Owners, Agile PMO, and **Lean Portfolio Management (LPM)** competencies in addition to their close relationship with ART.

Enterprise architects align and balance the architecture vision at the portfolio level in line with business and IT strategies to define portfolio canvas and lean budget guardrails. Enterprise architects closely operate with business executives and Epic Owners – who own and coordinate business initiatives. Enterprise architects act as advisors to the LPM, ensuring portfolio solution sets are evolved continuously in the right direction toward meeting the business's strategic objectives aligned with the architecture vision and principles. LPM consists of a senior group of business and IT members responsible for ensuring the portfolio is aligned and funded to meet their strategic needs. They are also responsible for ensuring that portfolio initiatives are well coordinated to achieve business milestones and that their strategic investments are safeguarded with lean governance. Enterprise architects work in close coordination with LPM in all these activities.

Enterprise architects continuously evolve portfolio solutions by regularly sensing opportunities to improve the portfolio's health and being ready to support business growth. Some of the key responsibilities of enterprise architects are the following:

- Working closely with ARTs by attending inspect and adapt sessions, and solution demos to exchange feedback

- Attending architect sync meetings with solution architects to receive feedback on current initiatives and the system landscape

- Being champions in communicating the business's strategic vision, purpose, and objectives to development and operations teams

- Guiding ARTs in developing and evolving architecture runway strategies by linking appropriate enabler Epics across ARTs

- Promoting technology strategies such as DevOps, continuous delivery, and so on

Enterprise architects work with a lean-Agile mindset, consistently using system thinking with economic sense. They also lead or participate in different architecture communities of practice. SAFe recognizes other architecture practices such as business architecture and information architecture as supporting mechanisms. Enterprise architects use portfolio Kanban and portfolio backlog to transparently monitor the progress of the enabler work in the delivery pipeline.

The enterprise architect role in Disciplined Agile

DA encourages a collaborative approach for enterprise architects to work with teams. It recognizes EA as an important enabler of Agile software delivery at an enterprise scale. It also observes that EA enables Agile teams to focus on creating value by reusing architecture patterns and solution components. EA, in large organizations with teams of teams, also brings higher degrees of consistency.

Enterprise architects participate in envisioning initial architecture, collaborate with business and IT stakeholders, and evolve architecture assets. DA recommends that enterprise architects, as active members of the team, help identify reuse opportunities and technical debts. They also create high-level architecture views, models, guidelines, and roadmaps to guide development teams.

DA recommends enterprise architects focus more on evolutionary collaboration over blueprinting, communication over perfection, enablement over inspection, high-level models over detailed documentation, and lean guidance over bureaucratic processes.

The enterprise architect role in Large-Scale Scrum

LeSS believes more in teams collectively taking architecture decisions and, therefore, there is no mention of enterprise architects. It is expected to have a broader vision and knowledge of EA available within the team.

Our next section will strengthen our understanding with some practical examples from Snow in the Desert based on SAFe.

Enterprise architects at Snow in the Desert

As we saw in the previous chapter, Snow in the Desert's IT software delivery operates under a single portfolio manned by one enterprise architect. In the traditional world, enterprise architects are often organized horizontally, aligned with various architecture disciplines such as business architecture, information architecture, application architecture, and technology architecture. The horizontal slicing impacts the solution's systemic view since each architect restricts their vision and steers the solution with a narrow frame of mind. Like a layered architecture, horizontally aligned organizations by design create abstractions between EA layers, which raises the complexity and deteriorates the quality.

In modern organizations driving business agility, try to vertically align enterprise architects to business portfolios to deliver better value. The vertically aligned design encourages better ownership and positive growth of the portfolio and significantly streamlines solution designs.

The profile of an enterprise architect

Traditionally, enterprise architects are far from hands-on technology explorations. This substantially challenges the ability of enterprise architects to sell technology innovations to businesses as well as delivering maximum possible value at scale with the right set of technologies and architecture patterns. At Snow in the Desert, the leadership made a conscious decision to appoint an enterprise architect who is well balanced with a profile as shown in the following diagram:

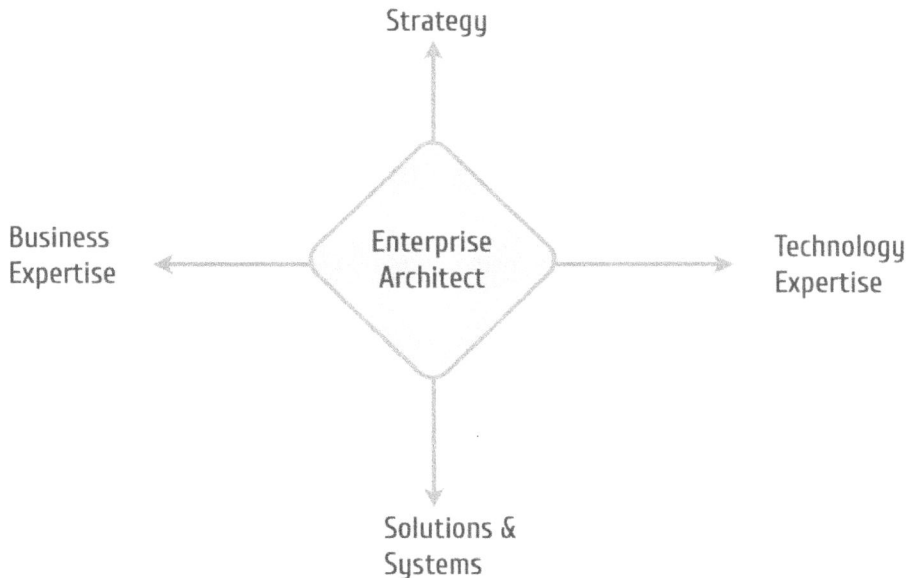

Figure 4.20 – Profile of enterprise architect

There is a rationale behind this choice at Snow in the Desert. While business domain knowledge is invaluable to have deeply engaged conversations with the business, a curious individual with a learning mindset earns a lot of respect within the developer community. Mastery in abstracting at the strategy level and zooming into details when needed is a leading sign of Agile enterprise architects. A chief architect style profile, who can think holistically and is good with technology, is a far better match for modern Agile enterprise architects.

Understanding the portfolio flow

An enterprise architect at Snow in the Desert is responsible for engaging closely with all four business units – front office, field operations, back office, and vehicle and asset management, to understand pain points, strategic directions, and business objectives. The enterprise architect uses a fixed weekly cadence to collaborate with business executives to realize strategic themes, review technology needs, and feedback. The enterprise architect captures strategies concisely, as shown in the following diagram:

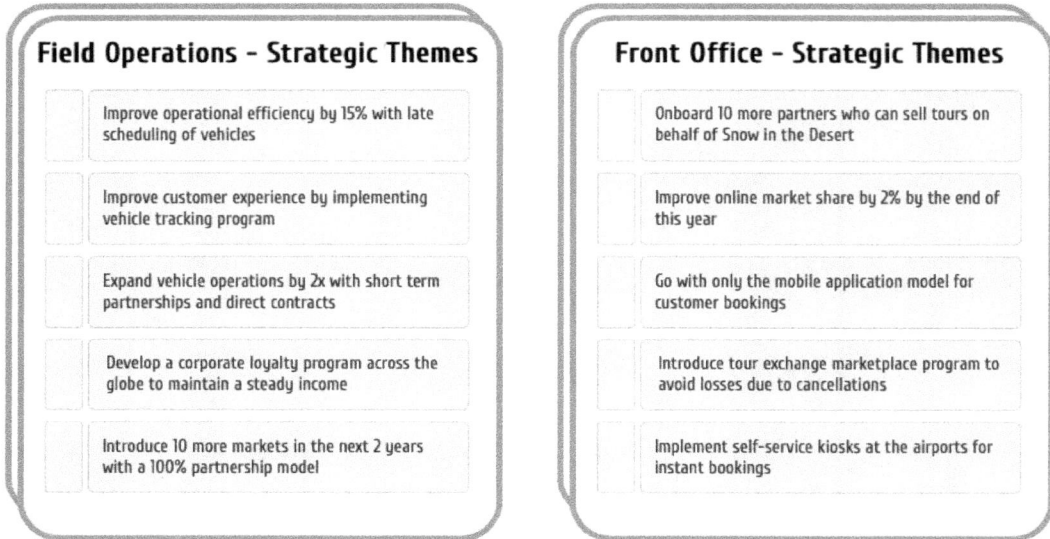

Field Operations – Strategic Themes	**Front Office – Strategic Themes**
Improve operational efficiency by 15% with late scheduling of vehicles	Onboard 10 more partners who can sell tours on behalf of Snow in the Desert
Improve customer experience by implementing vehicle tracking program	Improve online market share by 2% by the end of this year
Expand vehicle operations by 2x with short term partnerships and direct contracts	Go with only the mobile application model for customer bookings
Develop a corporate loyalty program across the globe to maintain a steady income	Introduce tour exchange marketplace program to avoid losses due to cancellations
Introduce 10 more markets in the next 2 years with a 100% partnership model	Implement self-service kiosks at the airports for instant bookings

Figure 4.21 – Strategic themes for Field Operations and Front Office

The preceding diagram shows strategic themes for field operations as well as for the front office business units. Strategic themes are near to medium term goals aligned with a long-term business vision. The enterprise architect carefully aligns the strategic business themes with IT strategic themes to create a harmonized architecture vision. The following diagram depicts the architecture vision for field operations:

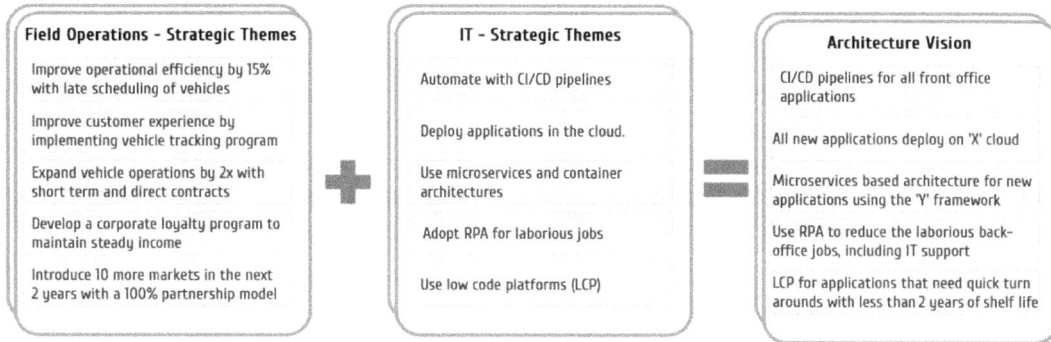

Field Operations - Strategic Themes	IT - Strategic Themes	Architecture Vision
Improve operational efficiency by 15% with late scheduling of vehicles	Automate with CI/CD pipelines	CI/CD pipelines for all front office applications
Improve customer experience by implementing vehicle tracking program	Deploy applications in the cloud.	All new applications deploy on 'X' cloud
Expand vehicle operations by 2x with short term and direct contracts	Use microservices and container architectures	Microservices based architecture for new applications using the 'Y' framework
Develop a corporate loyalty program to maintain steady income	Adopt RPA for laborious jobs	Use RPA to reduce the laborious back-office jobs, including IT support
Introduce 10 more markets in the next 2 years with a 100% partnership model	Use low code platforms (LCP)	LCP for applications that need quick turn arounds with less than 2 years of shelf life

Figure 4.22 – Architecture vision by combining business and IT strategy

The enterprise architect continuously engages senior business stakeholders to identify potential opportunities to leverage innovative technology solutions to overcome burning business problems. Such explorative ideas fed into the portfolio funnel trigger the portfolio Kanban flow similar to the one shown in *Figure 4.16*.

There may be many ideas coming in to the funnel from different parts of the business. One such idea based on the strategic themes of field operations is developing an **Automatic Vehicle Tracking System** (**AVTS**). These ideas, captured as Epics, are reviewed together with business and other IT portfolio stakeholders to build a solution hypothesis. Epics are then prioritized using the **Weighted Shortest Job First** (**WSJF**) method, using a relative scoring based on perceived business value. We will explore WSJF in *Chapter 6, Delivering Value with New Ways of Working*. This review process will spit out a priority list of Epics for further analysis to determine the investment size. To ensure these are processed without delay, a **Work In Progress** (**WIP**) method is used. We will explore more about WSJF and WIP in *Chapter 6, Delivering Value with New Ways of Working*.

At Snow in the Desert, the enterprise architect uses the EA repository as the primary means to perform a fit gap analysis against the current portfolio. At this stage, the enterprise architect looks for opportunities to reuse existing systems or services, impacting existing portfolios such as systems that need modernization to meet the intended business purpose, or simply decommissioning them. The enterprise architect also determines the dependency of the new Epic with other Epics in progress.

In the context of AVTS as an example, the enterprise architect collaborates with solution architects and relevant team members from the *Fulfilment ART* and *Fleet Ops ART* to collectively define the solution vision, guardrails, non-functional requirements, economic framework, solution context, and intentional architecture. We will discuss more on the solution aspects in the next chapter. In some cases, the enterprise architect discovers technical enablers required for delivering scalable and sustainable solutions. In this example of building an AVTS, a new **API Gateway** is required for providing access to tracking data.

With a clear solution vision, the enterprise architect helps teams break down the problem into small pieces at the earliest possible moment for value realization, starting with an MVP. The team estimates the solution, both the MVP cost and the full cost, for LPM approval. The enterprise architect develops an architecture roadmap that helps identify architecture dependencies, the right sequence of execution, and potential architecture risks. The following diagram captures the essence of a three-year architecture roadmap:

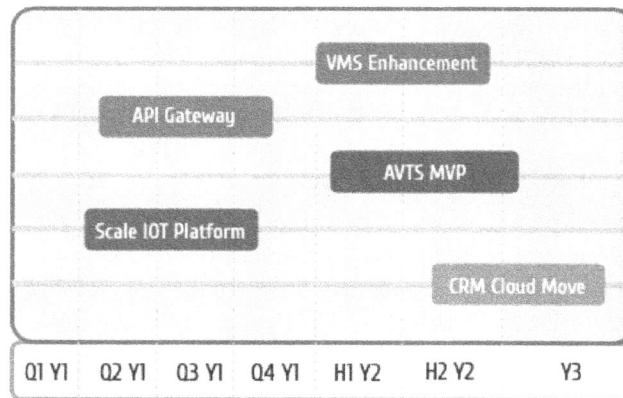

Figure 4.23 – Three-year view of the architecture roadmap

At Snow in the Desert, the portfolio backlog, portfolio Kanban, and the architecture roadmap are implemented using *JIRA advances roadmaps*. The architecture roadmap is nothing but a portfolio roadmap with additional filters to show architecturally significant initiatives.

In addition to ideas from the business, the enterprise architect continuously collaborates with solution architects and engineers across the ARTs to understand the challenges and improvements of existing live systems. Smaller improvements and technical debts are managed by the team based on their available capacity. If the ART has no sufficient capacity to include technical improvements, the enterprise architect pushes those demands into the portfolio funnel for review. **CRM Cloud Move** is one such example shown in the preceding diagram.

> **A day in the life of an enterprise architect**
>
> The day in the life of an enterprise architect poster at Snow in the Desert can be downloaded from the following GitHub location:
>
> `https://github.com/PacktPublishing/Becoming-an-Agile-Software-Architect/blob/master/Chapter4/EA-day-In-the-life.png`
>
> The different personas of the enterprise architect based on how things are organized at Snow in the Desert can be downloaded from the following GitHub location:
>
> `https://github.com/PacktPublishing/Becoming-an-Agile-Software-Architect/blob/master/Chapter4/EA-personas.png`

Summary

Traditional EA is strongly focused on adopting off-the-shelf EA frameworks. Framework-driven EA is dogmatically documentation-centric, forcing enterprise architects to produce a multitude of fine-grained models and documents with a longer lead time to realize value. Most of the organizations embarking on this route fail miserably to deliver their promises and value to enterprises. Therefore, framework-centric EA is no longer best suited for organizations driving agility.

Agile enterprise architects use Agile practices to operate EA. They stick to the five key principles – lean-Agile, value-driven, data-driven, collaboration, and intertwined IT delivery processes. They break down issues into smaller chunks, use iterations, and focus on delivering value with the shortest possible lead time. Agile enterprise architects balance between providing customer value faster and in a sustainable way. They use the DIMD approach for data-driven decisions with a lean-Agile mindset. They collaboratively engage business, IT leadership, IT operations, and development teams with an enabler mindset to build trust-based relationships. Enterprise architects follow a dual operating system model by focusing on three critical areas – technology innovations, system improvements, and delivering solutions. Agile enterprise architects use the EA repository as a supporting mechanism instead of the conventional repository-first approach. They rely on automation for data collection and incrementally enrich the repository contents.

Enterprise architects work very closely with solution architects and teams to avoid wastage due to any handoffs. They consistently meet, discuss, and share their ideas and understandings using cadence-based ceremonies with an open mindset. We will explore more of the solution architect's roles and responsibilities in the next chapter.

Further reading

- **The History of Enterprise Architecture – An Evidence-Based-Review by Svyatoslav Kotusev**: `https://www.researchgate.net/publication/308936998_The_History_of_Enterprise_Architecture_An_Evidence-Based_Review`

- **Cut and Paste of EA Framework**: `https://blogs.gartner.com/james-mcgovern/2017/04/17/cut-and-paste-enterprise-architecture/`

- **Collaborative Enterprise Architecture, Enriching EA with Lean, Agile, and Enterprise 2.0 Practices**: `https://books.google.ae/books/about/Collaborative_Enterprise_Architecture.html?id=dgegYa2qOokC&source=kp_book_description&redir_esc=y`

- **Investment by Horizon – The Alchemy of Growth**: `http://growthalchemy.com/introduction/the-alchemy-of-growth/`

5

Agile Solution Architect – Designing Continuously Evolving Systems

"As an architect, you design for the present, with an awareness of the past,
for a future which is essentially unknown."

– Norman Foster (designed the world's tallest bridge – Millau Viaduct)

In the previous chapter, we covered the important role of an enterprise architect in connecting strategy to code. This chapter will focus on one of the most critical roles in delivering reliable solutions in agile software delivery – the solution architect role.

In agile software development, the solution architect immeasurably focuses on rapidly delivering solutions with built-in quality that meet customers' near-term needs with the shortest consistent velocity, optimal cost, and maximum possible value. Enabling continuous flow by relentlessly eliminating flow barriers is critical for the success of sustainable delivery. Providing holistic vision, decomposing solutions into byte-size chunks, analyzing trade-offs, envisioning architecture, and anticipating decisions needed are critical responsibilities of solution architects. Non-intrusive means of technical alignment by actively participating in inspect and adapt ceremonies, enabling rapid feedback cycles with flow automation, and using influence over authority for guiding toward right designs are the mechanisms to ensuring architecture is built according to the vision. Being part of agile teams, developing a sense of care, empowering teams for decentralized decisions, promoting a growth mindset, fostering technical agility, and nurturing craftsmanship are integral to a solution architect's leadership qualities.

This chapter will go deeper into an agile solution architect's objectives and duties, and this will be reinforced with examples from Snow in the Desert. We will also examine how different agile scaling frameworks are handled solution architects. As we did in the previous chapter, we will also explore a framework for measuring a solution architect's success.

In this chapter, we're going to cover the following main topics:

- Solution architects – the busy bees of agile teams

- Maximizing value and eliminating flow barriers

- Understanding the duties of solution architects

- Measuring the success and maturity model

This chapter focuses on the **Solution Architect** focal point of **The Agile Architect's Lens**:

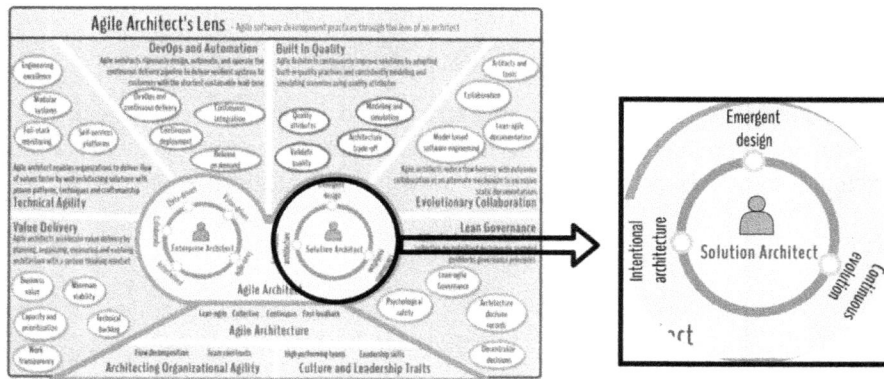

Figure 5.1 – The Solution Architect – focal point

Let's get started!

Technical requirements

The *day in the life of a solution architect* poster can be downloaded from the following link:

```
https://github.com/PacktPublishing/Becoming-an-Agile-Software-
Architect/blob/master/Chapter5/SA-day-in-the-life.png.
```

The different personas of a solution architect captured in a poster can be downloaded from the following link:

```
https://github.com/PacktPublishing/Becoming-an-Agile-Software-
Architect/blob/master/Chapter5/SA-personas.png
```

Solution architects – the busy bees of agile teams

Until the early 2000s, the solution architect role was not widely recognized as a formal discipline in software development. IASA Global positions solution architect, in *The Architecture Journal,* as a role that sits between the enterprise architect and the technical architect, and is responsible for strategy alignment, solution and technology decisions, and inter-team collaborations, and is *the go-to person for any technical conflicts, implementation issues, or decisions.* The position of the solution architect is illustrated in the following diagram:

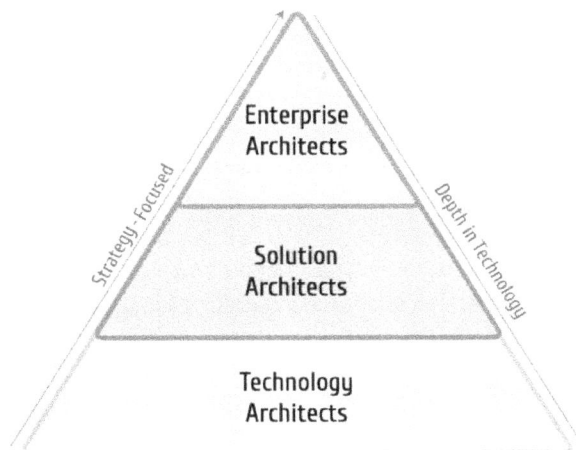

Figure 5.2 – Position of the solution architect, as defined by IASA

To explain the solution architect's responsibilities, let's say a customer in a restaurant is requesting a particular dish for dinner from the dine-in menu. In this context, the enterprise architect is responsible for the dinner menu that provides the best blend of indulging dining experience, the solution architect is responsible for the recipe that offers the best dish, and finally, engineers are accountable for finding the best quality ingredients and cooking the dish.

In another example of building a car, the Product Owner is responsible for its functional features. The solution architect is responsible for choosing the right engine, how the engine connects to the gearbox, and so on. Solution architects choose an engine type with a perfect balance between meeting customer expectations, long-term operability, and profit margins. The design of the engine is an active part of development.

Mike Cohn, one of the founders of Scrum Alliance, observed that if we consider "*the scrum teams as the car, then the Product Owner is the driver making sure the car goes in the right direction. The Scrum Master is the chief mechanic to keep the car well-tuned and performing well.*" Adding further, the Solution Architect is the co-driver for navigating technical aspects of the solution.

In summary, the solution architect is an equally important role as Product Owner and Scrum Master in agile software development. The next section will further explain the importance of solution architects.

A solution architect's mindset – the periscope in action

As we discussed in *Chapter 3, Agile Architects – The Linchpin to Success*, an architect's mindset is similar to a **periscope**. The periscope is used for seeing things that are not visible otherwise. It is commonly termed *Architectural Thinking*.

A friend of mine shared an interesting experience of observing a conversation between a business user and a very passionate, strong frontend developer. The business user wanted to add a new text field to search for customer data from an existing user interface that had been exposed to an external agency. The immediate reaction from the developer was to add a *typeahead* text box. After typing in the first three letters, the text box will start listing any matching customer names. The developer's approach was to get customer data from a central customer database that had been exposed over a REST endpoint. The business user really liked this idea because of its usability.

If you really dig deeper a bit – *are we okay with regulatory compliance of the proposed solution, such as GDPR? How much additional load is the new service going to introduce to the central customer service? Is the agency supposed to see customer names?*

Code is the starting position for many developers – specifically, code that is familiar to the developer. At times, developers miss the purpose and value of what the customer is asking. On the other hand, a seasoned solution architect's starting position is always the purpose, value, big picture, and trade-offs before they decide on the right solution.

As we discussed in *Chapter 3, Agile Architects – The Linchpin to Success,* developers going after technology lead to the *Big Ball of Mud.* However, it is natural for solution architects to think beyond technology, such as supportability, operations cost, technology longevity, license impact, reuse, performance acceptability, uninterrupted service, and so on.

Lean practitioner Mary Poppendieck mentioned, "*The biggest cause of failure in software-intensive systems is not technical failure; it's building the wrong thing.*" Committing designs without any clear understanding may lead to building the wrong things. Almost along the same lines, author Alan McSweeney observed that successful project delivery is when the "*right solutions are implemented successfully.*"

The following diagram captures four key parameters architects must consider for successfully delivering value:

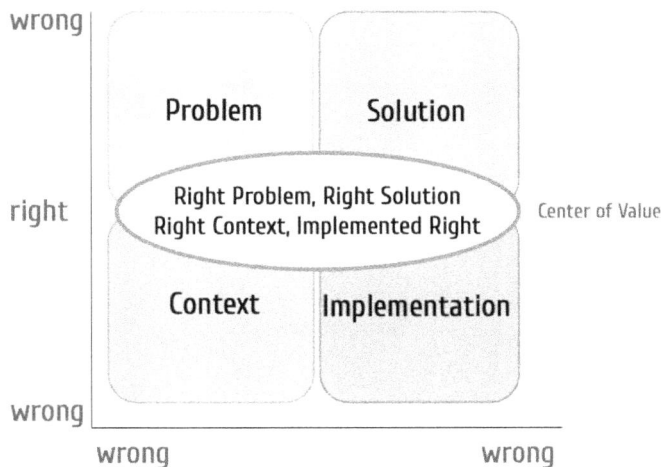

Figure 5.3 – Center of value

As shown in the preceding diagram, we have the center of value. Maximum value is given to the customer when the right problems are solved in the right context with the right solutions, when they're implemented correctly.

To understand the problem and context correctly, solution architects must also connect with the strategy, which is primarily the duty of an enterprise architect. In the next section, we will explore more about the overlapping roles of enterprise and solution architects.

The overlapping roles of enterprise and solution architects

In the previous chapter, we saw enterprise architects connect strategy to code by translating business strategies into an executable IT architecture. Furthermore, they help teams develop the right code that realizes the architecture's vision by closely collaborating with developers. As we explained in *Chapter 3, Agile Architects – The Linchpin to Success*, both enterprise architects as well as solution architects frequently travel up and down the enterprise tower.

Both enterprise architects and solution architects cross paths many times a day in their journeys, as illustrated in the following diagram:

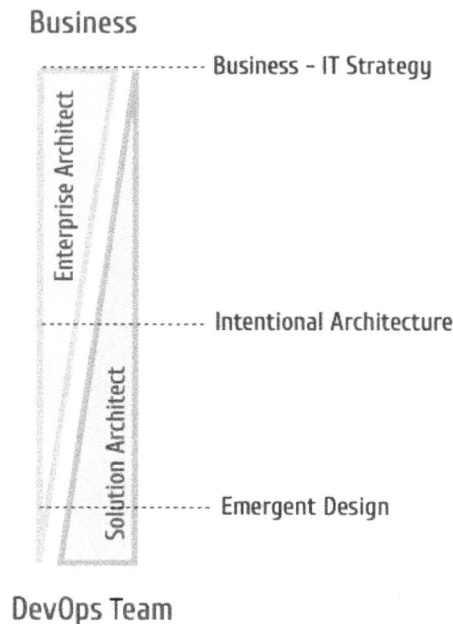

Business

.... Business – IT Strategy

Enterprise Architect

.... Intentional Architecture

Solution Architect

.... Emergent Design

DevOps Team

Figure 5.4 – Overlapping roles of enterprise architects and solution architects

While enterprise architects connect top to bottom with a heavy base at the top, solution architects connect bottom to top with a heavy base at the bottom. The size of the segment of the triangle indicates the amount of time they spend there. Solution architects may also cover an enterprise architect's duties in smaller organizations.

In summary, even though there is no explicit architect role defined by agile development frameworks, solution architects play a critical role in understanding the problem and context before offering the right solution. They also engage at the strategic level to ensure the business's purpose and objectives are clearly understood. While solution architects do many things, like a busy bee, their focus ultimately boils down to a few objectives. In the next section, we will examine those objectives further.

Maximizing value and eliminating flow barriers

Agile and Lean software development practices actively eliminate flow barriers such as phase-gated checks, handoffs, manual validations, deployments, and infrequent releases existing in traditional software development. Furthermore, agile software development practices need to consistently demonstrate the continuous, uninterrupted flow of values to rapidly respond to changing customer needs. As a key player in agile software delivery, solution architects ruthlessly focus on improving the pace at which value is delivered and substantially provide enhancing benefits to both customers and enterprises.

There are five areas where solution architects must spend their energy to maximize business value and faster flow. These are captured in the following diagram:

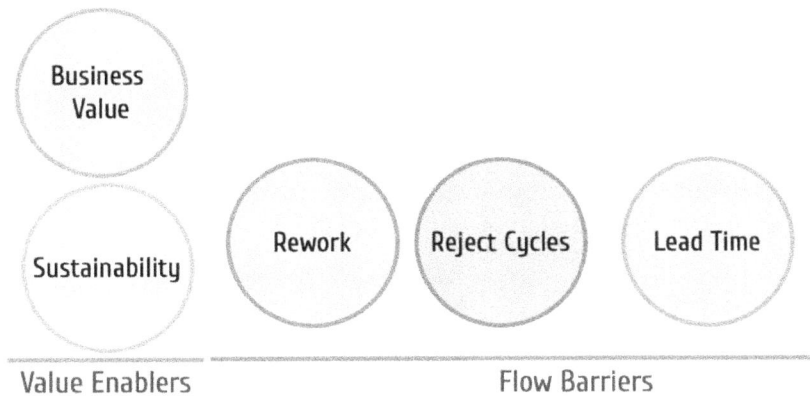

Figure 5.5 – Solution architect's key objectives

As we can see, solution architects maximize business value by quickly delivering the one fit for the purpose and sustainable solutions. They also focus on eradicating flow stoppers such as reject cycles, excessive lead time, and the cost of rework of the solutions being delivered. We'll look at these in more detail in the following subsections.

Starting with business value

To maximize the flow of value, architects need to ensure every piece of the architecture work delivers value to customers frequently. Complex architecture and design, future proofing based on imaginary, nonexistent requirements, and using technologies for the sake of using them are unnecessary architecture work that slows down real value to customers.

Simon Sinek, an industry-known motivational speaker, explained that you should always start with why because *"people don't buy what you do; they buy why you do it."* In agile teams, the Product Owner always starts with *why* a particular feature is requested and what problems it will solve. Solution architects need to understand the value of technical features. We will see more of the business value aspect in *Chapter 6, Delivering Value with New Ways of Working.*

Focusing on sustainable quality

An excellent article published in the Journal of Systems and Software, *Software Sustainability: Research and Practice from a Software Architecture Viewpoint*, explains that software sustainability is the ability of a software system to be resilient to changes by evolving its structure. This article defines many software metrics that are useful for measuring sustainability at the architecture and architecture knowledge levels across maintenance and evolution.

While focusing on enabling a faster flow of delivery, solution architects need to have a healthy balance between speed and sustainability, as we discussed in *Chapter 2, Agile Architecture – The Foundation of Agile Delivery*. The most important aspect of sustainability is to avoid accumulating technical debts, which increases software maintenance and operations liabilities, decays speed of delivery, and causes code smells, architecture erosion, and deterioration of software quality.

Reducing reject cycles

It is the responsibility of solution architects to curtail late surprises due to incomplete, unclear, and inaccurate architecture shared with teams. These are called *reject cycles*. Reject cycles significantly impact the flow rate.

Scaled Agile Framework (**SAFe**) uses the concept percent **complete and accurate** as a good measure for reject cycles. Percent complete and accurate can be measured by asking downstream customers what percentage of the time they receive work that's *usable as is.*

Solution architects focus on minimizing reject cycles by adopting customer-centric design thinking, always considering the field of usage, keeping the definition of ready in mind, using an architecture test case-first approach, employing sufficient modeling and simulations, making collective decisions, and using spikes effectively.

Minimizing lead time

Lead time in the continuous flow context is the elapsed time for moving a work item for the next action. The lead time is measured from the time the solution architect draws down a backlog item for emergent design to the time it is ready for development, as per the definition of ready.

Careful planning is required to avoid excessive and unplanned lead times. Identifying architecture decisions needed for a set of backlog items, planning and prioritizing in advance and considering the development timeframes, transparently sharing those decisions across all stakeholders, and continuously advancing decision backlogs by eliminating assumptions are essential in minimizing lead time.

Reducing rework

Excessive rework can significantly impact the velocity of value delivery. Rework and refactoring are measured against the effort to redevelop software and result in non-software costs such as technology cost, infrastructure cost, upskilling cost, and so on.

Adopting and implementing evolutionary solution architecture practices is critical for reducing rework. Early detection, correction, and refactoring of defects, continuously addressing tech debts, keeping the architecture simple, making high-quality decisions, using the last responsible moment for choosing the right option, and promoting continuous delivery pipelines help reduce rework.

To achieve these objectives, solution architects perform several activities throughout the life cycle of the solution. In the next section, we will look at the different duties that are performed by solution architects.

The duties of solution architects

Solution architects engage customers, business stakeholders, agile teams, external suppliers, and various IT teams to define and evolve quality technical solutions with an economic view of helping value at scale being realized faster.

Solution architects' duties are not distinctly different in agile software development, but the way of delivering architecture in collaboration with teams is significantly different from traditional practices. Agile solution architects and their duties are also not different across various scaling agile frameworks.

SAFe defines two architect roles in the solution space – Solution Architect and system architect. The roles and responsibilities of both solution architects and system architects are the same, with a difference in operating scope. Solution architects are part of large solution trains and are responsible for stitching end-to-end solutions together across multiple **Agile Release Trains (ARTs)**. Solution architects and system architects with product management and release train management functions form the *troika* at the solution train and ART level.

A solution architect in **Disciplined Agile (DA)**, also called an *architecture owner*, is someone on the team who's responsible for leading the architecture's evolution. DA also recommends having everyone in the team take responsibility for the architecture. The four key activities specified by DA concerning solution architects are envisioning the initial architecture, working with the development teams, communicating the architecture to stakeholders, and updating architecture work products.

There is no explicit solution architect role in **Large-Scale Scrum (LeSS)**. Instead, LeSS assumes a master programmer in the team performs solution architect duties.

Developing an intentional architecture, preparing for an emergent design, and enabling the continuous evolution of solutions are the three fundamental duties of solution architects in agile software development. These three aspects can be seen in the following diagram:

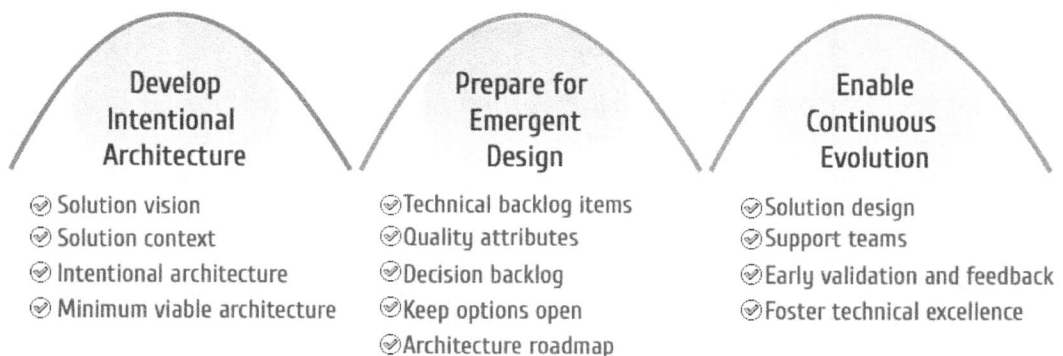

Develop Intentional Architecture	Prepare for Emergent Design	Enable Continuous Evolution
⊘ Solution vision	⊘ Technical backlog items	⊘ Solution design
⊘ Solution context	⊘ Quality attributes	⊘ Support teams
⊘ Intentional architecture	⊘ Decision backlog	⊘ Early validation and feedback
⊘ Minimum viable architecture	⊘ Keep options open	⊘ Foster technical excellence
	⊘ Architecture roadmap	

Figure 5.6 – Key duties of Agile for the solution architect

While the intentional architecture is primarily applied to new solution demands, preparing emergent design and enabling continuous evolution are part of the solution architect's everyday job. We will explain these three key duties in more detail in the following subsections.

Developing an intentional architecture

A software delivery cycle kicks off either when an enterprise discovers a new opportunity to drive innovation or to fix a bleeding business problem with a potential technology change. Both cases trigger *solution exploration* – a path to identifying the right-fit solution that satisfies those demands. Solution architects collaborate with enterprise architects, thus significantly contributing to developing solution vision, solution context, and an intentional architecture by supporting critical early architecture decisions such as buy versus build.

The key steps under the solution exploration phase can be seen in the following diagram:

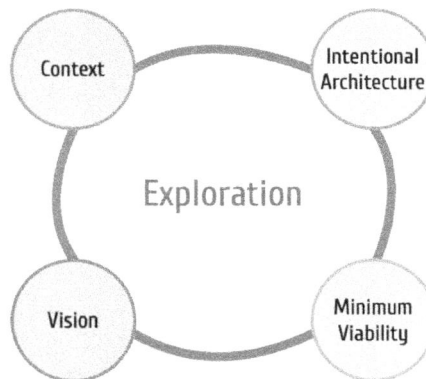

Figure 5.7 – Different steps under the solution exploration phase

With a clear understanding of the business's strategic objectives, needs, and constraints, solution architects intensively collaborate with other stakeholders to build upon the business idea. In the exploration phase, they gather market insights and trends, seek opportunities to leverage currently operational systems, envision the desired architecture, and simulate spikes to identify potential candidate solutions.

Bringing a shared understanding with the solution vision

The solution vision is an important outcome that captures the solution's operational purpose, users, value, and scale of use. While the Product Owner is responsible for building the solution vision, solution architects provide technical expertise and quality constraints for the intended solution.

SAFe uses the term solution vision to create a shared understanding across various stakeholders. Solution vision is a key element that gives clarity and direction for all stakeholders linked to the solution development. As shown in the following diagram, solution vision captures the most critical aspects, such as the purpose of the solution, the consumers, and their benefits.

It is important to keep the vision as simple as possible so that stakeholders can understand it easily. In the previous chapter, we introduced the scenario of building an **Automatic Vehicle Tracking System** (**AVTS**) solution. We will continue to use AVTS as a reference scenario in this chapter. The following diagram captures the AVTS solution's vision:

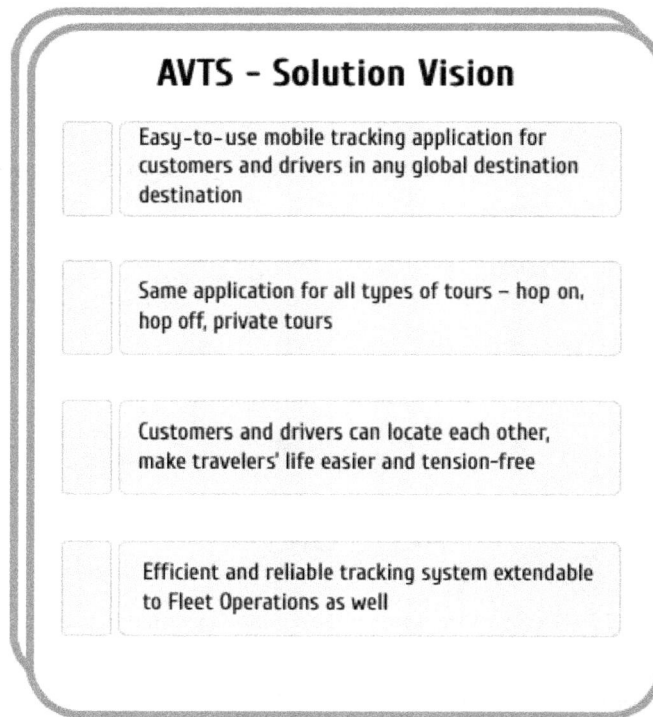

AVTS - Solution Vision

Easy-to-use mobile tracking application for customers and drivers in any global destination destination

Same application for all types of tours – hop on, hop off, private tours

Customers and drivers can locate each other, make travelers' life easier and tension-free

Efficient and reliable tracking system extendable to Fleet Operations as well

Figure 5.8 – AVTS solution vision

The vision shown in the preceding diagram covers consumers, the purpose of the solution, and the benefits of AVTS in a few sentences. In this case, AVTS Epic owners work in close collaboration with the enterprise architect, solution architects, and product management from the *Fulfilment ART* and *Fleet Ops ART* to collectively define a solution vision. Next, we will review the solution context.

Setting expectations on the operating environment with the solution context

The solution context brings about clarity and, therefore, sets the expectation of the solution's operating context, such as their deployment environment, users, geographical locations, end user devices, integration points, and so on. In the IEEE journal article *The System Context Architectural Viewpoint*, Eoin Woods defines system context as a combination of the **system** under consideration, external **entities** such as dependent systems, the environment and people, and **connections** such as the interface, protocols, and **connectors** that link the system to its external entities.

The solution context provides an important visualization for teams to understand the big picture view of the solution. The AVTS solution context is shown in the following diagram:

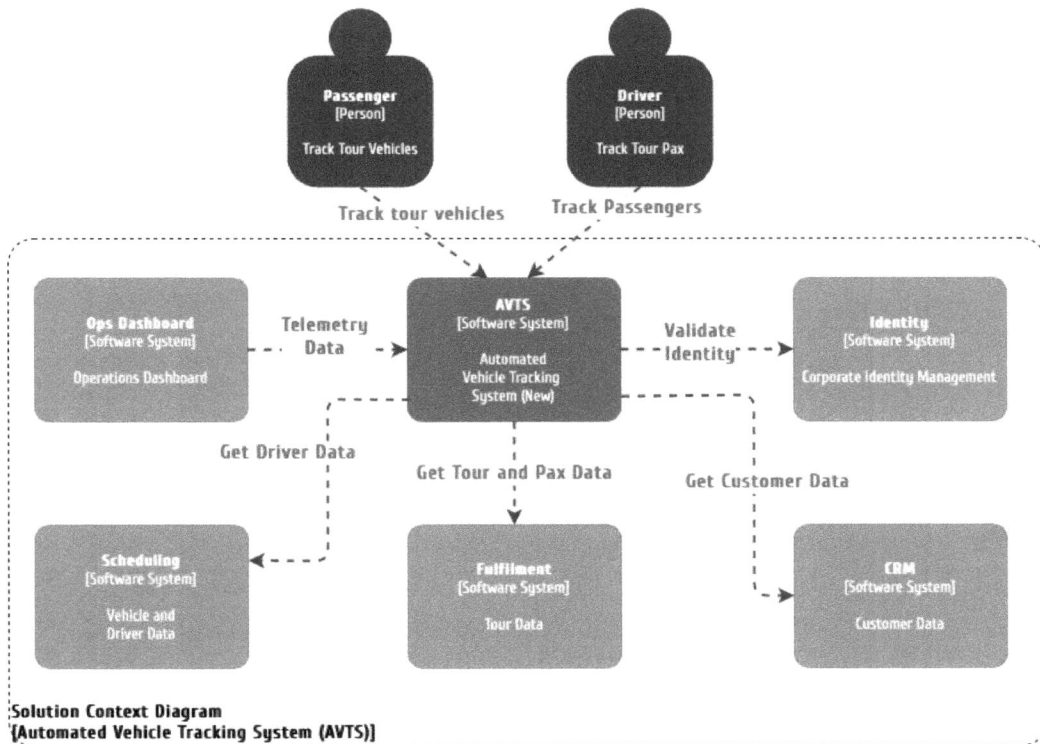

Figure 5.9 – AVTS solution context

As shown in the preceding diagram, the solution context captures organization boundaries, users, interactions, and integration points. This diagram is based on the C4 model, which we will discuss more in *Chapter 10, Lean Documentation through Collaboration*.

Guiding solution designs with intentional architecture

Solution architects develop an intentional architecture as a conceptual view of the solution combined with a set of guidelines, principles, and decisions. Some of these decisions and principles are non-negotiable as they are strongly linked to the business case. Therefore, any deviations may invalidate the business case. Some parts of the intentional architecture may evolve this new information. The development of the solution vision, context, and intentional architecture are not sequential; they influence each other and evolve together during the exploration phase.

The intentional architecture for the AVTS scenario captures the overall conceptual view of the solution, as well as its set of principles and guidelines, as shown in the following diagram:

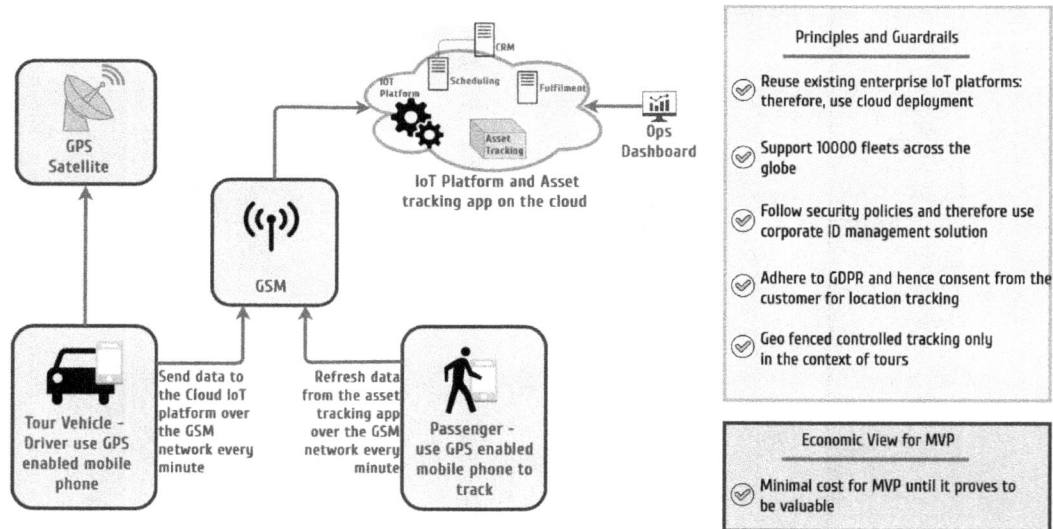

Figure 5.10 – AVTS intentional architecture viewpoint

This diagram contains three components: the conceptual architecture, its principles and guardrails, and its economic constraints. The internal architecture is simple enough to convey the overall architecture to both IT and business stakeholders.

Realizing early value with Minimum Viable Product (MVP)

Early and faster delivery of value without substantial investment is critical in agile software development. Solution architects contribute significantly to identifying the MVP's scope for the potential solution by considering business value, architecture viability, and optimal cost of delivery. At Snow in the Desert, it is a mandatory policy that everything must start with an MVP and must prove its value before large investments are made.

As recommended by SAFe, development starts with a *Lean startup* to detect any risks and realize early value. At Snow in the Desert, solution architects, the enterprise architect, and product management work together to define the MVP scope. A team led by solution architects defines the **Minimum Viable Architecture** (**MVA**) for the MVP. We will discuss this later as part of the solution design.

Based on the organization's strategic funding mechanisms, a Lean business case may be prepared for approval at the end of this exploration phase. After the exploration phase, architects prepare for emergent design, which we will discuss in the next section.

Preparing for emergent design

The intentional architecture, business case, and MVP execution plan form the foundation for teams to start development sprints. Fixed cadence-based backlog refinement, led by the Product Owner, is a substantially significant activity that's done ahead of development to ensure the **Definition of Ready** (**DOR**) is met when teams are ready to kick off development.

The backlog refinement phase, from the lens of the solution architect, can be seen in the following diagram:

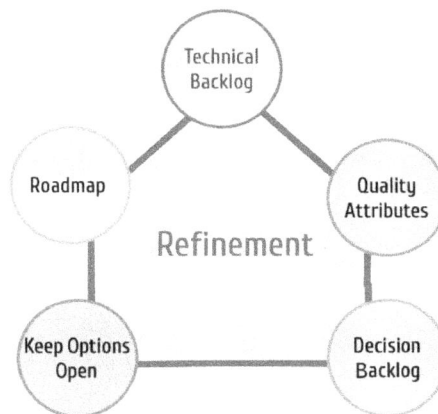

Figure 5.11 – Solution architect's role in backlog refinement

The elements in the refinement model mentioned in the preceding diagram are not sequential. The following section will explain this in more details.

Enabling sustainability with technical backlog items

Backlog refinement provides solution architects with the opportunity to identify technical components, platforms, and frameworks, such as introducing an API gateway platform, in order to build long-lasting, scalable solutions. Besides, solution architects also identify the work that's needed to enable quality for new solutions. They also focus on necessary work for improving the quality of existing solutions, such as performance improvements, remediating tech debts, upgrading the infrastructure, automation, and so on. For such technical backlog items, solution architects act as Product Owners, take responsibility for defining business value, write down requirements and acceptance criteria, and lead refinement activity.

SAFe defines four types of technical backlog items, called **enablers**, as follows:

- **Infrastructure**: New and upgrades of existing infrastructure components for development, build, test, and production environments.
- **Exploration**: Discover new solutions via industry research, repositioning and repurposing existing solutions, and prototyping innovative solutions.
- **Architecture**: Introduction of new technology components, architecture spikes, and software upgrades to evolve solutions and systems.
- **Compliance**: Build and validate local, enterprise, and industry compliance, which is mandatory for software solutions.

SAFe uses architecture runway to capture the enablers that are required for building business features and is continuously evolving as teams progress through iterations. These enablers are implemented just in time to meet the timelines of its dependent business features, thereby avoiding delays due to the non-availability of technical components.

In the AVTS MVP context, the following diagram captures MVP features and their enablers:

AVTS – MVP Features	AVTS – MVP Enablers
Real-time route tracking for passengers	Tracking API design
Alerts on the mobile application indicating distance and time	Tracking platform
Driver to track the passenger's location	Integration with existing systems
Passenger to share location with driver and request stop	IDAM integration
Support one city, hop off hop on tours	Scale IoT platform

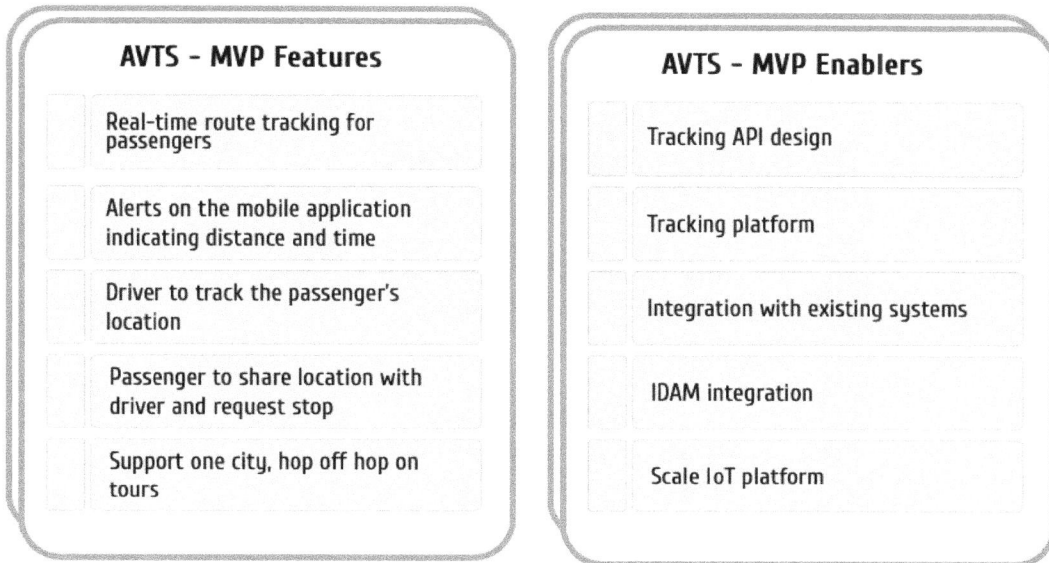

Figure 5.12 – AVTS MVP features and enablers

The enabler features are pushed into the backlog and are then captured in the architecture runway. The architecture runway is an integral part of the process at Snow in the Desert. Implemented in Jira, the architecture runway is consistently used as a planning and communication tool for transparently sharing architecture works. We will review the architecture runway for AVTS later in this chapter.

Promoting built-in quality with quality attributes

Quality attributes generally start with the needs of the business, such as the number of calls to be served in an hour in a call center scenario. Teams may discover and add further quality attributes as time progresses. It is important to understand the business implications of the quality attributes, such as revenue loss, dent in customer experience, increased run cost, and so on, since developing quality attributes costs money. Consistent use of the trade-off framework and economic view is important in making the right decisions that deliver quality.

In the context of AVTS, as an example, the following quality requirements are necessary for meeting customer satisfaction:

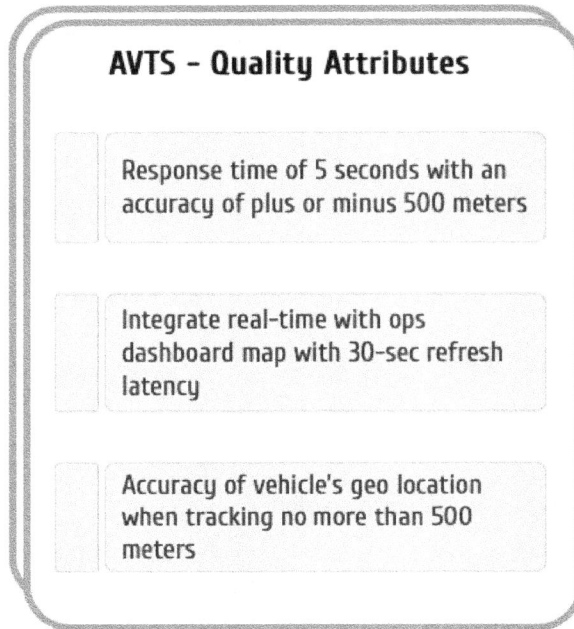

AVTS - Quality Attributes

Response time of 5 seconds with an accuracy of plus or minus 500 meters

Integrate real-time with ops dashboard map with 30-sec refresh latency

Accuracy of vehicle's geo location when tracking no more than 500 meters

Figure 5.13 – AVTS quality attributes for AVTS

Solution architects promote automation of architectures and quality attributes test cases in the continuous delivery pipeline. We will explore quality attributes in more detail in *Chapter 9, Architecting for Quality with Quality Attributes*.

Planning evolution with a decision backlog

Identifying architecture decision points early enough helps us understand and plan timelines for those decisions, to avoid rework. Using different viewpoints such as business, infrastructure, security, data, and so on helps us identify what decisions are required. Solution architects use the divide and conquer technique to break down complex problems into smaller issues without losing their holistic vision.

The following diagram shows a list of architectural decisions in the context of AVTS MVP:

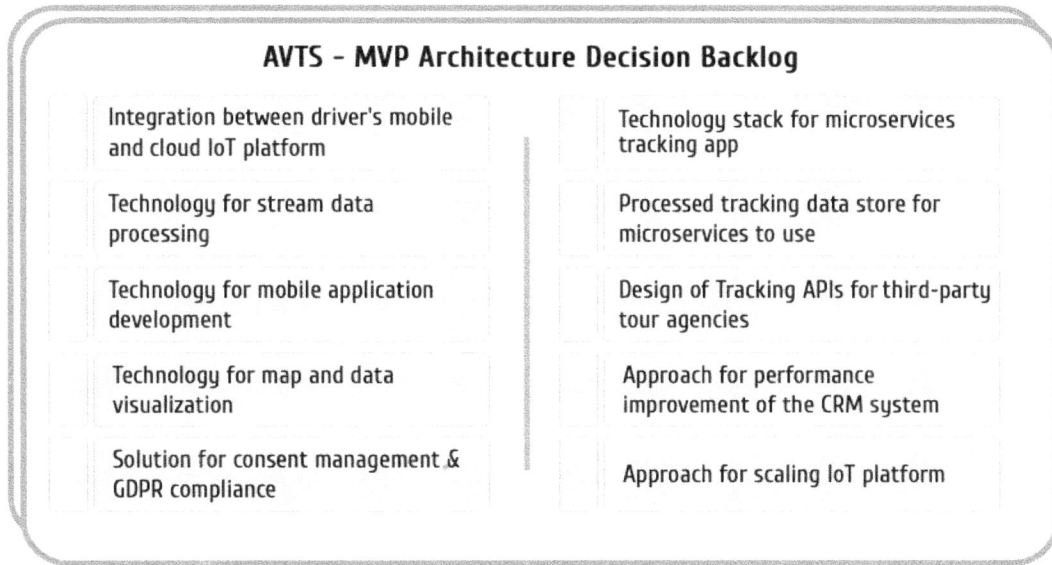

AVTS – MVP Architecture Decision Backlog

Integration between driver's mobile and cloud IoT platform

Technology for stream data processing

Technology for mobile application development

Technology for map and data visualization

Solution for consent management & GDPR compliance

Technology stack for microservices tracking app

Processed tracking data store for microservices to use

Design of Tracking APIs for third-party tour agencies

Approach for performance improvement of the CRM system

Approach for scaling IoT platform

Figure 5.14 – AVTS MVP architecture decision backlog

At Snow in the Desert, architecture decision backlogs are maintained at the ART level in Jira to track architecture decisions, similar to technical backlog items. Solution architects are mandatory participants in the feature backlog refinement sessions, and they are led by product management to ensure that all the architecture decisions that are required are captured.

Once these architectural decisions have been captured, the solution architects sequence those decisions by identifying the **Last Responsible Moment** (**LRM**) for each decision by keeping options open.

Delaying decisions by keeping options open

Solution architects capture facts, assumptions, and hypotheses for each solution option for every decision backlog item. Continuously analyzing option sets is required for eliminating unknowns, which helps in narrowing down the available options. Eliminating options is typically done by collecting facts, doing market research, monitoring architecture spikes, modeling, and performing simulations against each option.

One of the agile solution architect's key objectives is to delay decisions until sufficient information is available for decision making. However, a few decisions are still required upfront. Two upfront decisions that need to be made in the AVTS scenario are concerned with the device to be used on vehicles and whether to buy or build the AVTS solution. These decisions fundamentally impact the solution's delivery from the first day.

Solution architects at Snow in the Desert use set-based design to come up with a definite solution. We will explore set-based design in more detail in *Chapter 6, Delivering Value with New Ways of Working*.

Two options were identified for collecting signals from tour vehicles – use GPS tracking devices fitted onboard vehicles or use drivers' mobile phones. Both possibilities were explored in parallel and evaluated against the trade-off framework. Since the economic guardrail suggests minimizing investment, the architects settled for using drivers' mobile phones.

For the second decision, two options were considered – build in-house or use a third-party service. The team meticulously evaluated both options and decided to do it in-house since most of the infrastructure, including the IoT platform and data platform, is already available. Snow in the Desert has the excellent ability to develop mobile applications at a low cost, and they have expertise in integrating CRM and the Fulfilment system. Beyond vehicle tracking, an in-house solution offers the opportunity to extend for the enterprise's future needs, such as tracking for route optimization. Overall, building in-house was deemed commercially more viable and efficient. We will review these decisions in the context of architecture trade-off analysis in more detail in *Chapter 9, Architecting for Quality with Quality Attributes*.

The rest of the decisions are not required urgently and hence are sequenced for later decision making by exploring potential options further.

Visualizing progress with a roadmap

A typical architecture roadmap is a time series Gantt chart with a time horizon no more than 6 to 9 months ahead of the current sprint. Anything beyond this timeframe generates unnecessary work without value and can cause disconnection from the realities of development. A roadmap captures a mix of architecture decisions, technical enablers, significant upgrades to existing systems, and improvements.

The following diagram shows the architecture runway for AVTS MVP, which spans across three **Program Increments (PIs)**:

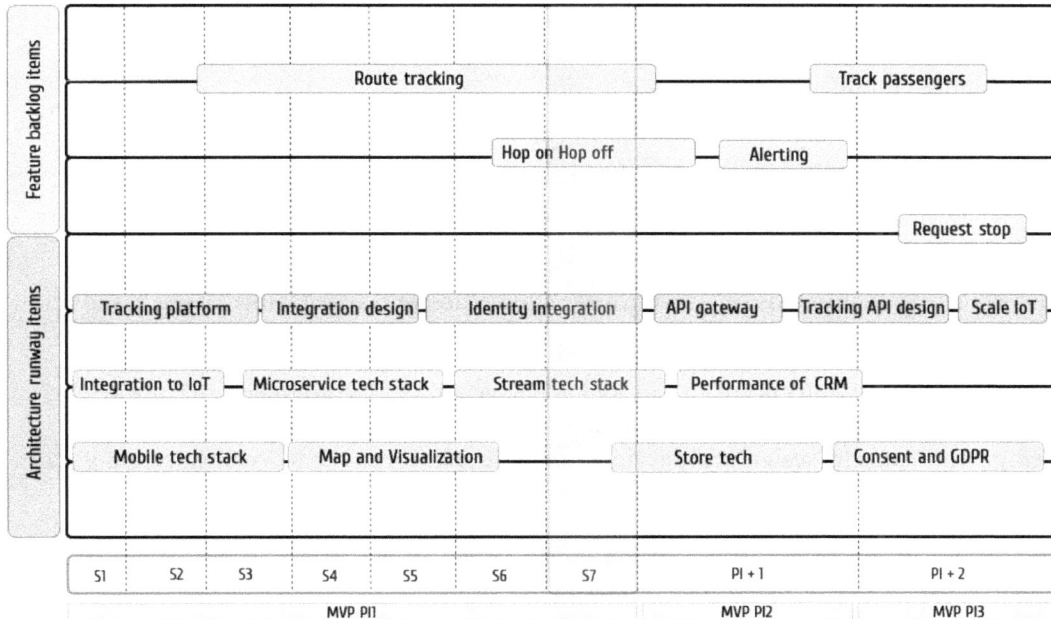

Figure 5.15 – AVTS architecture runway

The architecture runway at Snow in the Desert is implemented using the Jira advanced roadmaps plugin. It is a filtered view of the PI roadmap, with the necessary tags implemented for architecture runway items.

As shown in the preceding diagram, MVP PI1, MVP PI2, and MVP PI3 are three PIs for AVTS MVP. PI is a concept used by SAFe for scoping work for the next 6-7 sprints. Often, the seventh sprint is more for innovation and improvements. A smart, clear plan for three PIs helps architects work on architecture enablers and decisions ahead of development.

Every PI starts with a PI planning session with three segments – pre-planning, planning, and post-planning – and this is typically conducted over 2 days. Solution architects play an incredibly important role in the PI planning process. Pre-planning helps to build a plan for the next PI based on the capacity of the available ART resources.

During the PI planning phase, the solution and system architects are responsible for the following:

- Presenting the architecture vision, elaborating on Non-Functional Requirements (**NFRs**), discovering dependencies, and sharing implementation patterns and best practices for development

- Ensuring team alignment and risk mitigations by hovering across team breakouts and answering any solution architecture questions

- Reviewing, reflecting, communicating, and sharing feedback on the PI event as part of the post-PI ceremonies

- Adjusting solution vision, intentional architecture, and architecture scope based on feedback from PI planning

PI planning helps architects create a locked-down version of the architecture runway for the next set of sprints. With this clarity, solution architects work closely with the development teams to design and develop evolvable solutions. The following section goes more in-depth into solution design and development aspects.

Enabling continuous evolution

Collaboration for decision making and collective ownership are fundamentals of decentralized decision making in agile software development. Solution architects continuously reinforce the vision, purpose, and usage context of the solution to all concerned stakeholders to ensure they contribute with a sense of purpose.

The following diagram captures various aspects of the solution architect's collaboration and contribution in the context of solution design, development, and support:

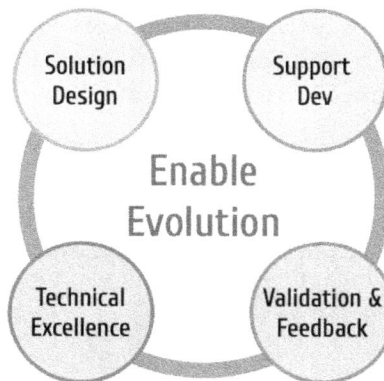

Figure 5.16 – Solution architect's collaboration points in design, development, and support

These collaboration points will be explained in the following subsections.

Collaborating for solution design

To achieve collective decision making without handoffs, a collaborative solution design workshop ahead of development but after backlog refinement is significant. Solution architects use intentional architecture, a customer-centric mindset, and design thinking to drive, align, and protect the overall architecture vision. In some cases, being educated about technologies and architecture techniques is necessary, in the context of the design workshop, for level settings. Solution architects lead fact-based discussions, prototyping, and spikes to effectively make decisions in case of conflicts.

The concept of a solution design workshop comes from LeSS. LeSS stresses the importance of design workshops being similar to requirements workshops and backlog refinement. In design workshops, all interested parties come together, usually at the beginning of every iteration. Solution design workshops are timeboxed for 2 hours to 2 days, creating opportunities for cross-pollination and knowledge sharing and reducing wastage and handoffs. Massive whiteboards are used for solution modeling. Keeping these whiteboard models on the wall throughout the iteration helps teams have inspiring conversations as they embark on the development journey. Multi-team design workshops are used for designing inter-team architecture dependencies, which occur once every few iterations. Design workshops focus on modeling designs for achieving the next desired state.

The idea of **Model-Based Systems Engineering** (**MBSE**), as recommended by SAFe, emphasizes the importance of modeling to achieve rapid learning and validation of the design. DA also recommends focusing on model-based designs as a fundamental mechanism for discovering faults early. As we mentioned earlier, Snow in the Desert uses a multi-team design workshop to design the solution collectively in order to develop design models.

The following diagram is an example output of a design workshop. It's a C4 container model that captures the MVA in the context of AVTS:

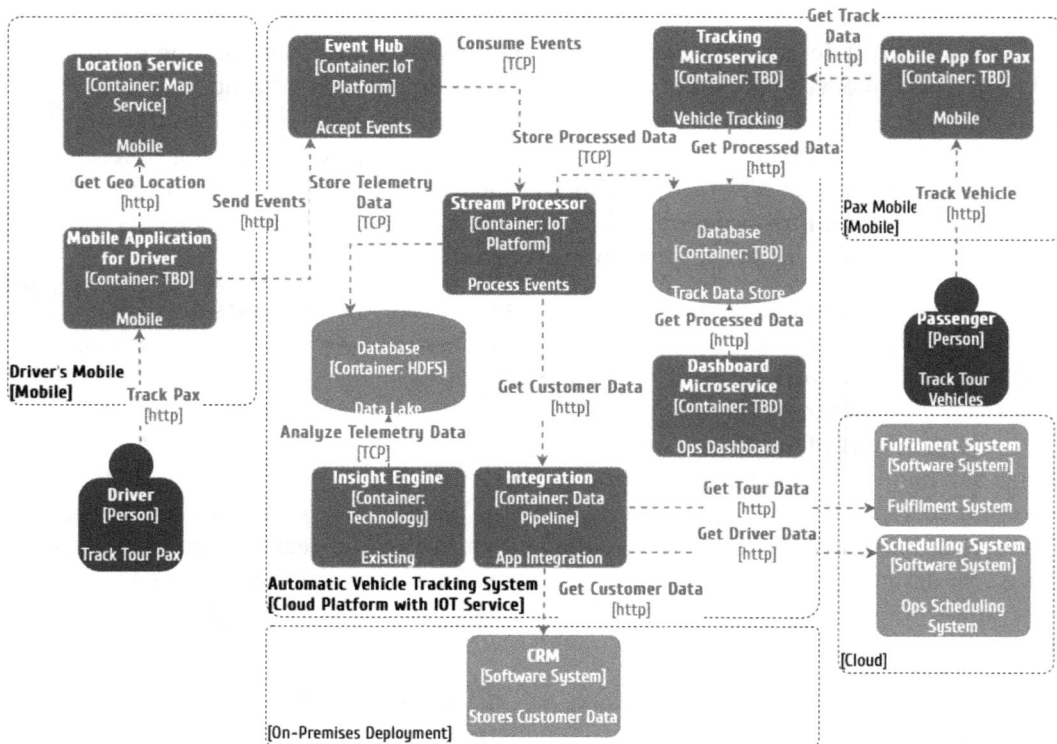

Figure 5.17 – AVTS MVA

The MVA shown in the preceding diagram has some technology decisions already embedded, but more decisions have been taken, all of which are deferred to the last possible moment. One of the other outcomes of the design workshop is having **Architecture Decision Records** (**ADRs**), which we will discuss in *Chapter 11, Architect as an Enabler in Lean-Agile Governance*. The MVA helps teams prioritize features for the initial sprints.

Supporting teams in development

Solution architects develop and share architecture patterns as code instead of large-scale documentation for higher productivity and consistency across development teams and team members. During sprints, solution architects reserve a fixed percentage of capacity for supporting teams in decision making, reviewing code, attending team ceremonies, and optionally taking part in coding. Solution architects work with teams to understand and resolve any architecture and technical impediments at the earliest possible moment.

LeSS proposes the concept of *tiger teams*. A tiger team is formed at the beginning of a project or when a large piece of architecture work is required. A tiger team is a combination of great programmer and architects put together in a collocated workspace. Once the tiger team completes their critical tasks, they go back and join feature teams and help them bootstrap faster. LeSS also recommends having a **System Architecture Documentation (SAD)** workshop at the end of the tiger team phase to disseminate the knowledge acquired during the initial architecture phase to wider teams. SAD workshops are generally conducted after the first cut coding stage is completed. The SAD workshop looks back at the finished system and then models it for communication, which helps them learn about the current architecture and collectively identify improvement areas.

Aligning with early validation and feedback

Demos and retrospections are the right avenues if you wish to align solutions against their intentional architecture and vision. Frequent demos and continuous delivery pipelines help the architecture to be validated in an unobstructive manner without compromising or hurting their autonomy and empowerment. Getting early feedback and detecting defects both offer a safety net for teams and therefore enable fearless innovation with psychological safety.

Solution architects attend ceremonies such as iteration planning reviews and demos. They effectively use these forums as a feedback mechanism. *Architect Sync* is one of those mechanisms in SAFe for architects that's used across multiple ARTs to collaborate with enterprise architects so that they can discuss and align solutions. Solution architects take part in a number of ceremonies, such as backlog refinement sessions, PI planning events, ART sync, architect sync, inspect and adapt, demos, retros, stand-ups, and so on, to ensure they stay connected with the team and flow of work.

Solution architects at Snow in the Desert use health dashboards at the ART and solution level to visualize architecture effectiveness against the five pillars of the **Well-Architected Framework (WAF)**, which has been adopted from AWS. Dashboard visualizations are done through integrating with live monitoring solutions, performing periodic solution assessments, and continually monitoring the mitigation statuses of tech debts.

Fostering technical excellence

Technical alignment is key across teams for developing a coherent solution that's aligned with a shared vision. Solution architects shepherd teams toward achieving this common purpose by aligning on technologies, tools, design patterns, and coding styles. Upskilling and cross-skilling team members are vital for improving the craftsmanship of developers. Solution architects lead with a growth mindset to guide and mentor teams and help cross-skill and upskill on today's technology needs and the future.

The continuous evolution of solutions needs the collective effort of all team members, and solution architects hold these efforts together. Solution architects succeed in agile development projects if they diligently perform these duties. Moving on to the next section, we will explore measuring success in more detail.

Measuring the success of solution architects

Similar to the measurement framework, which is used for measuring enterprise architects, another Lean and simple framework-based approach is recommended for measuring solution architects' success. This measurement framework is aligned with the five objectives we discussed earlier in the *Maximizing value and eliminating flow barriers* section. Solution architects, together with their key stakeholders, review the architect's performance on a relative scale in a retrospective meeting-style environment.

The following figure shows the measurement of solution architects on a five-point scale:

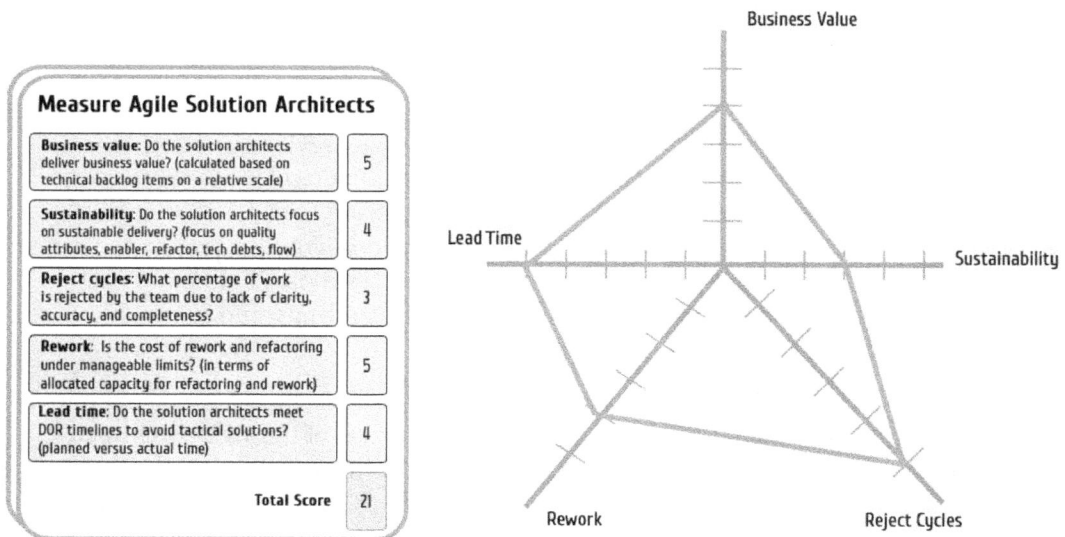

Figure 5.18 – Framework for measuring the success of solution architects

As shown in the preceding figure, for each parameter, the solution architects collaborate with other stakeholders to decide on a score between 1 and 5. The spider chart on the right of the preceding figure shows a visual view of the outcome.

Once the overall score is calculated by summing individual scores, you must use the maturity model shown in the following diagram to reflect the current maturity state of the solution architects:

5 (20-25)	Fly	Solution architects consistently maintain a healthy portfolio
21		
4 (15-20)	Run	Solution architects deliver value and dramatically improve the flow
3 (10-15)	Walk	Solution architects focus on sustainable solutions balancing agility
2 (5-10)	Crawl	Solution architects are part of the team, focus on the urgency of now
1 (0-5)	Pre-Crawl	Solution architects do not exist even for large projects

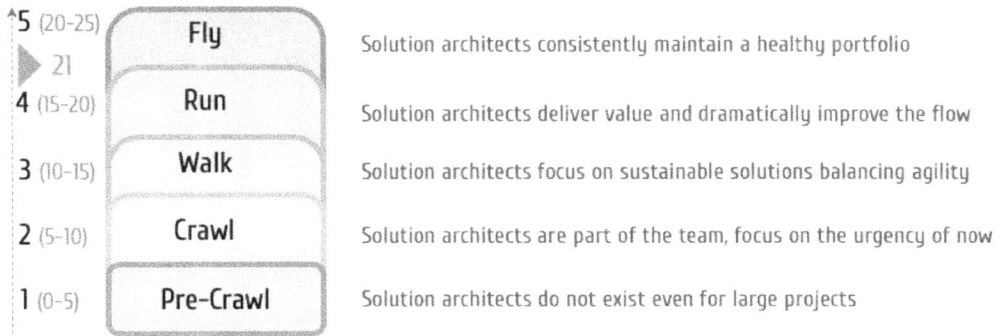

Figure 5.19 – Agile solution architect's maturity model

As we discussed in the previous chapter, frequent measurement of the scores, collaboratively with all stakeholders using a postcard-based approach, is advisable.

Summary

Solution architects in agile teams are just as important as the Product Owner. Solution architects play a technical leadership role in delivering high-quality, sustainable solutions with an optimal cost that meets the customer's needs faster without compromising quality. They ensure the right problems in the right context are solved in the right way and guide teams in implementing it correctly.

The solution architects in agile teams operate with a focused objective of increasing business value and sustainability. They also identify and eliminate flow stoppers by minimizing rework, lead time, and reject cycles. Solution architects work on exploring new solutions by defining intentional architecture, together with solution vision and context. They work with the Product Owner in refining backlog items to identify technical needs, quality attributes, and architecture decisions. These technical enablers and architecture decision backlog items are placed on a 3- to 9-month roadmap based on the LRM. Solution architects collaborate in design workshops, support development, look for early validation with rapid feedback cycles, tirelessly work on improving live systems, and foster technical leadership. Finally, solution architects are frequently and repeatedly measured based on business value, sustainability, rework, lead time, and reject cycles.

Enterprise architects and solution architects work hand in hand, cross over in their elevator journey, and share responsibilities. Both roles are critical for the success of agile software delivery. In the next chapter, we will look at the tips and techniques agile architects can implement to demonstrate value.

Further reading

To learn more about what was covered in this chapter, please take a look at the following links:

- **Different architect roles**: `https://www.iasa.se/wp-content/uploads/2009/08/TAJ15.pdf`

- **Thinking beyond technology**: `https://www.slideshare.net/alanmcsweeney/why-solutions-fail-and-the-business-value-of-solution-architecture`

- **Software sustainability**: *Research and Practice from a Software Architecture Viewpoint:* `https://www.researchgate.net/publication/321940604_Software_Sustainability_Research_and_Practice_from_a_Software_Architecture_Viewpoint`

- **The System Context Architectural Viewpoint**: `https://www.researchgate.net/publication/224605931_The_System_Context_Architectural_Viewpoint`

Section 3: Essential Knowledge to Become a Successful Agile Architect

In this part of the book, you will get a good grasp of some practical challenges beyond roles and responsibilities. It also explains some tricks and tips around how to address these challenges.

This section contains the following chapters:

6
Delivering Value with New Ways of Working

"In pure architecture the smallest detail should have a meaning or serve a purpose."

– Augustus W. N. Pugin (designed Big Ben)

So far, we have covered the concept of **agile architecture** and the duties of agile architects. This chapter will focus on the techniques needed to ensure value is delivered in every aspect of architecture work.

The highest priority in agile software development is to satisfy customer needs through the early and continuous delivery of value. The Lean principle **Muda** states that you should eliminate waste by avoiding non-value-added activities to deliver value to the customer. Architecture is no different; every architecture activity has to deliver value. Continuous and timely designed architecture solutions help in avoiding rework and improve quality and sustainable delivery. The on-time delivery of technical backlog items is only possible through the strict prioritization of functional and technical backlog items. Deferring decisions by keeping options open presents more opportunities to make the right decisions. These decisions need to be pragmatic by understanding and balancing the context of the customer at that specific point in time. The right scoping of MVP considering risk, value, lead time, and cost helps the business be in a comfortable position for large-scale investment decisions.

This chapter will examine the importance of agile architecture's business value and how architects can ensure they focus and deliver value with a Lean mindset. We will go through a handful of essential techniques useful for agile architects to be successful.

In this chapter, we're going to cover the following main topics:

- Understanding business value
- Linking architecture activities to the business backlogs
- Determining the business value of architecture
- Allocating capacity for prioritization
- Making the work transparent
- Looking ahead of development
- Working with a pragmatic mindset
- Keeping options open
- Delivering early value with MVA
- Managing tech debt

This chapter focuses on the **Value Delivery** focal point of the **Agile Architect's Lens**, shown in the following figure:

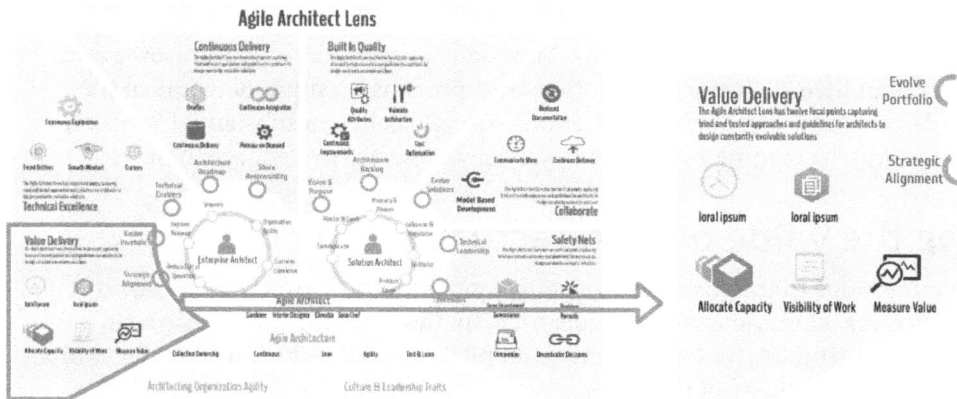

Figure 6.1 – Value Delivery – focal point

Throughout this chapter, we will cover examples from Snow in the Desert to reinforce our learning. We will also cover perspectives from different agile scaling frameworks.

Technical requirements

Additional materials related to this chapter are available at the following GitHub link for download: https://github.com/PacktPublishing/Becoming-an-Agile-Software-Architect/tree/master/Chapter6.

Understanding business value

In this section, we will learn about business value and understand that demonstrating the value of architecture is far from easy.

Delivering business value faster and more often is the most important aspect of agile software development. One of the Manifesto signatories for agile software development, Jim Highsmith, defines business value as *when a working software is delivered with features that customers accept, at the highest level of quality, and within acceptable cost and time limits.* Business value is created as a response to actual customer demands.

Agile organizations progressively shift their focus from shareholder value to ways to delight customers continuously and rapidly. Most agile organizations do not view profit as a direct result of their delivery; instead, they treat profit as a natural outcome of the continuous delivery of value. Such organizations strongly believe a substantial focus on profit leads to short-term thinking, sub-optimal deliveries, and low engagement.

Showing the value of architecture is difficult

While delivering value to the customer is significantly important, overly focusing on value by eradicating all non-customer-requested activities from the flow is also a sin. Such organizations tend to give low priority to architecture and technical activities, which eventually impacts the quality of the product and the sustenance of the delivery flow. Agilists in such organizations use **The Big Lebowski Dude's law** to justify their actions. The Dude's law is defined as $V = W/H$, where **Value (V)** is equal to **Why (W)** divided by **How (H)**. If W is what the customer expects, V increases when time spent on H is minimized. Since the customer does not directly request many architecture activities, they are classified as overheads contributing to H in such organizations.

Sustainability is a crucial aspect that often goes unnoticed and needs a lot of architecture and design effort. In the paper *Lean Primer*, Craig Larman and Bas Vodde reiterate the importance of maintaining *sustainability while delivering value fast*. The authors consider *sustainable shortest lead time, best quality and value (to people and society), most customer delight, lowest cost, high morale, and safety* as part of the sustainable delivery principle. The authors further provide evidence by stating "*Toyota strives to reduce cycle times, but not through cutting corners, reducing quality, or at an unsustainable or unsafe pace; rather, by relentless continuous improvement*". The result of ignoring sustainability is adopting tactical solutions and sub-optimal designs and a lack of quality considerations. Even though shortcuts increase the velocity of delivery initially, they substantially impact the sustainability of the velocity of the flow over time, as discussed in *Chapter 2, Agile Architecture – The Foundation of Agile Delivery*.

The real value of architects is their ability to deliver sustainable quality solutions. However, in many agile organizations, architects fail due to their inability to show value. Businesses worry only about product features, not architecture. Since architecture work is invisible to customers, architects need to spend time and energy *selling* their work. They have to use business language and economic terms to sell architecture.

As shown in the following diagram, architecture values need to be translated in terms of risk, cost, customer satisfaction, and competitive advantage:

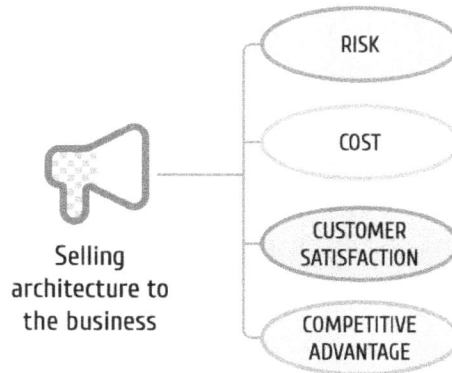

Figure 6.2 – Business language for selling architecture

The paper *Business Value of Solution Architecture* by Raymond Slot, Guido Dedene, and Rik Maes presents a quantitative measure of solution architecture's business value in software development projects. One of the parameters they used for measurement is customer satisfaction. They observed that *customer satisfaction is directly correlated to the experience levels of the architect – a lower experience level results in lower customer satisfaction.*

The rest of this chapter focuses on techniques that are useful for demonstrating the value of architecture work. The next section inspects how to make architecture backlog items more relevant.

Linking architecture activities to business backlogs

Architects need to adhere to the discipline of consistently using backlogs as the sole source of work assignments. The simplest way for architects to get their work noticed is to have a single backlog for both functional and technical backlog items following the same cadence of the team. Once architecture backlog items are captured within the same product backlog, it is easy to bring visibility by linking them using meaningful stereotypes such as **blocked by**, **related**, and **depends**.

There are three types of scenarios that need different approaches for selling to non-technical stakeholders, as explained here:

- **Direct**: The business understands these types of technical backlog items as there is a direct correlation with a business backlog item. For example, an application user login needs **single sign-on** or to design an Order API for third-party access.

- **Derived**: These technical backlog items are created to support business features. Business stakeholders understand these technical backlog items since they can correlate with business needs, for example, a hardware upgrade to support an upcoming software release, design for GDPR compliance, or platform selection for one of the business features.

- **Indirect**: These technical backlog items are harder to link directly to a business backlog item as they do not directly support business needs. A less technically sound business may not understand the value of such backlog items unless architects demonstrate their value by selling using business language. Examples are a **Continuous Integration and Continuous Delivery (CI/CD)** pipeline, implementing architecture styles and patterns such as microservices, and so on. These backlog items have to be linked to higher-level backlogs such as **Epics**.

In the previous chapter, we saw MVP features and related technical features for an **Automatic Vehicle Tracking System (AVTS)** at Snow in the Desert. Architects at Snow in the Desert diligently link all technical backlog items to business backlog items. They also use a relative scoring mechanism for calculating the estimated business value of enabler functions, as shown in the following diagram:

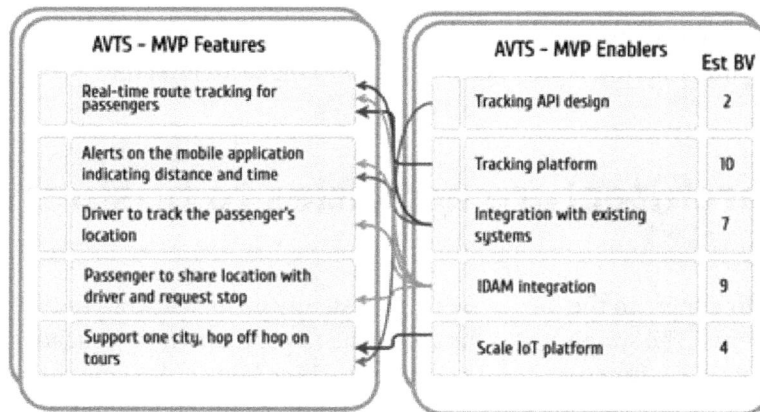

Figure 6.3 – Linking technical features and business value

As captured in the preceding diagram, all technical backlog items are linked to one or more business backlog items. For example, the IoT platform has to be scaled to implement the tracking solution in one city, for hop on hop off. Hence there is a dependency between these two.

Linking technical backlog to business backlog items is important in communicating why a piece of architecture work is undertaken. The next section explains determining the business value of technical backlog items.

Determining the business value of architecture

Since there is no practice in many organizations to measure each backlog item's value, it is harder to accurately gauge actual business value. The usual way is to measure the benefits of the overall initiative in terms of revenue increment, reduction in operational cost, customer happiness, acquisition, and retention.

Agile development uses **perceived business value** primarily for prioritization and measuring team efficacy. The perceived business value can be identified in many ways in close collaboration between the product owner and the business. Business value is generally mapped to profit- and non-profit-related options such as increased revenue, profit margins, better customer care, higher-quality outputs, reduced risks, employee satisfaction, reduced overhead, operations cost, increased service delivery, and so on.

There are complicated methods available for measuring estimated business value, such as cost-benefit analysis, cash flow analysis, net present value, and so on. However, using such sophisticated methods in agile software development is time-consuming and lacks accuracy and guarantee. Therefore, in many agile practices, a simple relative scale-based approach is adopted for defining anticipated business value.

This approach is illustrated in the following diagram:

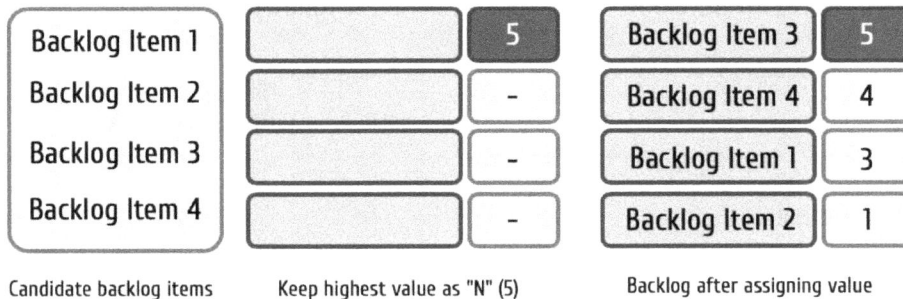

| Candidate backlog items | Keep highest value as "N" (5) | Backlog after assigning value |

Figure 6.4 – Assigning business value to backlog items

As shown in the preceding diagram, for a given set of backlog items, we identify one backlog item with the highest business value and assign a number – in this case, 5. The rest of the features are then relatively valued based on the highest-value backlog item. The same approach is also applicable for determining the value of technical backlog items.

Scaled Agile Framework (SAFe) recommends using a relative scale of 1 to 10 for perceived business value. Business value is calculated by closely engaging business owners at the beginning of every **Program Increment (PI)**. The actual perceived value is calculated at the end of the PI to measure the team's performance.

As shown in *Figure 6.3*, the tracking platform is the most important technical enabler and therefore given the highest business value score. Tracking APIs are for third parties to connect and hence aren't the most important MVP technical feature and so are awarded the lowest score.

The following section examines how to use business value for prioritization together with capacity allocation.

Allocating capacity for prioritization

This section will explain how to use capacity allocation to ensure technical backlog items are also prioritized with business backlog items.

Developing long-lasting, healthy systems requires well-balanced backlog distribution between functional and technical backlog items. Inclining more toward functionality increases operational overheads, and dents an organization's ability to respond to customer needs rapidly. Philippe Kruchten used the **zipper metaphor** to emphasize the importance of weaving functional and architectural activities together in the IEEE paper *Agility and Architecture: Can They Coexist?*. He observed *interleaving architecture elements in the plan is essential to avoid accidental architecture.*

In agile teams, often friction exists between the product owner and architects while prioritizing features for development, as illustrated in the following diagram:

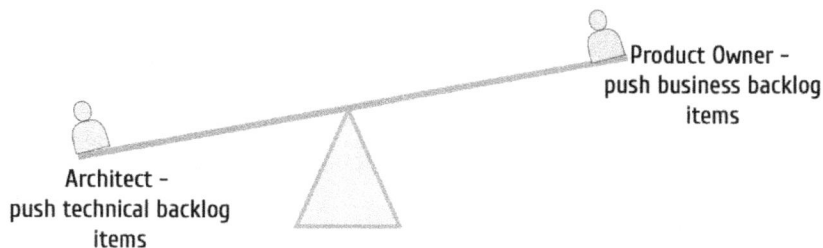

Figure 6.5 – Friction between architecture and product

As you can see, less technically sound product owners always try to push business backlog items ahead of technical backlog items. One way to address this issue is to create capacity allocation upfront for each type of activity, as shown in the following diagram:

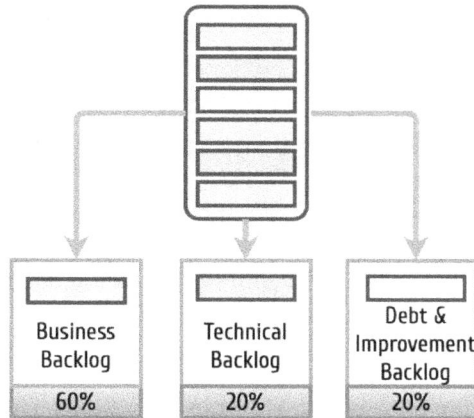

Figure 6.6 – Capacity allocation

As depicted in the preceding diagram, 20% of the team's capacity is reserved for technical backlog items to support business initiatives and another 20% for addressing tech debts. The capacity allocation provides a broad-level guardrail for prioritization, giving teams the flexibility to change across sprints. It provides grounds for architects to have meaningful and healthy prioritization discussions. SAFe suggests capacity allocation at the ART level as well as the team level.

Snow in the Desert uses capacity allocation techniques for balancing investments. The following diagram shows the capacity allocation model per ART for the upcoming PI:

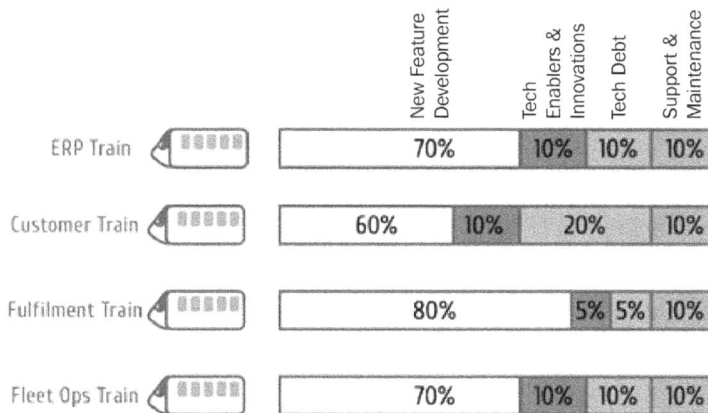

Figure 6.7 – Capacity allocation for ARTs

It is important to note that capacity allocation may be different for different trains, as shown in the diagram, based on the train's backlog and context. Different capacity categories are generally maintained at the organization level for consistency and tracking purposes.

As Mark Schwartz stated, *value should be delivered as quickly as possible—in small increments—and features should be prioritized based on the amount of value they deliver.* While business backlog prioritization is generally based on business value, technical backlog may use different strategies, as shown in the following diagram:

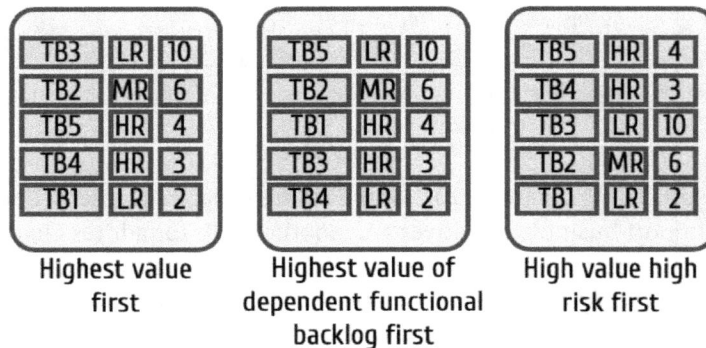

TB3	LR	10
TB2	MR	6
TB5	HR	4
TB4	HR	3
TB1	LR	2

Highest value first

TB5	LR	10
TB2	MR	6
TB1	HR	4
TB3	HR	3
TB4	LR	2

Highest value of dependent functional backlog first

TB5	HR	4
TB4	HR	3
TB3	LR	10
TB2	MR	6
TB1	LR	2

High value high risk first

Figure 6.8 – Prioritization strategies for technical backlog items

In the preceding diagram, **TB** stands for **technical backlog**. LR, MR, and HR represent **low risk**, **medium risk**, and **high risk**, respectively. The numbers represent perceived business value, with 10 as the highest value.

The following points explain these three strategies in detail:

- **Highest value first**: In scenarios with no strong link with another functional backlog item, the technical backlog item with the highest estimated business value with the lowest estimated development cost will be selected first.

- **The highest value of dependent functional backlog first**: In this case, technical backlog items with the highest-valued dependent business backlog item will be picked up early, with the view that those business backlog items are likely to be picked up first for development.

- **High value, high risk**: In this approach, technical backlog items with the highest risk with relatively high value will be taken up first. The risk in this context may be a business risk or risk of rework, risk of higher maintenance costs, and so on.

SAFe uses **Weighted Shortest Job First** (**WSJF**) for backlog prioritization. **Disciplined Agile** (**DA**) recommends using a combination of business value, risk, due date, dependency, and operational emergency as the primary criteria for backlog prioritization. Product owners in **Large-Scale Scrum** (**LeSS**) provide direction on how products need to be evolved by prioritizing the backlog with a relentless focus on delivering customer value. The prioritization is based on profit drivers, strategic customers, business risks, and so on.

Once technical backlog items are prioritized, it is important to show progress transparently. The next section will share techniques for making the work transparent.

Making the work transparent

Transparency and **openness** are fundamental principles of agile software development frameworks as they help to demonstrate the value of the work done by individuals and teams at all levels. Work transparency is crucial for architects to show their work consistently to all stakeholders.

The following points explore several ways to make architecture work more transparent:

- **Kanban boards**, burn-ups, and burn-down charts are consistently used by agile teams to show their work. Seamlessly integrating architecture work into these boards is an excellent mechanism to demonstrate architecture activities.

- An architecture roadmap showing the time sequence of decisions and enablers is a handy tool for evidencing what activities are in hand for architects. Roadmaps can exist at the initiative level as well as at the team level.

- Application health dashboards capturing the current state and next desired state are excellent ways to show how continuous improvements related to backlog items stabilize the application's health.

- An architecture maturity dashboard is a way to bubble up issues and challenges related to ways of working, such as the urgency of now, non-compliance to the architecture vision, aging of tech debts, architecture risks, and so on.

- A **DevOps health radar** for application automation captures the state of automation and continuous delivery for solutions under development.

Show walls, using physical walls, are useful for better communication as well as selling and explaining architecture work. Use hallways and team rooms for showcasing architecture models so that they live in the stakeholders' minds in the longer term.

SAFe recommends kanban and roadmaps at the portfolio, solution, and ART levels to enable transparency of work. Similar to SAFe, work transparency is achieved in DA using techniques such as task boards, kanban, and velocity burndown charts.

The next section highlights the importance of moving away from the urgency of now to working ahead of development cycles.

Looking ahead of development

Developing solutions without a view of what is coming in the near- to mid-term causes several issues. Due to the urgency of now, a solution's architecture decays progressively, and a move toward accidental architecture results in higher operational costs and reduced agility. Just-in-time decisions and delayed decisions increase the adoption of tactical solutions. As a result, velocity goes up before deteriorating due to higher degrees of rework. Tactical solutions impact the technical quality and predictability of the release process. Solution qualities such as stability, reliability, performance, and so on also decreases as time progresses.

As discussed in the horse and buggy metaphor in *Chapter 3, Agile Architects – The Linchpin to Success*, architects often work on future iterations alongside the product owner to ensure backlog items have enough clarity at the time of the development sprint. Preparing the **Definition of Ready** (**DOR**) on time ahead of development improves velocity and predictability. Looking ahead is important for architects to develop solutions for approaches and options by sufficiently allocating time for explorations to solve complex architecture and design challenges. DA calls this look-ahead planning and modeling.

While architects are running ahead, it is also important to stay connected with the development team. For example, for balance, architects work 70% of the time looking ahead, paving the way for development teams, whereas 30% of their capacity is reserved for working with the team to address current challenges.

In an increasingly complex and turbulent environment, the perfect architecture may not give the best value to the customer's needs. Therefore, architects from time to time need to make adjustments to the vision and roadmap based on newly discovered constraints with a pragmatic mindset. The next section covers this aspect in more detail.

Working with a pragmatic mindset

As mentioned by Gene Kranz, **NASA**'s mission director for **Apollo 13**, *I don't care about what anything was designed to do. – I care about what I can do.* The customer's mindset is always the usability of the system, not the complexities behind the solution. Therefore, architecture decisions have to be balanced between long-term sustainability and the customer's near-term needs.

This section will explore the importance of pragmatism when working with architecture to deliver quick value to the business.

Understanding the last responsible moment

Deferring decisions is a good thing to do in agile software delivery, which means in addition to architects thinking *why?*, they must also think *when?*. In simplistic terms, a decision is required when customers need a particular functionality. For example, an MVP can run without high availability and is only needed when the business is ready to scale the solution to more markets. Therefore, the decision on high availability can be deferred closer to that milestone. DA recommends deferring commitments by scheduling irreversible decisions to the **Last Responsible Moment** (**LRM**).

The following graph shows the LRM and its various impacts:

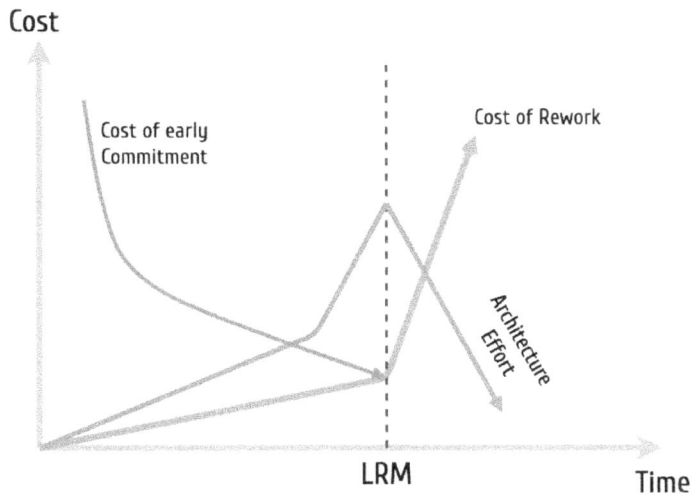

Figure 6.9 – LRM

As shown in the preceding diagram, the cost of early commitment goes down towards the LRM, whereas the rework cost goes sharply up beyond the LRM due to the adoption of wrong or tactical solutions. Understanding the LRM can be tricky. In most cases, the LRM is calculated based on approximation. The architecture work kicks off quite a bit ahead, as shown in the diagram, which reduces the risk associated with the estimation of the LRM.

Using eventual integrity

At times, businesses need certain functionalities quicker than anticipated due to sudden and unpredictable changes in customer and market demands. Besides, failing to resolve technical challenges on time also leads to unexpected situations. These scenarios lead to unexpected re-prioritization by compromising sustainability for time to market.

To support such scenarios, architects have to re-evaluate proposed solutions and offer alternate solution options, often tactical in nature, to support immediate business needs. Such decisions lead to tech debts but for genuine reasons. In such cases, architects compromise on eventual integrity by transparently establishing a roadmap to rectify accumulated debts.

Using a risk- and cost-driven approach

George Fairbanks proposed *architecture efforts should be commensurate with the risk of failure* as a key principle. He continued with an analogy stating *if security risks are not a concern then spend no time on security design. On the other hand, when performance is a project threatening risk, work on it until you resolve that issue.*

Risk- and Cost-Driven Architecture (RCDA) is an approach proposed by the CGI (cgi.com). This approach is described in the IEEE conference paper *Architecting as a Risk- and Cost Management Discipline* by Eltjo R. Poort. The author recommends considering *the cost of architecture concern as a sum of the cost of development and cost associated with the risk of failure.*

In the RCDA approach, everything is translated to a risk-first approach. Architects take a piece of activity only when there is a risk associated with that concern. Ignoring such decisions may increase the cost of delivery. This approach is useful as a means for prioritizing the architecture decision backlog.

Ensuring anti-viscosity

Viscosity is one of the symptoms of poor architecture in object-oriented programming proposed by Robert C. Martin in *Design Principles and Design Patterns* along with a few other principles, such as **rigidity**, **fragility**, and **immobility**.

There are two forms of viscosity: **design** and **environment**. The viscosity of design is considered high when the solutions deviate from the intended design as a result of developers finding an easier approach than the original design to solve the same problem. The viscosity of the environment arises when developers cut corners due to sub-optimal environments, such as the slowness of machines. High viscosity negatively impacts the long-term sustainability of solutions.

Collaborating with team members by exploring and debating alternate solution choices is important in agile software development.

In this section, we learned about the pragmatic mindset. We also learned about the LRM along with using eventual integrity. We also understood using a risk- and cost-driven approach and maintaining anti-viscosity.

The next section will expand on approaches for keeping options open.

Keeping options open

Architecture work starts far ahead of the LRM, as shown in *Figure 6.8*. During this period, architects can use multiple approaches to explore solutions for decision making.

This section will examine a few techniques to refine the decision backlog to arrive at a final solution by analyzing options. In some cases, we have to eliminate options early, but in other instances, techniques that keep as many options open as possible are more appropriate.

Using a hypothesis-based solution

A **Hypothesis-Based Solution** (**HBS**) is a structured mechanism for exploring, analyzing, and finalizing solutions based on point-based design. The HBS approach is particularly useful in cases where the architect is confident about a particular solution, based on experience and expertise, even before all the requirements are known.

In such cases, architects start with one potential solution and keep exploring that solution further with an open mind until they discover negative cases to disprove the proposal. However, architects commit to a final decision only at the LRM. For example, an experienced e-commerce architect can easily design an e-commerce solution long before many requirements are known. This approach substantially reduces the efforts required for exploring too many options. It also presents an opportunity to prove the solution earlier through spikes, model-based simulations, and expert engagements.

The following diagram captures the HBS approach:

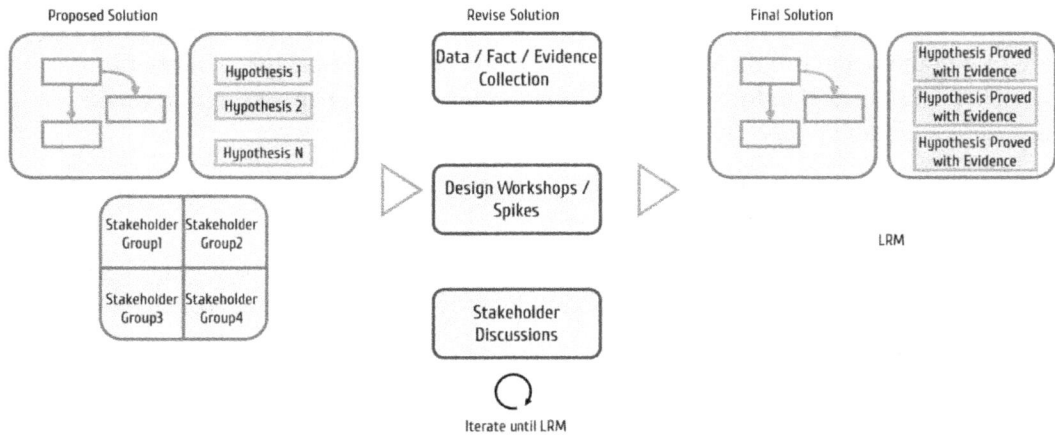

Figure 6.10 – HBS

As you can see, HBS starts with a potential solution based on several hypotheses. As time progresses, these hypotheses are validated by important stakeholders who add value by offering challenges and facts. Architects make decisions once all or most of the hypotheses are proven with evidence.

Using real options theory

Real options theory appeared in the early 1980s in the finance domain as a tool for making investment decisions. The three critical factors in real options theory are as shown in the following diagram:

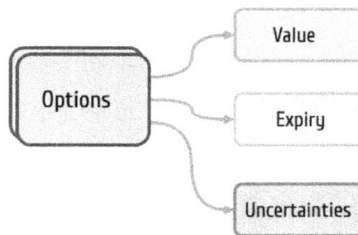

Figure 6.11 – Real options theory

There are benefits or value associated with every option at hand, such as customer satisfaction. There is an expiry for every option, such as a delivery milestone. Beyond that point, the option is no longer viable. There are also uncertainties attached to every option. Options with uncertainties are not the right choice and have to be eliminated.

Using set-based concurrent engineering

The term **Set-Based Concurrent Engineering** (**SBCE**) was introduced by Ward and is documented in the paper *Toyota's Principles of Set-Based Concurrent Engineering*. SBCE, or in short **set-based design**, is the name used for Toyota's method of managing product development processes. In the set-based design approach, alternative solutions are generated, explored, and evaluated before shortlisting the best solution for every design.

Set-based design allows architects to explore multiple options concurrently before reaching a final decision by applying trade-offs based on an economic framework. It also helps architects explain why a decision cannot be taken and what additional information is required. At the LRM, when the final decision is taken, outstanding assumptions are converted to risk statements.

The following diagram captures the set-based design approach:

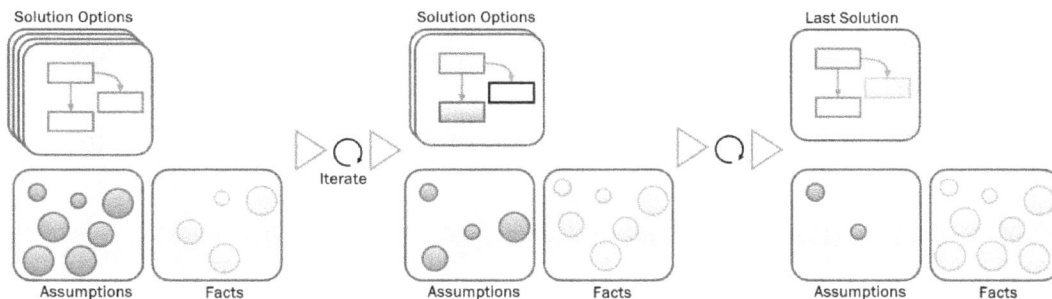

Figure 6.12 – Set-based design

As shown in the preceding diagram, as iterations progress, assumptions go down and facts go up. While architects get thinking time for most decisions, a few critical decisions have to be taken upfront, especially when innovations start with an MVP cycle.

DA suggests proof of technology spikes and set-based design for evaluating two possible solution options. SAFe also uses set-based design for decision making by exploring multiple options in parallel until all the required information is available.

At Snow in the Desert, architects use both set-based design and HBS for solution options analysis. For example, in the mobile application technology stack, architects used HBS because Snow in the Desert has deep expertise in **React Native**. Therefore, architects want to use React Native as the hypothesis for mobile application development. The team still keeps an open mind and welcomes challenges to their hypothesis. On the other hand, as shown in the following diagram, set-based design is used for technology selection for the streaming solution:

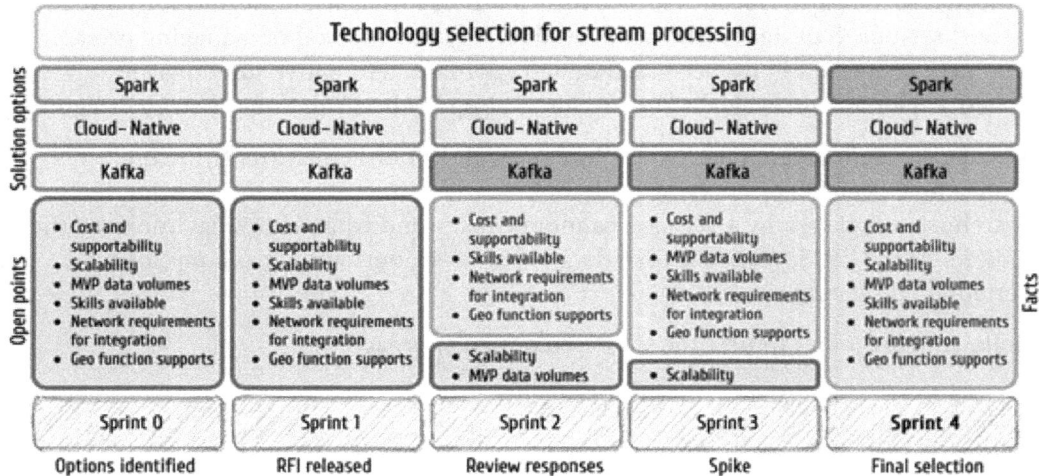

Figure 6.13 – Set-based design example

As depicted in the preceding diagram, open points are replaced with facts that eliminate options as time progresses. For example, Kafka is eliminated due to additional costs to the company in terms of skills and support. The other two solutions are already part of the Snow in the Desert ecosystem, and skills are immediately available.

In this section, we learned about keeping options open until the LRM to take the opportunity to gain as much information as possible before committing to a decision. In the next section, we will cover the architectural aspects of MVP.

Delivering early value with MVA

Early validation, feedback, and learning cycles are critical practices in agile software delivery. Developing an MVP helps enterprises test new innovations in production and receive customer feedback early enough to reduce risk before committing to large investments.

Architects play a critical role in defining the technical scope of MVP as well as designing the **Minimum Viable Architecture** (**MVA**). As a rule of thumb, high business value features are delivered first to ensure the business gets maximum value upfront. When designing the MVP scope, features are carefully chosen based on a combination of risk and value, as shown in the following diagram:

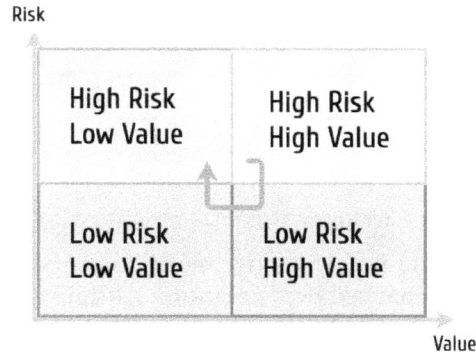

Figure 6.14 – Risk- and value-based prioritization for MVP

The preceding diagram is adapted from Ken W. Collier's view on the prioritization approach for business features. However, this approach is not particularly viable for architecture unless the business solution is centered around technology. For example, if the solution is based on **blockchain**, it is prudent that the blockchain must be proved. In contrast, if businesses want to build a **staff rostering system**, technology may not be verified as part of the MVP. In such scenarios, MVA is calculated as an option that delivers maximum value with minimal development cost, lead time, and risk, as shown in the following figure:

$$MVA = \frac{Value}{Cost + Lead\ time + Risk}$$

Figure 6.15 – Calculating MVA

This formula is an extension of the **VRC model** for upfront design, discussed in *Chapter 2, Agile Architecture – The Foundation of Agile Delivery*, by adding lead time as an extra parameter to the formula to factor the faster time to market aspect of MVP.

The following diagram captures the two possible approaches for designing MVA:

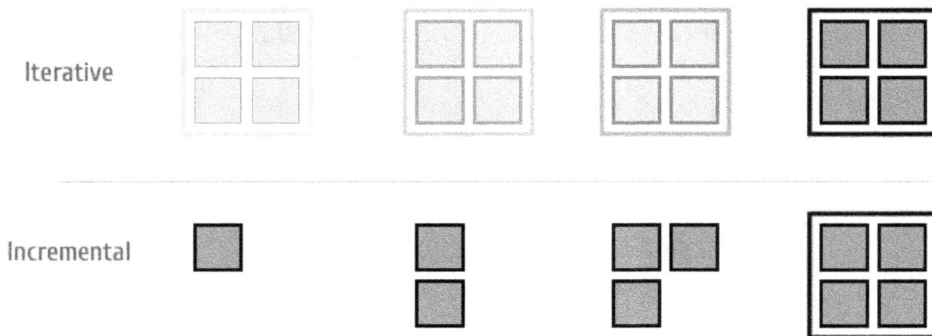

Iterative

Incremental

Figure 6.16 – Iterative versus incremental development

As shown in the preceding diagram, the iterative model is a version of the best solution that is built upfront and then enhanced over iterations. Alistair Cockburn describes this as the **walking skeleton**. On the other hand, incremental development only includes certain parts of the system as a starting point and incrementally adds additional parts. In agile software delivery, it is always recommended to use the iterative model for MVA.

SAFe adopted a Lean startup cycle to build, measure, and learn the viability of new innovations. So far, we have reviewed many techniques to improve the visibility and value of an architect's work. In the next section, we will discuss the important aspects of tech debt.

Managing tech debt

Ward Cunningham, who introduced the term **tech debts**, used a bank analogy to explain tech debts, which appeared in the *OOPSLA 92 Experience Report–The WyCash Portfolio Management System*. He observed that *creating debt is like not repaying a loan. One or two missed payments is fine, but longer than that will lead to an irrecoverable situation.*

There are **genuine** tech debts and **manufactured** tech debts. Genuine tech debts are to support the business to meet sudden customer needs. On the other hand, manufactured debts are due to architecture erosion, lack of knowledge of people, lack of motivation, lack of discipline, the urgency of now, the pressure of delivery, lack of intent to do the right thing, lack of focus on architecture and design, and so on.

Robert C. Martin observed in his article *A Mess is not a Technical Debt* that in acceptable tech debt situations, *decisions are taken in an informed way by properly analyzing trade-offs with a true intention to deliver value to customers faster in specific situations.*

In agile, tech debts cannot be avoided altogether; however, continuously refactoring and remediating debt is critical in developing sustainable software systems. The following graph captures various aspects of tech debt:

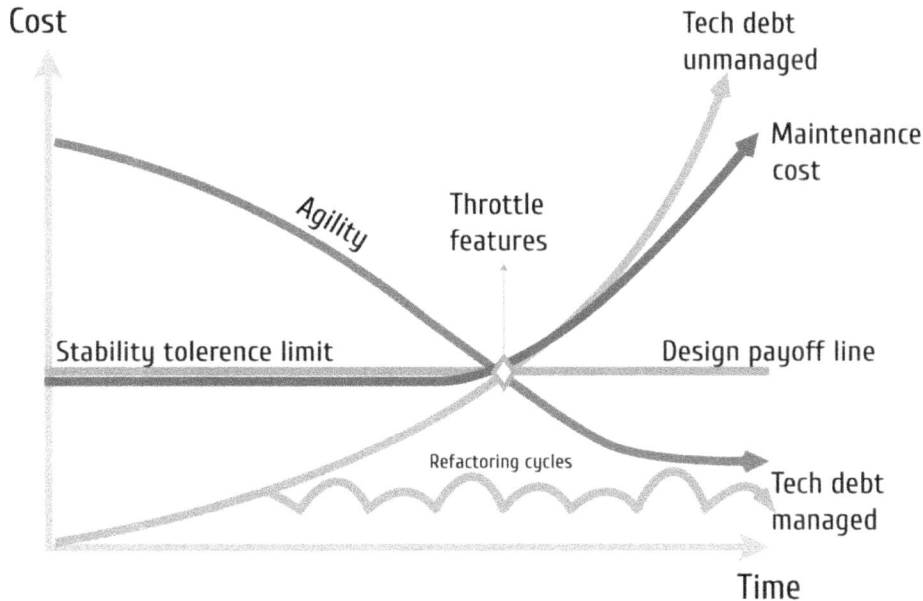

Figure 6.17 – Tech debt curve and stability tolerance limit

Well-managed tech debt with continuous refactoring cycles helps to contain the cost of maintenance well below the stability tolerance limit. The maintenance cost goes sharply up once it crosses the stability tolerance limit. To improve the stability of the system, beyond the stability tolerance limit, functional features will be throttled for mitigating tech debt. Martin Fowler called the **design pay-off line**, in his article *Design Stamina Hypothesis*, the point beyond which features cannot be traded-off for quality. Unmanaged tech debt beyond the stability tolerance limit also severely compromises the business agility.

The eight laws of Lehman appeared in *The Evolution of the Laws of Software Evolution: A Discussion Based on a Systematic Literature Review*, which defines a number of laws balancing the progressive evolution of systems against factors that deteriorate software quality. The laws of continuing change, increasing complexity, and declining quality are the most important ones related to tech debt.

As observed by Carola Lilienthal, the steps to avoid tech debt are the need for the participation of architects in team discussions regularly to avoid architecture erosion, using the right architecture styles and patterns, adopting domain-driven design, implementing automated testing to support refactoring, continuously analyzing architecture with the help of tools, reviewing architecture, and training.

Snow in the Desert pays special attention to tech debt and refactoring with specific capacity allocation. The simplified process flow for managing tech debts is captured in the following diagram:

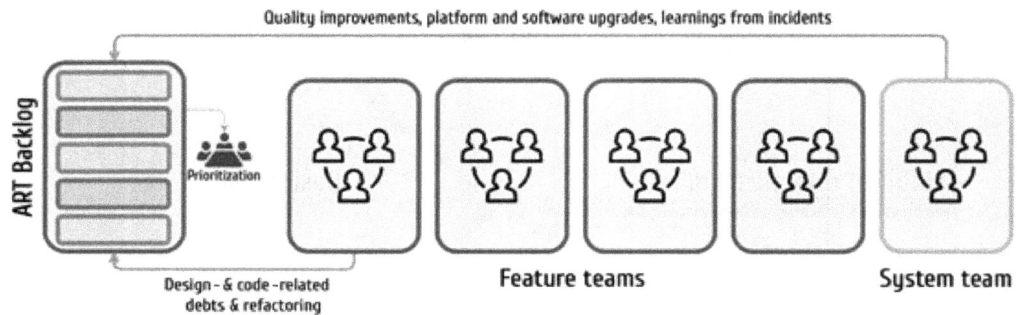

Figure 6.18 – Tech debt process

As you can see, the system team is a special team in every ART. One of their responsibilities is to continuously discover tech debts based on application monitoring, profiling, and assessing systems using automated and manual tools. Improvement items are fed into the backlog as and when they are discovered. For example, an incident may result in adding new monitoring requirements to the backlog.

Several self-assessment models are used at Snow in the Desert to measure various aspects of value delivery. These can be downloaded from the following GitHub location:

```
https://github.com/PacktPublishing/Becoming-an-Agile-Software-
Architect/blob/master/Chapter6/Self-Assessment-Models.png
```

Summary

Even though the architect role is not standard, like Product Owner and Scrum Master in agile development, it is still critical for many essential activities. However, it is the architect's responsibility to showcase their work's value in a transparent way, such as by using business language for selling architecture ideas.

In this chapter, we have learned that one of the key objectives of agile software development is to satisfy customers' needs early and frequently. The purist view of demand-based pull systems often challenges architecture activities. However, robust evolutionary architecture is important for delivering quality solutions with a sustainable flow in reality.

Then, we learned about the importance of architects demonstrating value and useful techniques. Architecture in agile needs rigor and work transparency, and must manifest value. Using shared product backlogs and linking technical and business backlogs help architects sell their work. Determining the business value of architecture backlog items, capacity allocation, and effective prioritization is a mechanism to ensure the right balance between functionality and quality. Architects add real value when they look ahead and spend their fair share of capacity on upcoming backlog items. They need to approach solutions with a pragmatic mindset by delaying decisions to the LRM, accepting deviations with eventual integrity in mind, and appropriately adopting RCDA. Architects defer decisions by keeping options open using set-based design, HBS, or real options theory. High-risk, high-value architecture and business backlog items are candidates for MVP with careful planning to balance the cost, value, lead time, and risk. Lastly, we looked at the tech debts aspect. Continuous management of tech debt is essential for long-lasting solutions with the optimal cost of delivery and maintenance.

This chapter has covered several useful techniques for architects to demonstrate their value in agile software development. However, process excellence is just not enough to deliver value. In the next chapter, we will explore the technical excellence dimension by analyzing the right architecture patterns and practices to adopt.

Further reading

- **The Big Lebowski Dude's law**: https://digitalcommons.pace.edu/cgi/viewcontent.cgi?article=1938&context=plr

- **Lean Primer**: https://www.leanprimer.com/downloads/lean_primer.pdf

- **Business value of solution architecture**: https://silo.tips/download/business-value-of-solution-architecture

- **Zipper model**: https://www.researchgate.net/publication/224118841_Agility_and_Architecture_Can_They_Coexist

- **LRM:** https://iglcstorage.blob.core.windows.net/papers/attachment-b9415f20-aa6a-4f80-a7d7-ea3f06529ec3.pdf
- **RCDA:** https://www.researchgate.net/publication/252019959_Architecting_as_a_Risk_and_Cost_Management_Discipline
- **Viscosity:** http://www.cvc.uab.es/shared/teach/a21291/temes/object_oriented_design/materials_adicionals/principles_and_patterns.pdf
- **Real options:** https://www.infoq.com/articles/real-options-enhance-agility/
- **Set-based design:** https://sloanreview.mit.edu/article/toyotas-principles-of-setbased-concurrent-engineering/?gclid=CjwKCAjw0On8BRAgEiwAincsHGCqacQF5lcfdm9FjPLUBu_PhwuLsrAH_inf-0hb3vXz8Ru0i_89GRoCbdMQAvD_BwE
- **Tech debt:** http://c2.com/doc/oopsla92.html
- **Tech debt:** https://sites.google.com/site/unclebobconsultingllc/a-mess-is-not-a-technical-debt
- **Lehman's law:** https://www.researchgate.net/publication/262297736_The_Evolution_of_the_Laws_of_Software_Evolution_A_Discussion_Based_on_a_Systematic_Literature_Review
- **Design stamina hypothesis:** https://martinfowler.com/bliki/DesignStaminaHypothesis.html

7
Technical Agility with Patterns and Techniques

It is possible to fly without motors, but not without knowledge and skill.

-Wilbur Wright, American inventor and aviator

The previous chapter focused on process excellence in delivering value with the shortest sustainable lead time without compromising quality. This chapter will explore the technical excellence needed to successfully deliver quality solutions without losing sustainability and lead time.

In the fast-moving world of technologies delivering high-quality solutions, architects must be at the forefront of technology advancements by curiously learning and continuously applying niche technologies. Adopting the right technologies and techniques for the right use cases enhances technical agility, quality, and customer satisfaction. Organizations have to maximize technical agility by investing in cultivating individual craftsmanship. Achieving technical agility is primarily done by adopting patterns and techniques as well as nurturing a DevOps culture. Evolutionary architecture to respond rapidly to customer requirements, modularizing applications to reduce coupling, and adopting modern technologies are necessary enablers for technical agility. While greenfield systems are easy to manage, many organizations continue to deal with legacy systems due to their long tail of tech debts. Re-platforming those systems to an acceptable technical level with incremental modernization is essential to continue deriving value. Teams deliver better-quality solutions when they follow good coding practices with a strong focus on testability and full stack telemetry.

Architects have the inherent responsibility to enable technical agility that brings stated benefits in delivering high-quality solutions faster. Therefore, we must learn the art of promoting technical agility by consistently applying patterns and techniques. This chapter sets the overall context of technical excellence and then zooms into the aspect of technical agility with architecture patterns and techniques. We will discuss many areas that architects have to focus on in order to achieve technical agility. We will also discuss legacy modernization techniques for graceful upgrades.

In this chapter, we're going to cover the following main topics:

- Amplifying agility with technical excellence
- Building technical agility with patterns and techniques
- Architecting for change
- Developing good code with engineering excellence
- Understanding enterprise integration
- Developing for testability
- Treating infrastructure like software with the cloud
- Monitoring everything with full stack telemetry

This chapter focuses on the **Technical Agility** focal point of **The Agile Architect's Lens**:

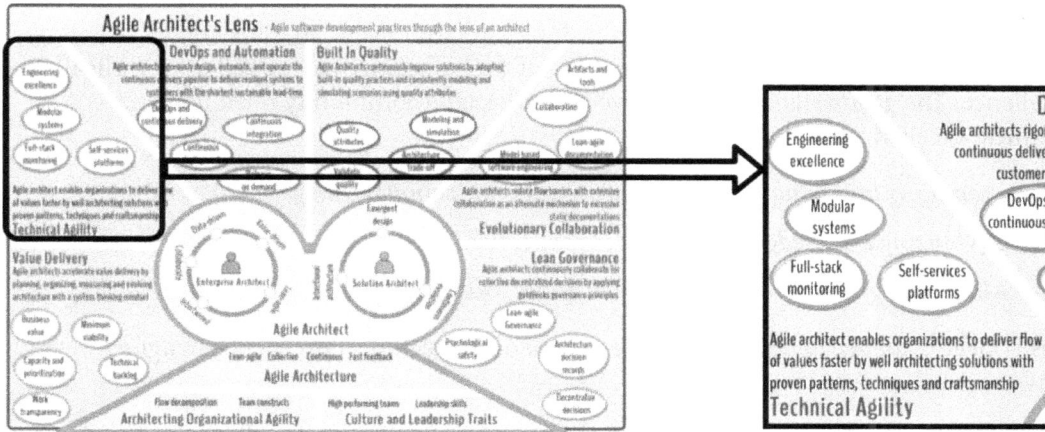

Figure 7.1 – Technical Agility – Focal point

Amplifying agility with technical excellence

Technical excellence is a term widely used beyond software engineering to refer to the rigorous and outstanding qualities and capabilities of individuals and organizations in a particular field of expertise.

In the paper *Technical Excellence: A Requirement for Good Engineering*, Paul S Gill and William W. Vaughan explain technical excellence at NASA as *technical thoroughness and rigor applied to software engineering*. This paper also observes that *technical excellence has a personal accounting aspect as well as an organizational responsibility*. Technical excellence refers to the state of an organization where people have the culture of favorably adopting technologies by going beyond their call of duty to deliver anti-fragile, reliable, and durable solutions at optimal cost. As proposed by Jim Highsmith, one of the manifesto signatories, technical excellence *is measured as the ability to deliver solutions for the customer's current needs and its adaptability for future needs.*

One of the manifesto principles of agile software development is that continuous attention to technical excellence and good design enhances agility. 10 years after publishing the manifesto, the signatories met again, at the Snowbird 2011 event, to reflect on its adoption status. One of the key messages that came out of the gathering was the call for technical excellence. The group signatories observed that the agile community must demand technical excellence by doing the following:

- *Promoting individual change and leading organizational change*
- *Organizing knowledge and improving education*
- *Maximizing value creation across the entire process*

This is clear evidence demonstrating the pertinence of technical excellence in agile software development.

Large-Scale Scrum (LeSS) observes that the maturity of technical excellence restricts an organization's degree of agility. The key characteristics of a learning organization competing and thriving to attain agility, with technical excellence as a critical enabler, include building an adaptive culture, software craftsmanship, technical mastery, and continuous learning of technology advancements. As an example, Snow in the Desert implemented an *internal academy* to foster continued learning by partnering with other enterprises and universities. Rewards, recognition, and leader boards are implemented to keep the motivation of employees high.

Well-engineered products flawlessly adapt to changes with minimal impact on serviceability, speed of delivery, and cost. Often, maintenance costs are regarded as a way to measure technical excellence. Obsession over the best quality, with a strong inclination toward minimizing operational overheads, is a positive sign of technical excellence.

The influence of technical excellence in the software life cycle is illustrated in the following diagram:

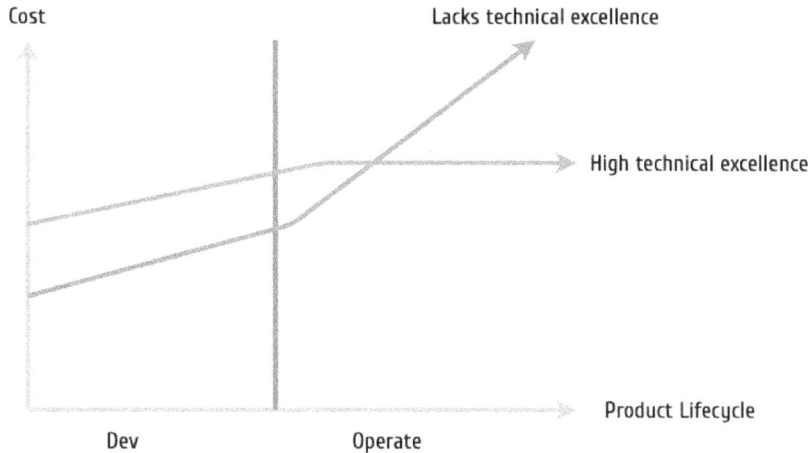

Figure 7.2 – Impact of technical excellence on software delivery

As shown in the preceding diagram, the development cost may be marginally high when organizations focus on technical excellence. In the long run, maintenance costs offset the initial development cost.

Architects play an incredibly significant role in enabling and promoting technical excellence through adopting the right technologies and techniques by providing technical thought leadership and responding to complex problems with simple, sustainable solutions. Agile architects need to critically and continually reflect and improve their technical knowledge in order to remain relevant.

As shown in the following diagram, technical excellence has two aspects:

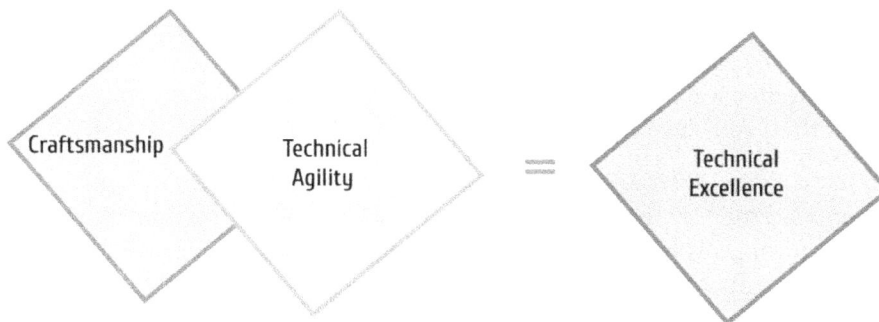

Figure 7.3 – Components of technical excellence

Software craftsmanship is associated with individuals, whereas technical agility is at an organization and team level. This is explained further in the following sections.

Adopting software craftsmanship

Intrinsically motivated individuals are significant assets to any organization for delivering well-engineered software products. Focusing on the quality of the solution is a habit for such individuals more than their duty. Regardless of processes, instilling craftsmanship can dramatically improve delivery success, quality, and maintainability of the software when individuals assume a sense of ownership.

The most formal documentation on software craftsmanship is the *manifesto for software craftsmanship* defined in 2009. It states that practicing and educating on the craft of professional software development steadily adds value to business change by delivering well-crafted working software. Such practitioners foster a community of professionals with a productive customer partnership beyond collaboration.

Adding value by delivering a well-crafted software solution is one of the key highlights of this manifesto. Software craftsmanship requires personal commitment for self-improvement, a curiosity to learn, and striving for better quality. As a master programmer, software craftsmanship is core to an architect's role. Software craftsmanship is the state of an individual with a growth mindset when they do the following:

- Relentlessly stretch their comfort zone by continuously striving to acquire new skills
- Fearlessly adopt good technical practices
- Systematically apply learning with passion, rigor, and thoroughness
- Selflessly share learning for the betterment of the people around

Beyond merely getting their job done, these aspiring and disciplined individuals care about the code they write and invest time in avoiding code depreciation. They use simple designs and clean and concise code, and they automate mundane tasks. They take pride in delivering bug-free code and always challenge the status quo for their own betterment. They invent innovative ways to perform the same job differently for their satisfaction and deliver better, faster, and cheaper solutions.

Software craftsmanship is possible for individuals to achieve only with immense encouragement and support from the organization. Organizations need to invest in individuals' technical skills by continuously amplifying learning and consistently applying reflective improvements.

Enhancing quality with technical agility

Organizations cannot accomplish agility by just adopting agile development and delivery practices. It requires the efficacious use of technologies as a cardinal tool to enable agility. For example, agile software development using in-house infrastructure management with a lead time of months for procurement and the provisioning of new servers induces flow barriers.

Teams excel in technical agility when contextually adopting modern technologies such as the cloud, microservices, and automation for accelerating the flow of delivery continuously and sustainably. High-performing teams enable agility by imprinting principles such as clean code, proven engineering practices, appropriate use of technical patterns, and adherence to object-oriented concepts and design techniques.

Focusing on technical agility helps organizations to harness current technologies and embrace new technologies quickly and easily. Technical agility helps avoid excessive reworking, improves productivity, and ensures a faster time to market, higher quality, better longevity, reduced operations cost, and improved organizational agility.

There are three segments in technical agility, as shown in the following diagram:

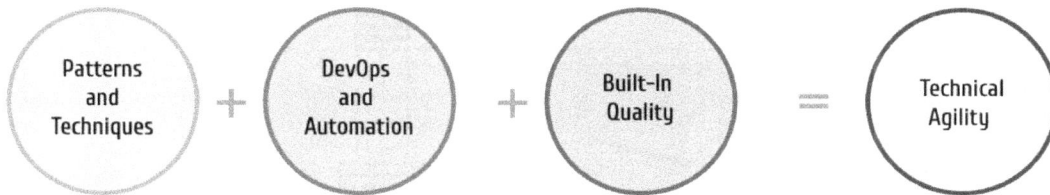

Figure 7.4 – Components of technical agility

The three segments are explained as follows:

- Patterns and techniques help deliver high-quality evolvable solutions.

- DevOps and automation support the continuous faster flow of values.

- Built-in quality supports high-performing, scalable, and sustainable solutions.

Scaled Agile Framework (**SAFe**) describes technical agility as one of the seven competencies. It consists of critical skills and competencies required for high-performing teams to develop and deliver high-quality solutions to customers.

So far, we have examined technical excellence, which consists of software craftsmanship and technical agility. We have also seen the importance and characteristics of technical agility. The following section introduces some of the principles of technical agility.

Building technical agility with patterns and techniques

Embracing new technologies for adding new features or incrementally enhancing live systems without disruptions necessitate an evolutionary architecture. Carefully choosing the right patterns and techniques enhances software evolution and, hence, technical agility.

Neal Ford from ThoughtWorks defines evolutionary architecture as an *architecture that supports guided, incremental changes across multiple dimensions*. Evolvability needs to be considered as a first-class citizen when targeting technical agility. A sound, evolvable architecture demonstrates patterns and techniques such as simplicity, well-formed structures, and modular, repetitive components.

Many real-world architectural marvels, such as the Taj Mahal and Tower Bridge, reflect these characteristics. These large architectural monuments were built using the principle of self-replicating geometry and symmetry of architectural elements, as illustrated in the following diagram:

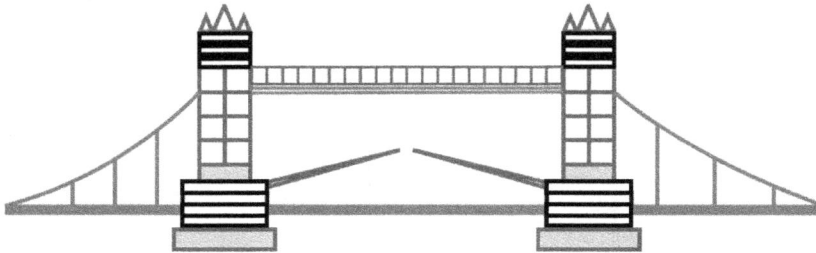

Figure 7.5 – Tower Bridge – the symmetry of architectural elements

As you can observe in the diagram, simple, standalone, and replicated components integrate with the overall structure in symmetric order. Architecture evolves by extending these repeated structures in multiple directions. In software engineering, simple, smaller parts of the system stand on their own and integrate to form larger systems. Repeating the same structural patterns differently to build larger systems is the way to build stable systems faster.

Accomplishing technical agility with evolutionary architecture patterns and techniques can be summarized as five core principles, as shown in the following diagram:

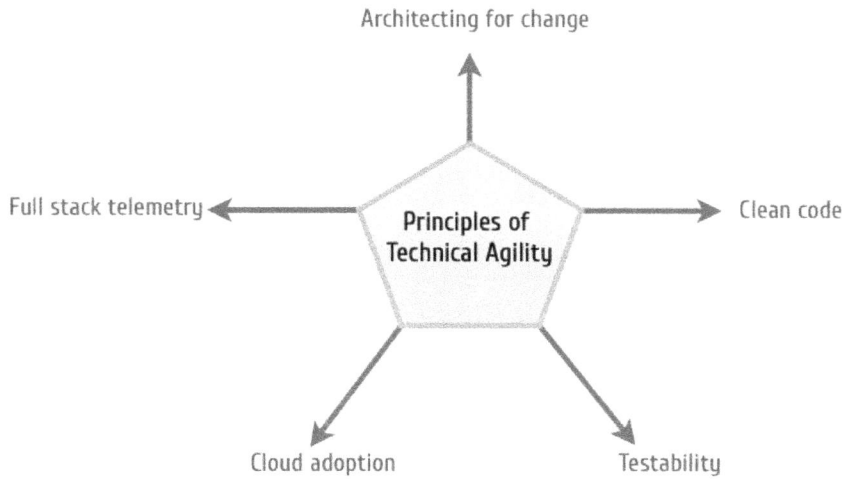

Figure 7.6 – Principles of technical agility

The principles of technical agility enable developers to accept changes without fear of failure. These five principles are explained further in the following sections.

Architecting for change

Architecting for change is an art. It requires a change-aware mindset irrespective of working with greenfield or legacy systems. This section explains the different patterns and techniques for enabling architecting for change.

Driving technical agility with simplicity

The **Keep It Simple, Stupid** (**KISS**) principle is one of the oldest principles for relating good design and engineering. This principle originated from an American aircraft engineer, Kelly Johnson, referring to the simplicity in designing military aircraft to be repaired with a limited set of tools in a war zone. This principle highlights that systems with simple designs work better than their complicated alternatives. Simplicity must be one of the key goals when architecting for change to support software evolution.

Simplicity is also reflected in **Occam's Razor**, or the **Law of Parsimony**, as a principle related to problem solving. This principle states that if there are multiple hypotheses for a solution, choose the one that has the fewest assumptions. Fewer assumptions brings better clarity and simplicity.

The following points highlight the consequences of complex design:

- Complex designs due to overengineering result in increased costs of change, evolvability, and maintainability.

- Complex designs due to feature creep lead to extra efforts in building features beyond what is needed for the solution.

- Complex designs due to a lack of refactoring lead to software bloats, which consume more resources and slow the system down.

- Complex designs with more moving parts, such as components, dependencies, configurations, technologies, and layers, lead to reduced productivity and an increased number of faults.

Adopting the clean architecture principles defined by Robert C. Martin helps to develop the most straightforward but effective form of layered architecture. Good engineering practices and an individual's maturity are essential for keeping the design as simple as possible.

Simplicity alone cannot deliver the necessary levels of technical agility. Isolation by design is one of the most critical design aspects to enable technical agility. The next section explains the different patterns and practices of isolation by design.

Evolving with isolation by design

Isolation by design is one of the critical principles of architecting for change. **Isolation by design** is defined as follows: *for software to be evolvable, tightly dependent components have to stay together, exhibiting high degrees of cohesion, whereas components that are likely to change need to be decoupled with loose integration patterns and techniques.*

Several patterns and techniques related to isolation by design are explained in the following sections.

Slicing architecture vertically

The traditional way of organizing architecture is by slicing an application horizontally with technology layers as the boundary. Product-based development is one of the standard practices in agile software development. Architecture needs vertical slicing along the boundaries of the product instead of horizontal slicing with technology as the currency. Doing so helps in significantly minimizing the blast radius of any system changes.

The following diagram captures both the horizontal slicing and vertical slicing approaches:

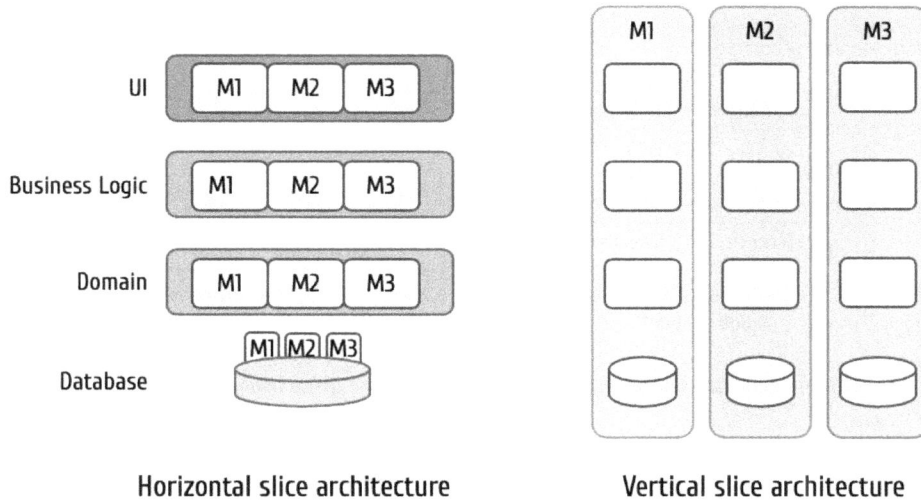

Horizontal slice architecture   Vertical slice architecture

Figure 7.7 – Horizontal and vertical slice architectures

M in the diagram stands for functional modules. As shown in the preceding diagram, vertical slicing is a miniature version of the whole system that helps accelerate feedback cycles with minimal investments. Technologies and layers are abstracted within the vertical slices to provide empowerment and autonomy for product-specific teams with full stack developers. The components within a vertical slice are high on cohesion, whereas across slices they are loosely connected. These vertical slices are progressively enhanced with vertical user stories without disrupting other slices.

Vertical slices not only reduce the blast radius, but also aid in substitutability and maintainability. They also help you to stay customer-focused, accelerate innovations, understand and organize teams better, achieve faster build cycles and structured integrations, improve testability, automate deployments, facilitate segmented delivery and release pipelines, and achieve better performance, and they are less laborious to operate.

Decomposing using domain-driven design

Domain-Driven Design (**DDD**) is an effective approach to vertically slice larger systems across functional and product boundaries. DDD is a development approach proposed by Eric Evans to help build evolvable systems by organizing complex logic and data models with *bounded contexts*. A bounded context is a mechanism for establishing functional boundaries between various parts of the system.

The following diagram gives an illustration of domain objects and bounded contexts:

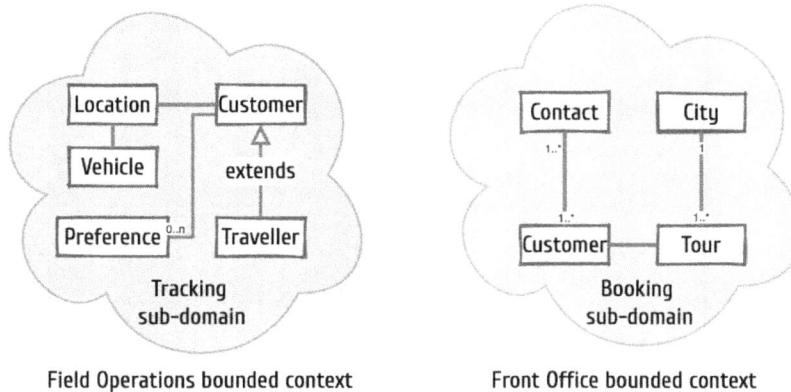

Figure 7.8 – DDD using bounded contexts

When introducing Snow in the Desert in *Chapter 2, Agile Architecture – The Foundation of Agile Delivery*, we saw four departments – Front Office, Back Office, Field Operations, and Vehicle and Asset Management. These are often treated as bounded contexts as they share domain objects, the same business language, and relationships. The domain decomposition usually starts with these higher-level bounded contexts, which are decomposable into more granular sub-domains.

The Field Operations domain can be further decomposed into the Fulfillment sub-domain, the Scheduling sub-domain, and the Planning sub-domain. The Fulfillment sub-domain can be further decomposed into the Tracking sub-domain and others. Similarly, the Front Office domain can be decomposed into more granular domains, such as the Booking sub-domain.

Figure 7.8 shows two bounded contexts from Snow in the Desert – Field Operations and Front Office, focusing on the Tracking and Booking sub-domains. As shown in the diagram, the Customer domain object is repeated in both bounded contexts, but may represent different meanings.

DDD uses domain objects, object-oriented principles, and business language to model complex problems. Unrelated domain objects and business language indicate that they belong to a different bounded context. Domain objects within a group are highly cohesive, whereas the objects across domain boundaries are loosely coupled. Typically, bounded contexts are owned by a line of business and aligned to a development team.

These models reflect the interactions and behaviors of the business in the real world. The model evolves by adding new relationships and objects as new scenarios are introduced. Since models use business language, stakeholders from business can easily validate and share feedback based on their experience ahead of the development.

A good domain model developed using object-oriented patterns and principles reduces breaking changes, improves modularity, and produces better-quality clean code.

Implementing vertical slices and DDD using microservices

Even though microservices are a relatively new concept, Dr. E. E. David, at the *NATO Software Engineering Conference* back in 1968, observed that the building of large systems needs to start small by developing and deploying a *small subsystem and then building on that.* He also observed that *systems need to be designed in modules that can be implemented, tested, and evolved independently.*

Microservices is a realization of Dr. E. E. David's vision. It is an architecture style that primarily helps in implementing vertical slices and bounded contexts. The microservice approach decomposes large applications into a series of smaller autonomous services minimally integrated with loosely coupled designs. The life cycle states from inception to operations of these services are relatively independent of each other.

The following diagram shows the concept of microservices as a combination of DDD and vertical slicing:

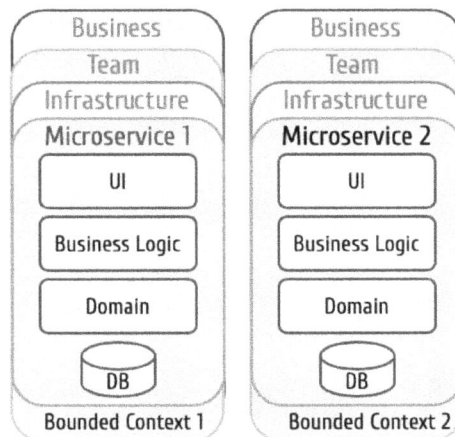

Figure 7.9 – Microservices architecture

As shown in the diagram, application components, infrastructure, development teams, and business can all be isolated from other services by providing greater degrees of autonomy.

Some of the key benefits of microservices in the context of agile software development are as follows:

- Organizing teams around service boundaries to reduce co-ordination efforts, improve autonomy, and better engage with business.

- Separating deployment pipelines and infrastructure for each service, thereby helping teams to determine the frequency of deployment and releases independently.

- Automating build, test, infrastructure provisioning, and deployment is then easier to achieve compared to large monolithic systems.

- By enabling polyglot technologies, teams are empowered to choose the right technologies for the right purpose to deliver maximum value at optimal cost.

- Enabling innovations with a faster test learn cycle, as the impact of the failure is restricted to a smaller set of features.

- Reduced cost of operations is achieved with selective autoscaling and service-based availability.

In the context of Snow in the Desert, two microservices were represented in the container diagram explained for AVTS in *Chapter 5, Agile Solution Architects – Designing Continuously Evolving Systems*. Extending that further, the following diagram shows the component view of **Tracking Microservice**:

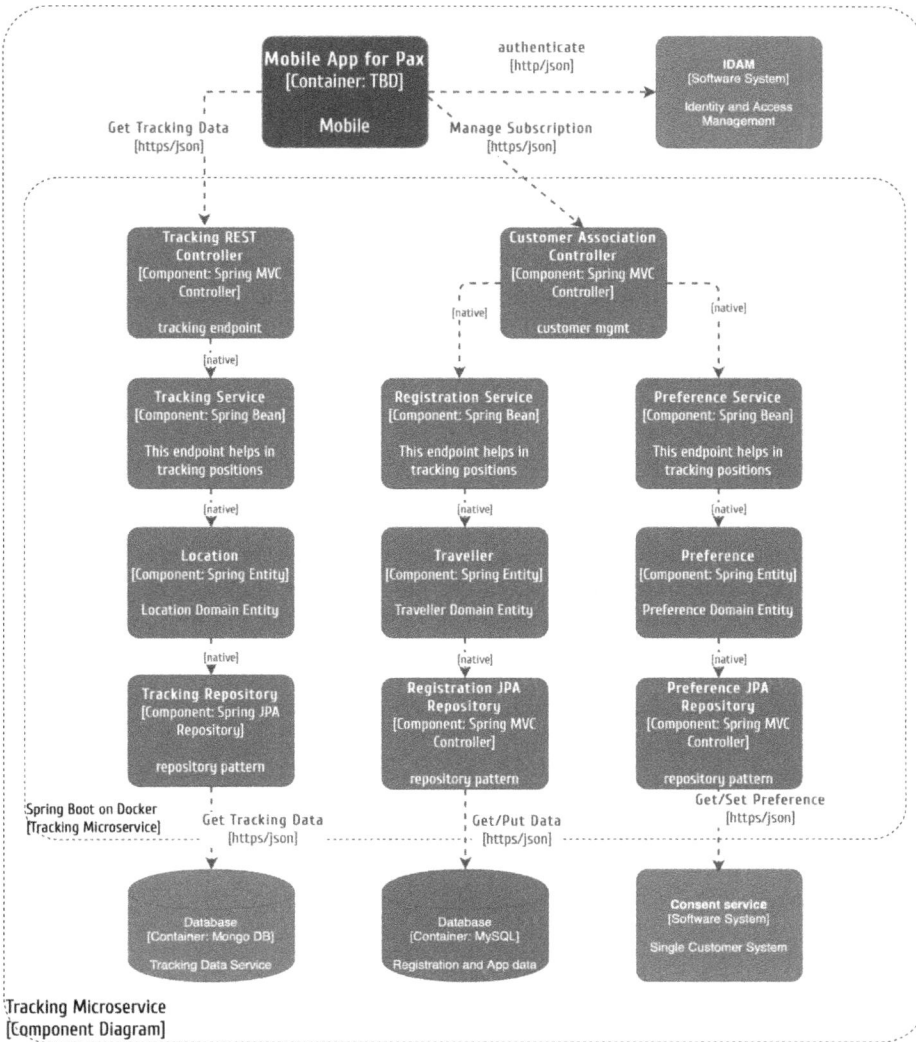

Figure 7.10 – Component diagram for a tracking service

As captured in the diagram, **Tracking Microservice** uses a clean and straightforward structure based on Java and the Spring Framework. The self-contained solution is deployed as a containerized Spring Boot application, and managed by a separate team with isolated continuous integration and deployment pipelines.

Isolation by design techniques are straightforward to apply in greenfield development scenarios. However, they are incredibly complex and pragmatic to use with legacy monolithic applications. The next section will explain some of the patterns that are useful when dealing with legacy applications.

Legacy modernization architecture

In a large enterprise with a long tail of legacy, architects have to deal with live legacy systems, which are hard to change, test, deploy, monitor, and scale. Most of these systems are incompatible with and non-responsive to modern techniques and technologies, such as containers, the cloud, infrastructure as code, and continuous delivery pipelines.

Legacy systems require considerable investment for large-scale modernizations. Many times, the business value of such large-scale improvements is outweighed by the cost of upgrades. Therefore, such approaches are considered infeasible. Instead, small incremental changes that move systems to the next immediate desired state is a better proposition. A few patterns are handy for achieving graceful changes to legacy applications.

Strangler pattern

The strangler pattern is used when new changes are built outside the core legacy system or incrementally by moving functionalities from the core system to outside. Services built outside the legacy system follow modern architecture and engineering practices. The legacy system makes its way to retirement once all features are migrated to the new architecture.

The following diagram captures different phases of the strangler pattern:

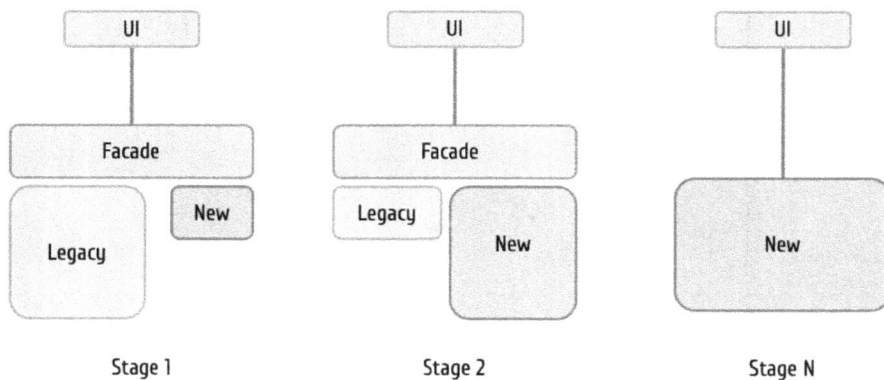

Figure 7.11 – Strangler pattern for legacy modernization

As shown in the diagram, the façade between the UI and backend helps avoid disruptions to customers. The strangler pattern helps in reducing the migration risk and interruption to customer services. However, this incremental migration is a bit trickier for frontend applications as users may experience a sub-optimal user experience during the migration period.

Stalling migration halfway is one of the common challenges associated with the strangler pattern, which leads to a migration flux. Migration flux is a complicated situation with old and new systems working in conjunction for a more extended period without further migration. Full commitments from the sponsors and an aggressive migration plan are essential for avoiding migration flux.

Anti-corruption layer

The anti-corruption layer is another pattern that is useful for legacy migration. In this case, the old systems continue to have the old way of working, and the new system serves new functionalities.

The following diagram captures the anti-corruption layer pattern:

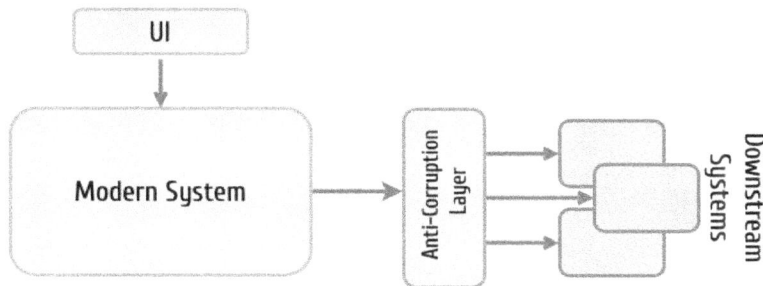

Figure 7.12 – Anti-corruption pattern

As shown in the diagram, downstream systems that rely on the old system may take time to migrate completely so as to be compatible with the new system. The anti-corruption layer acts as a message translating bridge between the old and new systems. With this approach, the new system is not polluted with legacy semantics and avoids significant investment in the old system to make it compatible with new semantics. Similar to the strangler pattern, the anti-corruption layer is only a transitionary stage. It is essential to change downstream systems to comply with new semantics to avoid unnecessary complexity and maintenance overheads.

So far, several tactics helpful for architecting for change to support technical agility have been explored. The next section will examine the engineering excellence aspects of technical agility.

Developing good code with engineering excellence

The rich and adequately modeled domain of DDD is an excellent starting point and a strong foundation for building evolvable systems. Architects identify opportunities, and foster, demonstrate, and continuously reinforce good technical practices. The technical quality of a team is directly connected to the technical competency of the leader.

The development of high-quality code demands that engineers adopt several engineering disciplines. The subsequent sections cover these essential practices, starting with principles, techniques, and patterns, which are the foundation.

Applying coding principles, techniques, and patterns

The development of mission-critical resilient systems with longer estimated service periods needs to focus on, and invest in, good-quality code. This section explores design techniques and patterns for building such long-lasting systems.

Enabling techniques

Software design patterns and practices are fundamentally based on several enabling techniques of object-oriented programming. These enabling techniques include abstraction, encapsulation, information hiding, modularization, separation of concerns, coupling and cohesion, and the separation of the interface from implementation. Meticulously practicing enabling techniques helps to develop robust foundational design and code.

Foundation patterns

The **Gang of Four** (**GoF**) design patterns provide solutions to recurring problems in the most elegant and optimal way, with well-structured and proven code. In addition to the GoF patterns, principles such as **SOLID** and **Don't Repeat Yourself** (**DRY**) are also very useful in developing resilient systems. The *Design Principles and Design Patterns* paper by Robert C. Martin documents several design principles and patterns for developing good-quality code.

Enterprise application patterns

Enterprise-grade patterns have to solve more complex and challenging problems beyond what the GoF and other foundation patterns address. Enterprise patterns are solutions to repeatable complex application design problems, such as **Model View Controller** (**MVC**) for web application development, and Front Controller. These principles are related to the layering of application components, structuring domain logic and web applications, connecting to databases, handling sessions, and so on. Other higher-level design patterns, such as event sourcing, **Command Query Responsibility Segregation** (**CQRS**), and circuit breakers, are also useful for solving enterprise application architecture problems. *Microsoft Cloud Design Patterns* (`https://docs.microsoft.com/en-us/azure/architecture/patterns/`) is a good repository of application-level patterns that are quite useful for building modern cloud-based applications.

With Snow in the Desert, architecture and design patterns are developed and maintained in a Git repository as *patterns as code* for easy consumption.

Good coding practices

Good coding practices come with experience and craftsmanship. Great programmers define and continually follow enterprise coding guidelines, including naming and structuring standards, to produce consistent and maintainable code. While an average developer just handles the exception when they encounter an issue, a good programmer fixes the exception's root cause, even if it takes more time.

Too many control statements reduce readability and code performance, where a master programmer uses object-oriented constructs in place of control statements. **Cyclomatic complexity** is a general measure of the code's complexity by measuring the number of independent control flow paths in the source code. A lower value for cyclomatic complexity indicates well-structured code. *Clean code*, as proposed by Robert C. Martin, advocates writing code in a simple, structured, orderly, and elegant fashion with the right levels of abstraction and good readability.

Snow in the Desert continuously enhances coding and design guidelines and applies them automatically by integrating with development IDEs and continuous integration tools. Knowledge of coding practices and patterns is spread throughout learning sessions and code fests.

Continuous refactoring

Sustainable delivery of software in agile development requires continuous improvement of systems that require continuous refactoring. Refactoring is not optional; it is a must and has to be part of all team members' day jobs. Cumulative change by many developers over time invariably increases design and code complexity. Refactoring requires a different mindset in which developers need to anticipate code changes to accommodate new requirements, no matter how well the code is written initially.

Refactoring, by definition, is changing the structure of the code without altering the behavior of the solution. To ensure that behavior is intact, an automated regression test pack is essential. Refactoring is not only applicable across all layers, such as the UI and database, but also applicable at architecture, design, and code levels.

As mentioned in *Chapter 6, Delivering Value with New Ways of Working*, small but repeated refactoring is preferred in order to reduce complexity, tech debts, and code smells. Refactoring improves performance, reduces resource utilization, enhances readability, and promotes reusability.

Twelve-factor principles

Twelve-factor principles, made popular by Heroku, now part of Salesforce, is a methodology describing the characteristics expected from modern cloud-ready applications.

Developers building modern, cloud-native, scalable, and distributed applications have to adhere to several fundamental coding practices in addition to the good coding practices. These are a set of principles relating to coding, configuration, dependencies, and deployment. They help in decoupling application code from the cloud environment so as to insulate against infrastructure volatility. For example, internal and service URLs have to be separated strictly from code into, preferably, an externalized configuration file. By doing so, developers ensure that code does not change between deployment environments.

Now, let's understand what is meant by enterprise integration.

Understanding enterprise integration

Modern software development following agile practices needs to consider integration as a first-class citizen. Short-sightedness in integration may quickly turn into a flow impediment. The early and continuous testing of integrations is a good practice in agile software development to avoid this.

Designing and implementing robust integrations requires many considerations, including the following:

- The right level of coupling by choosing integration types and protocols, such as synchronous versus asynchronous and HTTP versus TCP/IP

- Non-functional requirements, such as fault tolerance, performance, data transfer size, cost of transmission, and scalability

- Appropriate data exchange and transformation mechanisms

- Implementing appropriate security controls

Designing reactive event-driven services can significantly minimize dependencies between subsystems of organically evolving systems. Organizing teams around bounded contexts restricts inter-team communications through a loosely coupled API-first approach. Using enterprise integration patterns such as publish-subscribe, message routing, message transformation, and message consumption solves the most common integration patterns.

The following example is from Snow in the Desert to demonstrate how trains and systems communicate with loosely coupled event-driven interfaces:

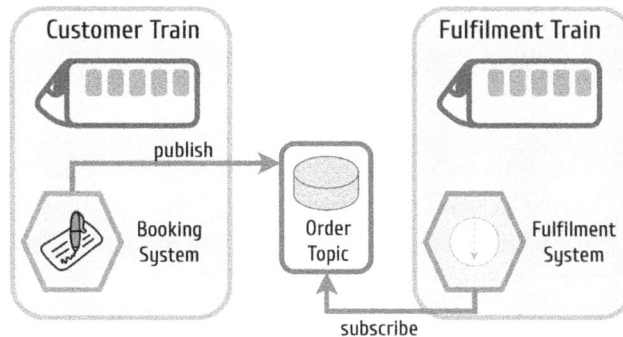

Figure 7.13 – Teams and systems communicate with event-driven interfaces

In the case of Snow in the Desert, the **Booking System**, which is part of the **Customer Train**, and the **Fulfilment System**, which is part of the **Fulfilment Train**, interact through events over a message bus. Upon completion of order creation, the **Booking System** publishes **Order** events to a topic to which the **Fulfillment System** is subscribed. In this case, teams agree on a contract for events exchanged between them. This way, both these systems and, hence, teams can work with minimal dependencies.

In summary, developing clean code requires the diligent application of design principles, patterns, and guidelines, together with continuous refactoring. The next section covers the testability aspect of technical agility.

Developing for testability

Agile software development focuses on rapid tests and learns cycles implemented with repeatable and automated testing. Therefore, designing systems for testability is critical. A fully automated regression test pack helps developers fearlessly make changes to code since automated testing can discover faults as early as possible in the delivery cycle before interrupting customers' services.

Testability refers to the ability to test application code, preferably in an automated form. Testability and automatability enable faster feedback cycle time, reduce rejections, and minimize the number of faults in production. Let's see how this works with the TDD approach.

Test-driven development

Test-Driven Development (TDD) is a software development approach that advocates writing test cases before code. The code is written against pre-created test cases, which represent scenarios of customers' needs. The *red-green-refactor* approach is typically used in TDD as follows:

1. First, write a test case that is failing because the code is not yet ready.

2. Then, write the most straightforward possible code to make the test pass.

3. Lastly, add functionality and refactor the code to make it more structured, efficient, and better.

TDD helps in avoiding overengineering by forcing you to write only the required code to meet test cases. It also helps in clearly meeting customer expectations and introducing new features since the test pack can detect faults. In most cases, developers apply TDD alongside **Behavior-Driven Development (BDD)**, which is an extension of the TDD concept.

Using contract-first design and a fake implementation of a *test double* enables parallel development for producers and consumers. It ensures that producers are developing just enough code to meet the contract and empowers producers to make behavioral changes by just testing against consumer contracts.

Architecting for testability and automation together with TDD helps guarantee the quality of delivery. The next section examines the infrastructure aspect of technical agility.

Treating infrastructure like software with the cloud

Enterprises cannot achieve business agility without adopting a cloud for application deployments. Agile software development, microservices (or loosely coupled architecture), DevOps, and the cloud are critical forces that help organizations realize business agility.

The traditional way of managing infrastructure is no longer feasible when the focus is on rapidly delivering value. There are several disadvantages associated with using traditional infrastructure models in agile software development:

- The traditional approach of labor-intensive and less responsive infrastructure management is considered a flow impediment in agile software development, impacting lead times, developer productivity, operational excellence, and knowledge acquisition challenges.

- In the traditional approach, architects and developers spend a fair share of their energy and time focusing on architecting, deploying, and maintaining infrastructures and software platforms instead of focusing on building business solutions.

There are several benefits in actively adopting the cloud instead of the traditional data center approach. Some of the key benefits are as follows:

- The cloud provides an easy-to-manage, software-friendly environment for enterprises instead of buying, owning, and maintaining hardware components and technical platforms.

- The cloud offers modern enterprise-grade services that are battle-tested for high scalability, availability, resiliency, fault tolerance, and performance at an optimal cost.

- Enterprises adopting the cloud realize improvements in terms of agility, productivity, and cost of operations.

- Embracing the cloud enables architecture evolution by supporting the incremental expansion of infrastructure and platform services without requiring capacity predictions.

- The cloud also allows teams to have frequent releases with a high degree of automation and unparalleled security and compliance levels.

- Cloud environments support the rapid development, testing, and launching of business applications by rapidly provisioning infrastructure and platform services with lower investments and upfront contractual commitments.

- Platform services such as database as a service, messaging as a service, containers, serverless computing, APIs, CI/CD pipelines, and application telemetry help bootstrap modern application development faster without spending a significant amount of time on plumbing needs.

Since cloud services are API-based, the adoption of the cloud gives teams more autonomy by reducing dependencies on central infrastructure teams. Development teams effortlessly provision, configure, and operate infrastructure as code by maintaining it in a version control repository just like any other code.

The last principle to be taken care of when building evolutionary architecture is full stack monitoring. The next section will cover the monitoring aspect of technical agility.

Monitoring everything with full stack telemetry

Proactively collecting data to determine possible faults helps to significantly improve **Mean Time To Recovery** (**MTTR**) and, therefore, minimize disruption to customer services. Telemetry is a rapid automated feedback mechanism for collecting data for measurement purposes from various data sources. Full stack instrumentation and telemetry are the cornerstones of effective monitoring. They connect infrastructure, networks, applications, and customer interactions to provide end-to-end visibility of operations.

Outcome-based monitoring, product-centric monitoring, and observability are best practices for monitoring modern systems that enable technical agility. The following sections cover these topics in detail, starting with outcome-based monitoring.

Outcome-based monitoring with progressive enhancement

Several metrics across applications, infrastructures, networks, platforms, interfaces, and security are required to maximize the system's serviceability with monitoring. The data collection includes logs, metrics, user activities, security, pipelines, and distributed systems tracing. While traditional monitoring is limited to monitoring the data collected, modern monitoring systems need to show the value of monitoring by rolling up as **Key Performance Indicators** (**KPIs**) at various levels, including business, IT service management, products, and infrastructure, as shown in the following diagram:

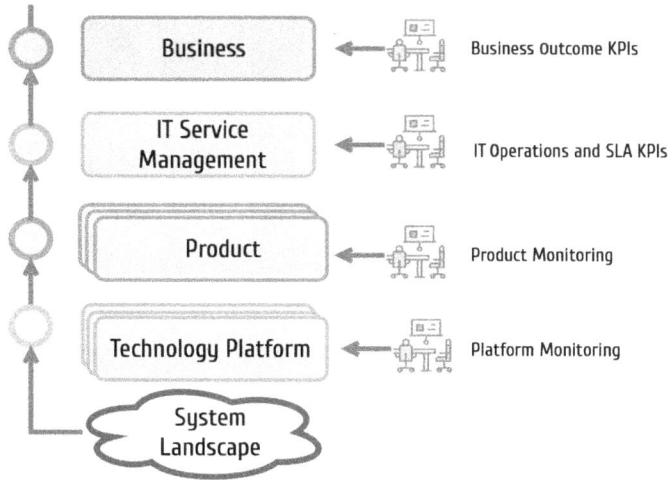

Figure 7.14 – Different levels of monitoring

KPIs and outcome-centric monitoring help in quantitatively determining the impact of an incident. For example, revenue loss due to an online e-commerce website being down for 30 minutes is a business KPI.

Well-established processes to progressively enhance monitoring capabilities help minimize interruptions to services and, therefore, improve customer satisfaction and reduced revenue loss. With such processes, every root cause analysis leads to not only fixing issues, but also enhancing monitoring to proactively detect and prevent similar incidents from occurring in the future.

Moving to product-centric monitoring

In the traditional approach, monitoring is fragmented. Current stack-based monitoring practices focus on isolated technical components and layers such as a database, server, and network. In a product-centric environment, development teams must be autonomous as much as possible.

A comparison of traditional monitoring and product-centric monitoring is captured in the following diagram:

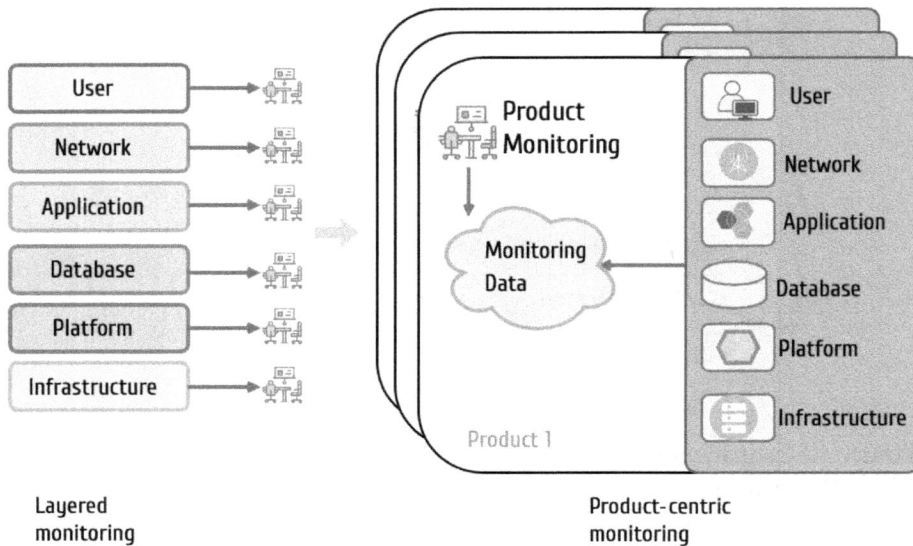

Figure 7.15 – Product-centric monitoring

While traditional monitoring requires effort to aggregate and create insights, product-centric monitoring does this by design. The approach must shift from horizontal to vertical full stack monitoring by offering end-to-end diagnostic information for development teams to minimize dependencies on others. Such product-centric teams use contextualized visual displays for instant access to metrics and logs collected for diagnosis.

Building observable systems

Modern systems must be like a transparent watch – the inner mechanisms are visible from the outside. Observability in software systems is a concept that helps to bring this transparency.

Observability helps to understand the inner state of a system, collecting signals that originated from its internal components. Observability starts with signals as opposed to the monitoring requirements and metrics.

The following diagram captures a typical process of monitoring observable systems:

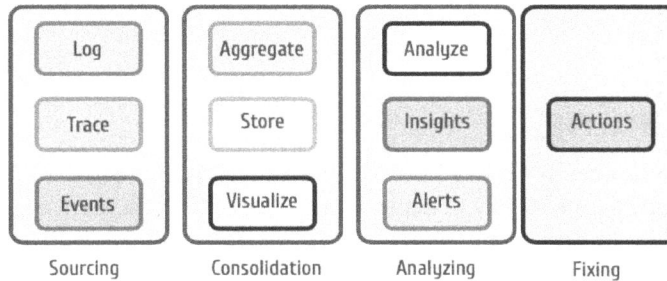

Figure 7.16 – Observability-based architecture

As shown in the diagram, the key patterns for signal collection for observability are metrics, tracing, and event logs. Aggregation, storing, and visualization are additional capabilities for full stack telemetry. Insights and alerts are generated based on the analysis of aggregated data. **AI Ops – Artificial Intelligence for IT Operations – is an** upcoming trend to identify and process patterns to create insights and automated actions. Structuring logs properly, adding instrumentation at multiple levels, emitting signals as events, and adopting distributed tracing are good design patterns for full stack telemetry.

Summary

Technical excellence brings rigor in quality development, which substantially improves organizational agility. Technical agility and software craftsmanship are underpinning forces enabling technical excellence. While technical agility implies good practices for architecture, development, testing, and deployments with built-in quality and automation, software craftsmanship is about an individual's ability to exhibit mastery in technology.

Learning organizations elevate technical agility by indefatigably adopting good design and coding practices and strenuously implementing DevOps and automation techniques. Technical agility fosters architecture evolution when architecture principles, design coding, testing, and operations are thoughtfully applied. Architecting for change is primarily attained by vertically slicing solutions, adopting DDD for decomposition, and implementing microservices-style architecture. Anti-corruption and strangler patterns are inevitably important for legacy migration. Clean design and leveraging clean coding techniques, principles, and patterns significantly promote quality and sustainability. Investing in writing test cases with TDD and developing automated regression packs helps developers fearlessly make code changes. Studiously adopting the cloud, platform services, and infrastructure as code is indispensable for developing modern applications. Full stack telemetry and observability across various environments enable the early detection of faults and minimize disruption to customer services.

This chapter focused on patterns and techniques – aspects of technical agility. The next chapter continues to focus on technical agility, deep diving into DevOps and automation aspects.

Further reading

- **Technical excellence**: https://www.researchgate.net/publication/228956099_Technical_Excellence_A_Requirement_for_Good_Engineering

- **Manifesto for software craftmanship**: https://manifesto.softwarecraftsmanship.org

- **Keep it simple**: https://www.interaction-design.org/literature/topics/keep-it-simple-stupid

- **Clean architecture**: https://blog.cleancoder.com/uncle-bob/2012/08/13/the-clean-architecture.html

- **Design principles and design patterns**: https://fi.ort.edu.uy/innovaportal/file/2032/1/design_principles.pdf

- **Cloud pattern repository**: https://docs.microsoft.com/en-us/azure/architecture/patterns/

- **Refactoring**: https://sourcemaking.com/refactoring

- **12-factor principles**: https://12factor.net

- **Software Engineering, NATO Science Committee Report**: https://www.scrummanager.net/files/nato1968e.pdf

- **Spring microservices**: Rajesh R V

8
DevOps and Continuous Delivery to Accelerate Flow

Lean is about constant ticking, not occasional kicking.

– Alex Miller, Professor of Management at the University of Tennessee

Chapter 7, Technical Agility with Patterns and Techniques, explained technical excellence and focused on patterns and techniques for attaining technical agility. This chapter focuses on DevOps and automation, which is the second segment of technical agility.

DevOps has gained significant momentum in the last few years and fueled success for many enterprises. Adopting a DevOps culture in software development and operations is increasingly necessary to deliver reliable systems faster through better collaboration and operational excellence. Continuously delivering value with a substantially high velocity with quality and flawless solutions demands a considerable amount of automation, similar to that in the manufacturing industry. Like a conveyor belt in manufacturing, in which goods pass through different stages, a fully automated delivery pipeline transfers source code immaculately through several pre-configured processing steps, such as building, validating, deploying, and releasing into a production environment, making it accessible to customers. Architecting to enable incremental builds, frequently repeated testing, zero-downtime deployments, and progressive releases to customers is incredibly important to successfully adopt DevOps and continuous delivery.

This chapter explains the importance of DevOps and DevSecOps culture and discusses other concrete practices of DevOps concepts. We will also discuss the idea of continuous delivery pipelines and explain the techniques and patterns relevant to continuous integration, automated deployment, and releasing on demand. We will also discuss the importance of security by design in delivering secure applications.

In this chapter, we're going to cover the following main topics:

- Embracing DevOps culture
- Enabling flow with continuous delivery
- Adopting continuous integration
- Automatically deploying to production
- Releasing on demand
- Securing systems by design

Thoughtfully integrating full stack telemetry into the continuous pipeline ensures the systems and services deployed are measurable against agility and stability. Delivering security-hardened and battle-tested systems against possible attack vectors without adversely affecting the speed of delivery needs careful architecture considerations for the early detection and rectification of security vulnerabilities, including the automatic patching of zero-day exploits.

This chapter focuses on the **DevOps and Continuous Delivery** focal point of the **Agile Architect Lens**:

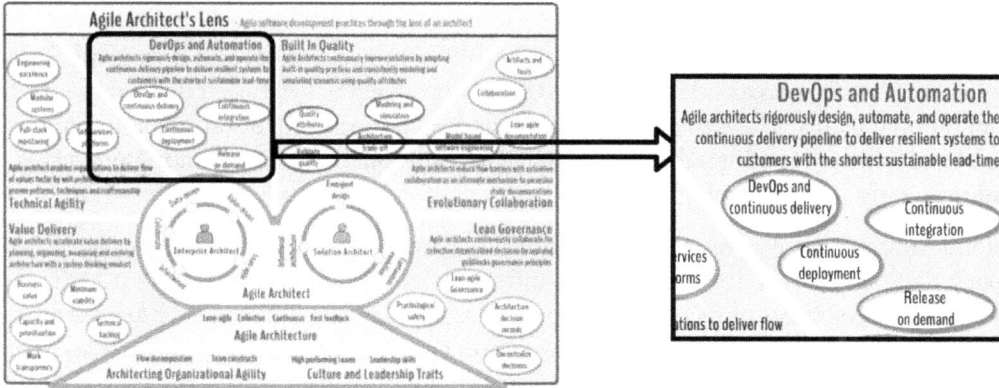

Figure 8.1 – The DevOps and continuous delivery focal point

Embracing DevOps culture

As articulated at `https://devops.com/`, DevOps was first introduced by Patrick Debois as a concept to get rid of the wall of confusion and lack of cohesion between development and operations teams. In traditional software delivery practices, developers used to throw code over the wall to the operations teams for deployment and operations. DevOps annihilated the walls of confusion by aligning development and operations teams' efforts, visions, and capabilities to produce valuable outcomes for businesses and customers.

The following diagram illustrates how DevOps avoids traditional handoffs:

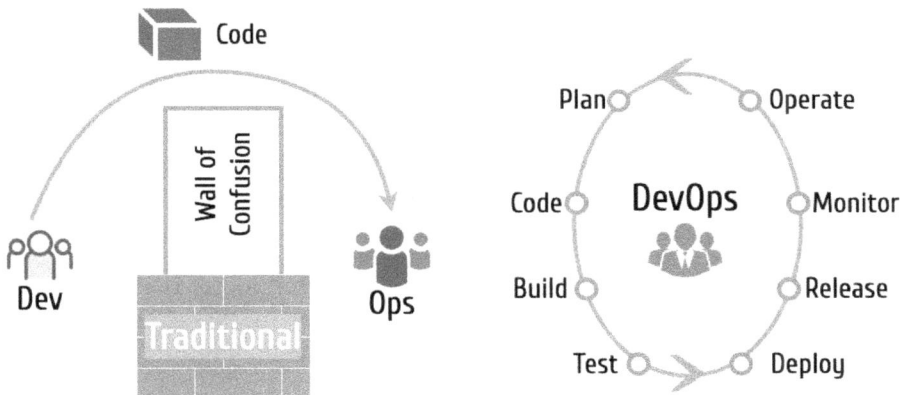

Figure 8.2 – Traditional approach of development and operations versus DevOps

With the **Wall of Confusion**, deployment teams without enough knowledge of the workload lose opportunities for optimization, resulting in suboptimal and erroneous deployments. DevOps breaks this wall by establishing strong ties between development and operational teams. The DevOps concept, anchored around collaboration, automation, and speed of delivery, is adopted by implementing team topologies, automating processes, and adopting modern technology practices.

The cloud infrastructure advancement is increasingly moving toward a software-based approach that enables developers to handle infrastructure like any other software. On the other hand, system admins are trying to acquire application knowledge to keep them relevant. DevOps needs a radically different culture by adopting lean-Agile principles to create a shared responsibility between development and operations teams to build, test, deploy, and operate systems that continuously deliver the maximum possible value to customers with a minimum lead time.

Jez Humble introduced the **CALMS** principles for DevOps. These five foundation principles are expanded and explained as follows:

- **Culture**: Faster delivery of high-quality secured solutions is the responsibility of everyone in the team with no siloed thinking or conflicts of objectives.

- **Automation**: Automating the delivery pipeline removes toil and makes the flow more predictable, reliable, and repetitive.

- **Lean**: Avoiding flow barriers by relentlessly focusing on eradicating wastage such as handoffs and wait times reduces reject cycles and improves lead times.

- **Measurement**: Continuously collecting signals from observable systems to build information radiators helps to determine areas for improvement.

- **Sharing**: Dev and ops teams communicate transparently and collaborate frequently with a sense of purpose.

Architects enable and foster DevOps culture with lean Agile leadership and enable a continuous flow of values by architecting modular, secure, and reliable solutions that are incrementally testable, deployable, and releasable.

Scaled Agile Framework (**SAFe**) uses a slightly different definition of **CALMS** by replacing **S** with **R** for **recovery**. Building automated recovery mechanisms to roll back and recover from unexpected deployment failures is substantially important for high-throughput deployments.

Over time, DevOps practitioners realized the importance of blending security into DevOps, which resulted in DevSecOps. The next section elaborates on the concept of DevSecOps.

Enhancing security with DevSecOps

Security is substantially important for any business since compromising security can lead to severe damage. A critical data breach can bring enterprises to a grinding halt. Hence, security cannot be traded for functional features, cost, and speed of delivery.

In large enterprises, cybersecurity is rightfully one of the most powerful departments empowered to stop projects from going live if not aligned with security guardrails. Security is by far the most critical non-functional aspect of any software system. Traditional stage-gate security validations only work well for systems and services with a slow release rate, and can adversely impact the lead time of projects having high-velocity releases.

DevSecOps is the next evolution of DevOps practice. DevSecOps introduces security as code into day-to-day programming and delivery practices by shifting security concerns to the left of the delivery cycle to earn time and cost savings. In DevSecOps, every team member, including cybersecurity members, has to have a shared responsibility to ensure they collaborate with less friction to develop secure applications without compromising the speed of delivery or customer satisfaction.

In Agile software development, security engineers are integral parts of the DevOps team. However, the federation of security engineers across every team is infeasible at scale since security engineers are scarce resources and cost more. Upskilling developers on secure coding practices and the automation of soft gate security validations to the extent possible is ineluctable to achieve consistent adoption of security policies and practices across the organization.

In DevSecOps, security is not just coding; it needs to be baked into the architecture and design as well. Fundamental principles of DevSecOps are the following:

- Security coding, testing, and fixing are done by the development team.
- Security needs and concerns are consistently added to the backlog.
- Security failures discovered are recorded in the backlog as technical debt.

Every member of the team assumes the responsibilities of developing secure applications, including architects. Architects play a vital role in architecting and designing security as a primary citizen.

Site reliability engineering is a practice overlapping with many DevOps practices. In the next section, we will take a closer look at site reliability engineering.

Crossing over with site reliability engineering

Site Reliability Engineering (SRE) is a software engineering approach practiced extensively at Google. SRE practice takes an engineering approach to deploying and operating reliable software systems by employing people with a strong engineering background with a deep understanding of the system's internal designs. Along with their design knowledge, their problem solving and analytical skills with operations experience make them a perfect candidate for managing system operations most efficiently and effectively. While development teams focus on building new features, SREs complement that with an operational and reliability view of the system.

While DevOps is an umbrella term for bringing about a collaborative culture between development and operations, SRE practice is a real implementation of DevOps philosophies. Since SREs are part of the development team and understand the application design much better, they help the development team to deploy and operate applications with operational excellence. While DevOps focuses on automation as the primary purpose, SREs focus more on highly available and reliable system delivery in which automation is just one element.

SREs take a risk-based approach to implementing technical aspects of the system, following a similar approach to the **Risk- and Cost-Driven Architecture (RCDA)** – if there is no risk, no improvement is required.

With Snow in the Desert, every **Agile Release Train (ART)** consists of a system team responsible for supporting feature teams with infrastructure, the automation of delivery pipelines, and validating system quality. The following diagram illustrates the concept of system teams:

Figure 8.3 – System and enabler teams at Snow in the Desert

The system team follows SRE practices, including infrastructure, security specialists, and other specialist support roles. As shown in the preceding diagram, there is also an enabler team for every ART. The enabler team is short-lived, joining the ART with a particular mission, such as operationalizing new toolsets for bootstrapping automated delivery pipelines. Once operationalized, they return. They work with the principle of **build-operate-transfer (BOT)**.

While DevOps has a broad spectrum of principles and philosophies, continuous delivery practices define much more tangible principles and practices to rapidly deliver value. The next section elaborates on continuous delivery patterns and practices.

Enabling flow with continuous delivery

One of the Agile Manifesto principles is to *satisfy the customer through early and continuous delivery of valuable software*. Adopting lean-Agile development practices alone is insufficient to accelerate the flow of work; build-test-deploy-release cycles must equally be lean and continuous. Continuous delivery is one of the underpinning practices in DevOps for automating build-test-deploy-release cycles.

CI and **CD** are common abbreviations used in the context of Agile software delivery and DevOps. CI stands for **Continuous Integration**, and is well understood as an evolution of engineering practice that automates building, testing, packaging, and deployment. On the contrary, CD is a confusing term. In general, *CD is interchangeably used for continuous delivery or continuous deployment*. The recent state of DevOps report *Accelerate: State of DevOps 2019* defines **Continuous Delivery** as **CD** and that is adopted in this book.

The following diagram illustrates various terminologies related to the automation aspect of DevOps:

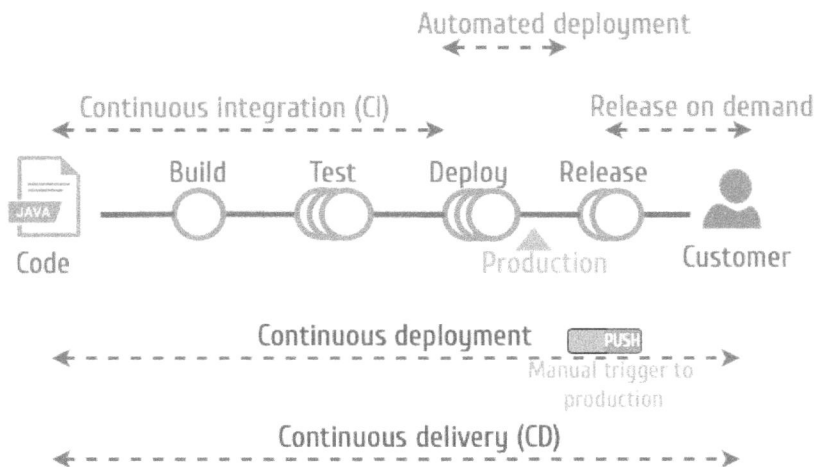

Figure 8.4 – Continuous integration, delivery, and deployment

As shown in the diagram, the end-to-end CD pipeline is broken down into three major components, as follows:

- **Continuous integration**: Run an integrated build, package the artifact, deploy it into various pre-production environments, and perform validations.

- **Automated deployment**: Packaged, validated, and hardened artifacts are deployed into the production environment with a dark launch.

- **Release on demand**: Make new features already deployed to production available to customers.

CI automates the first phase, covering building, testing, and, to some extent, deployment. Continuous delivery connects build, integrate, test, deploy, and release cycles in an end-to-end pipeline. At the end of every stage, work items are pushed to the subsequent stage automatically without manual interventions. On the other hand, continuous deployment replaces push with a manual pull for the last mile of moving work items to a production environment.

CD is a lean practice of automating the path to production from building, testing, configuring, deploying, and releasing with a minimal sustainable lead time. CD moves innovations, incremental code, configuration changes, and bug fixes to production in minutes by eliminating wastage, minimizing risk, and ensuring quality with almost no human touches post-development.

Disciplined Agile (**DA**) observes that CD is best suited when stakeholders need solutions on a frequent and incremental basis, the frequency of changes is often, and organizations have streamlined deployment practices and procedures and there is good maturity in terms of DevOps practices.

When CD is implemented properly, it can bring many benefits to Agile enterprises. Now let's explore some of the critical benefits of CD.

Benefiting from CD

Embracing CD brings several benefits in enabling a faster, uninterrupted flow of values, as captured in the following points:

- **Reduced lead time and costs**: Eliminating handoffs with automation enables a faster continuous flow of values, resulting in lower resource costs.

- **Predictable**: Fully automated CD pipelines enhance the predictability of system behavior in production.

- **Lower risk**: Frequent commits, incremental builds, continuous integration of code, and automated execution of test cases substantially reduce the risk of failure.

- **Better quality**: Automated tests with fast and frequent feedback cycles enhance the quality of products.

Together with Agile software development practices, CD substantially enhances organization agility. But to derive the best out of CD pipelines, systems have to be architected in a certain way. The next section explains this further.

Architecting for continuous delivery

CD goes far beyond the automation of builds, deployments, and releases of software systems. Architectural choices significantly impact the degree of automation possible and the true value proposition of automation. Patterns and practices discussed in the previous chapter, such as vertical slicing and microservices, are instrumental techniques for achieving small, independent, and incremental builds, deployments, and releases. The IEEE paper *Architecting Towards Continuous Delivery*, by Lianping Chen, observes that *in addition to traditional quality attributes, several other attributes are influential when architecting for CD, such as deployability, security, logability, modifiability, monitorability, and testability.*

In addition to architecting solutions for CD, architecting the pipeline itself is incredibly important. Some of the considerations are as follows:

- Thoughtfully invest time in designing the right-fit extendable CD pipeline by identifying and organizing steps specifically needed for a given system.

- A strong release management strategy, solid version control, and well-managed dependencies are critical.

- Determining bottlenecks by measuring efficiency and resolving root causes by continuously redesigning and adopting the right technologies are key to success.

- Integrating automated compliance checks for architecture, security, and other system quality attributes considerably improves the speed of delivery and quality.

- Fostering continuous collaboration between the various teams involved in the CD pipeline is one of the key responsibilities of an architect as a technical leader.

There is no one-size-fits-all solution for implementing a CD pipeline. Investing time in designing a suitable CD pipeline with progressive enhancement is essential for sustainable delivery.

Now that we have learned the importance of architecting for CD, we will now examine different parameters for measuring the success of continuous delivery.

Measuring the effectiveness of continuous delivery

Measuring the efficiency of CD is significant in determining areas of challenges and progressively refining it to achieve better results. **Software Delivery and Operational (SDO)** performance defined by the *Accelerate: State of DevOps 2019* report suggests four measurement parameters for the betterment of CD, as follows:

- **Lead time for changes**: The length of the pipeline time is measured as the time taken between a code commit and successfully running that feature in production.

- **Frequency of deployment**: Measured as the number of change releases of a particular system or service to its intended customers in a given timeframe.

- **Time to restore**: Measured as **Mean Time To Restore (MTTR)** after a failure that impacted users.

- **Change failure rate**: Measured as the percentage of successful versus failed deployments in production.

Measuring against these four parameters helps to understand two broad levels of KPIs – the throughput of the delivery process and the stability of the deployment. The report also highlights that these are not trade-off parameters, but are equally important outcomes.

It is straightforward to implement CD for greenfield applications but much harder for legacy applications. The next section throws some light on how to manage CD in legacy applications.

Implementing continuous delivery for legacy applications

CI and CD adoption are significantly painful with monolithic legacy applications. Even though the size, complexities, dependencies, and legacy nature of the technology stack hinder efforts to automate fully, it is still possible to substantially improve delivery throughput and stability by adopting relevant strategies of CD. In the research paper *An Empirical Study of Architecting for Continuous Delivery and Deployment*, the authors observe that monoliths and CD are not intrinsically oxymoronic. There are many effective approaches for embracing CD for legacy systems.

A few key strategies for adopting CD for legacy applications are as follows:

- Adopting a highly customized CD approach, tools, and technologies for automation
- Consistently testing interfaces by treating the system as a black box
- Adopting infrastructure as code for deployment automation
- Externalizing configurations from code
- Incrementally automating test cases for functional and non-functional requirements
- Adopting containerization strategy for applications where this facilitates deployment and maintenance

Continuously measuring effectiveness based on the lead time for changes, the frequency of deployments, the time to restore, and the change failure rate is helpful in determining further improvement areas. Gracefully adopting CD practices with a test, learn, and improve cycle is a better approach for legacy applications. Consistently measuring the complexity of code and progressively decoupling the architecture of various subsystems can further improve stability and throughput.

We have learned the purpose and benefits of CD in Agile software development. Subsequent sections explore different aspects of CD in detail, starting with CI.

Adopting continuous integration

CI, originated from extreme programming, is a natural extension of development practices for constantly building, integrating, and validating code automatically. Providing feedback to developers on the risk associated with the integration as early as possible helps to fix integration and quality problems with minimal impact on the speed of delivery. CI is the first step in the CD process. The traditional approach of infrequent manual builds and integrations is no longer viable for high-velocity developments and deployments. The manual build-integration-test-correction cycle significantly impacts the continuous flow of work.

A typical CI process flow is depicted in the following diagram:

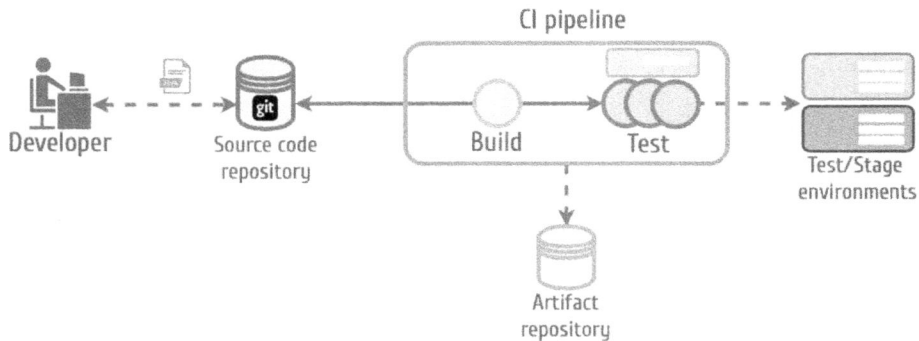

Figure 8.5 – Continuous integration pipeline

A **Developer** continuously commits code in a shared **Source code repository** that triggers the **CI pipeline**. Adopting pre-commits eliminates faulty code moving to the code repository. Quality assurance routines are automatically performed against the new build, ensuring the packaged artifact meets desired quality conditions. The **CI pipeline** also publishes the validated artifact to a central **Artifact repository**. The final stage of CI is to provision and deploy newly published packages to one or more pre-production environments. Some organizations use CI only for deploying artifacts to an integration server and leave other deployments to an automated deployment function. CI is based on the philosophy of build once, deploy many times to ensure that packaged artifacts are consistent across environments.

A well-architected CI pipeline offers the following benefits:

- The flexibility to execute tasks sequentially and in parallel
- Scalability to handle a large code base
- A low footprint and optimal resource utilization
- High performance and a shorter CI cycle time
- It's easy to provision and configure a pipeline for new systems and subsystems
- End-to-end monitorables and notifications
- Pluggability support to include additional processing steps

Teams also check-in configurations and **infrastructure as code** to the shared source code repository for the CI pipeline to use. Monitoring and securing pipelines is also critical for efficient CI implementation.

Large-Scale Scrum (**LeSS**) observes that CI is not just automating builds and running tests; it includes a set of practices such as **Test-Driven Development** (**TDD**), small incremental changes, frequent integrations, fast feedback cycles, and trunk-based development.

A good CI pipeline starts with better development and source code management practices. We will cover this in the next section.

Improving development and source code management

Enterprises can further scale the value of CI by adopting good development and source code management practices. Clean coding practices were explained in the previous chapter, together with pair programming, peer code reviews, and good development practices. Using IDE-based code analyzers, executing functional and performance unit test cases, testing integrations with test doubles, and committing code often adds further value.

A sophisticated, shared source control system such as **Git**, with an effective branching strategy, is essential for efficient source code management. There are two key repository management strategies that are popular – **Monorepo** and **Multirepo**, as shown in the diagram:

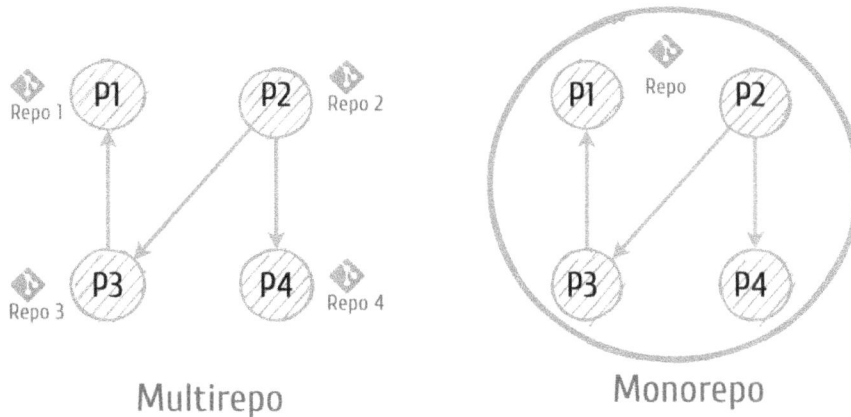

Figure 8.6 – Multirepo versus monorepo

In the case of **Monorepo**, multiple projects share the same repository by maintaining dependencies between each other. This approach offers a consistent developer experience with real-time change awareness as it can quickly identify whether changes break another project and reduce dependency management overheads. **Multirepo**, on the other hand, isolates project code with separate source code repositories with no build-time dependencies.

Many large organizations effectively make use of monorepo. However, considerable efforts are required to build tooling around monorepo to reduce associated complexities, such as too many branches and pull requests. It also needs higher degrees of team maturity to eliminate potential chaos. Complex enterprises and product companies with heterogeneous applications may choose multirepo over monorepo to manage their products better. Monorepo requires significant discipline as accidental mistakes can lead to unexpected errors in fellow developers' code. It can also cause adoption and performance challenges, such as slowness in downloading the full code base. There is no one-size-fits-all strategy; therefore, careful analysis is necessary when selecting a repository strategy.

Choosing the right branching strategy is also significantly important when implementing CI. There are two branching strategies, as shown in the following diagram:

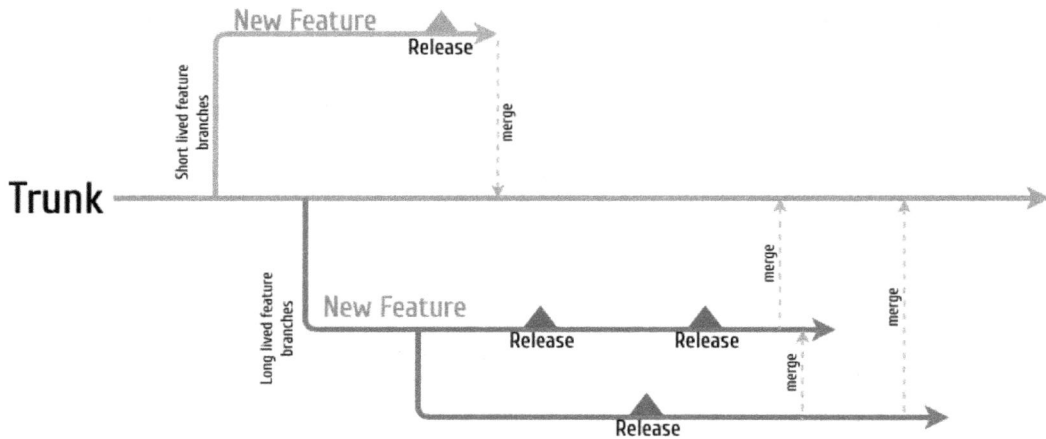

Figure 8.7 – Trunk-based and feature-based development

In your feature branch strategy or GitFlow, create a long-living branch for every new feature and merge it back to the trunk only when development is completed, as shown on the bottom side of the **Trunk** line in the preceding diagram. This approach induces minimal dependencies between different feature teams. However, feature branching strategy leads to complex and time-consuming merges using pull requests.

Trunk-Based Development (**TBD**) scenarios avoid long-living feature branches that complicate build systems over time. Instead, TBD recommends that developers commit directly to the trunk or use short-lived feature branches merged within a day or two with or without using pull requests.

There are advantages and disadvantages to both branching strategies. Feature branches are better if the release frequencies are longer, features are less dependent, and features are not released together. If the application is well modularized with minimal dependencies between modules and the release frequencies are very high, trunk-based development is better. Product maturity may also play a role in adopting the branching strategy. Teams might use TBD in the initial stages before migrating to feature branching as they gain stability and become established on the market.

Snow in the Desert uses monorepo as well as multirepo. All legacy applications are on multirepo, whereas new microservice developments use monorepo but are confined to certain business boundaries. Similarly, Snow in the Desert uses a combination of trunk-based development and feature branch strategy. Most old systems continue to use feature branching strategy with long-lived feature branches as releases are less frequent. Modern applications such as the B2C online frontend use trunk-based development.

In addition to good development practices, good testing practices are also equally critical for the success of CI, which we will look at next.

Automating quality assurance

Automated testing is a fundamental aspect of CI that enables repeated predictable releases that improve quality and stability. Automatic testing includes unit, functional, integration, security, and other types of testing. Repeatedly executing regression test suites increases developer confidence and protects from unstable code flowing into production.

The pipeline also automates architecture and quality tests such as performance, security, coupling, complexity, and dependency. Using environments that are correctly sized is essential to get realistic results. Failure trails and antifragility checks using approaches such as Chaos Monkey ensure the resiliency of systems. Test data management, writing independent test cases, and adopting service virtualization with test doubles such as mocks, stubs, and fakes are also essential considerations for automating tests.

Building a rigorous testing culture is important for the success of automated testing. Continuous automated testing frees up resources, improves consistency, and creates psychological safety.

So far, we have gone through CI, which is the first step of CD. The next section will elaborate on automated deployment.

Automatically deploying to production

Automated deployment is a mechanism for continuously deploying systems and services to production without downtime. An automated deployment pipeline kicks off when CI completes successfully and further deploys artifacts from the artifact repository to a target production environment silently – also called **dark mode**. Automated deployments use immutable infrastructure concepts with **infrastructure as code** for deployment.

Continuous deployment also includes provisioning and configuring required environments using infrastructure as code before deploying the system. By fully automating production deployments at scale bring an incredibly high degree of consistency, repeatability, and protection from disruptive failures. Some organizations extend their CI tools to perform automated deployments, whereas others use different tools integrated with CI to execute deployments. Smaller changes and high-frequency deployments such as feature-based micro deployments are best practices to handle the efficiency of automated production deployments. In the event of failures, automatic rollback routines revert and restore the deployments back to a stable state to provide uninterrupted services.

There are different techniques for deploying to production without impacting customers. One such technique is dark mode deployment, which we will discuss next.

Launching in dark mode for early feedback

A **dark launch** is a common technique used for deploying and testing applications in production silently. Dark mode deployments help development teams to deliver faster without compromising the stability of production systems.

In many organizations, the release frequencies may not be high, and businesses usually wait for the right time to release certain features. However, the development team can decouple releases from deployment and continue to deploy and test solutions in production using an automated deployment pipeline without making them available to customers. **Feature flags** and **blue-green** deployments are useful for dark launches without disrupting the production environment. Dark launches allow a subset of users, primarily development teams and business users, to access and validate new releases.

A feature flag is another technique for deploying applications silently to production. The next section will cover how developers can use feature flags.

Using feature flags to release features selectively

A **feature flag** is a mechanism for contextually determining the control flow of programs at runtime based on toggles to determine releasable and non-releasable features. Feature flags are the foundation for dark launches and progressive releases.

The following diagram shows the use of feature flags:

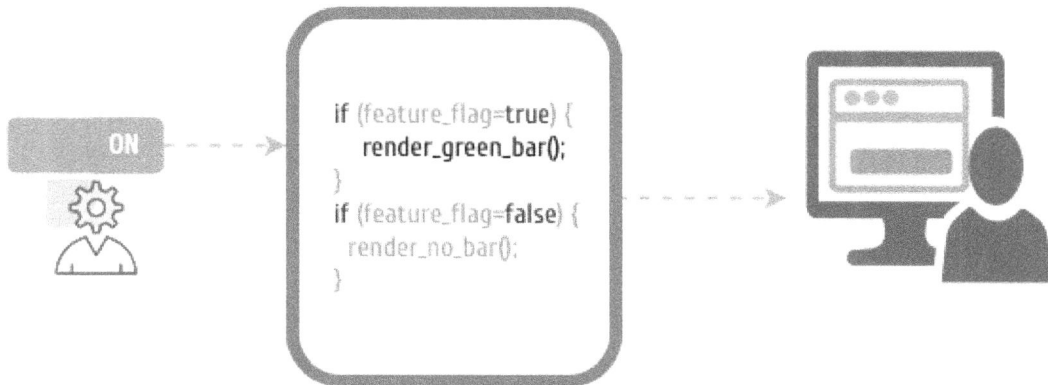

Figure 8.8 – Illustration of feature flags

As shown in the diagram, based on the toggle state, the flow of control goes to the new feature or uses existing features.

Enterprise-scale feature flag solutions provide a central dashboard to manage and monitor feature flags easily and efficiently with a high-performing architecture. The use of feature flags needs to be really controlled to avoid the proliferation of uncontrolled feature flags, which results in higher degrees of technical debt. One way to manage feature flags is to set an expiry date for each feature flag and attach test cases that validate and fail on expiry. Another approach for managing feature flags is adding compensating technical debt in the backlog to remove corresponding feature flags. Feature flags are also useful for managing the progressive enhancement of databases for deploying with a blue-green strategy for uninterrupted services.

Blue-green deployment is another pattern to deploy new releases rapidly to production without interrupting existing services, as shown in the following diagram:

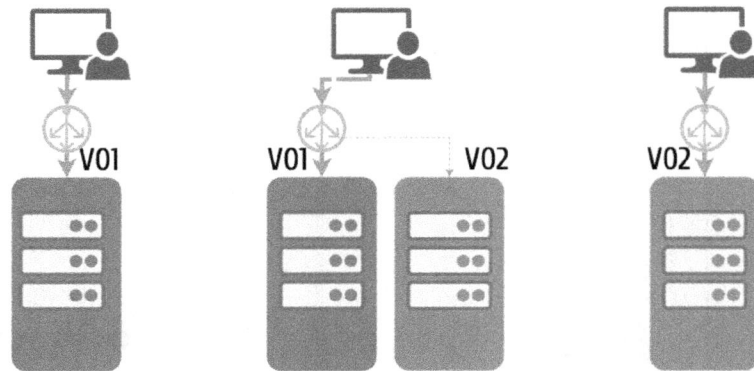

Figure 8.9 – Blue-green deployment

Blue-green presents two identical production environments where new code is deployed to a freshly provisioned green environment. The green environment is then used for sanity tests, chaos tests, system quality tests, user experience tests, and so on in production. Most of these tests are conducted automatically as part of the deployment pipeline. Full stack telemetry is used for monitoring failures and rectifying them before pushing users to green. Blue-green deployments also enable automatic recovery from unexpected failures.

Architecting systems for automatic and feature-based deployment is critical for releasing stable and reliable software to customers with rapid fix forward cycles. Implementing and managing feature flags at scale needs architecture thinking to avoid the diffusion of feature flags. Using the right architecture patterns and technologies such as statelessness, the cloud, and containers enables blue-green deployment.

So far, we have covered the first two phases of CD – continuous integration and automated deployment. The next section covers release on demand, which is the last phase of CD.

Releasing on demand

Release on demand is proposed by SAFe to decouple deployment from the actual release of software to customers. Continuous deployment pushes the software into production without making it widely available to customers for regular use. The time between deployment and release is the window of safety to ensure customers get stable features from the moment it is released.

In most cases, the time of release, what to release, and whom to release to are business decisions based on market events and milestones. Delineating deployment from release elevates the business confidence to go to market with release announcements as the features are already available and are tested in production.

During the window of safety period, the development team and business users test deployed features functionally and non-functionally, release them progressively to a smaller subset of users, collect their feedback, and make necessary refinements before releasing them to broader users.

Progressively releasing with canary releases

A canary release is one of the release strategies for the progressive rollout of new software to reduce the risk of early feedback from real customers. The following diagram explains the approach for a canary release:

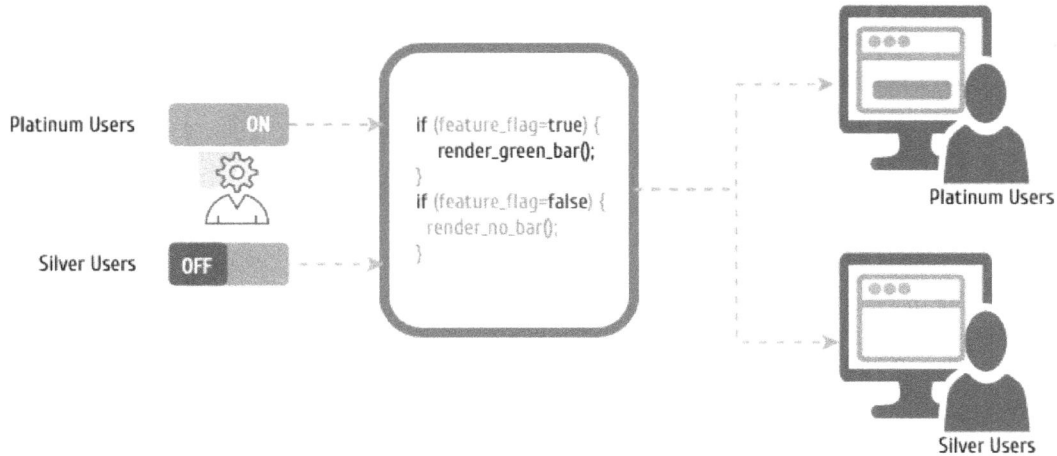

Figure 8.10 – Canary release for a selected user community

As shown in the diagram, traffic is routed to the new software, with an additional bar on the screen version, progressively to a subset of users selected based on candidate user selection strategies. The candidate user selection strategies could be random samples, internal users, premium users, market-based, and so on. Usually, collecting feedback is one of the key constraints when choosing the right canary user segment. Typically, the canary starts with a set of change agents, approximately 2-5% of the overall user base. There is a little difference between dark launches and canary releases as canary releases mainly target an informed community. In contrast, dark launches are generally restricted to internal users and might not be released to actual customers.

In addition to canary style releases, enterprises also use A/B testing for collecting early feedback on innovative changes before mass release. We will see more on A/B testing in the next section.

Collecting early feedback with A/B testing

A/B testing or **split testing** is another release strategy to get quick feedback primarily for experimental features, as shown in the following diagram:

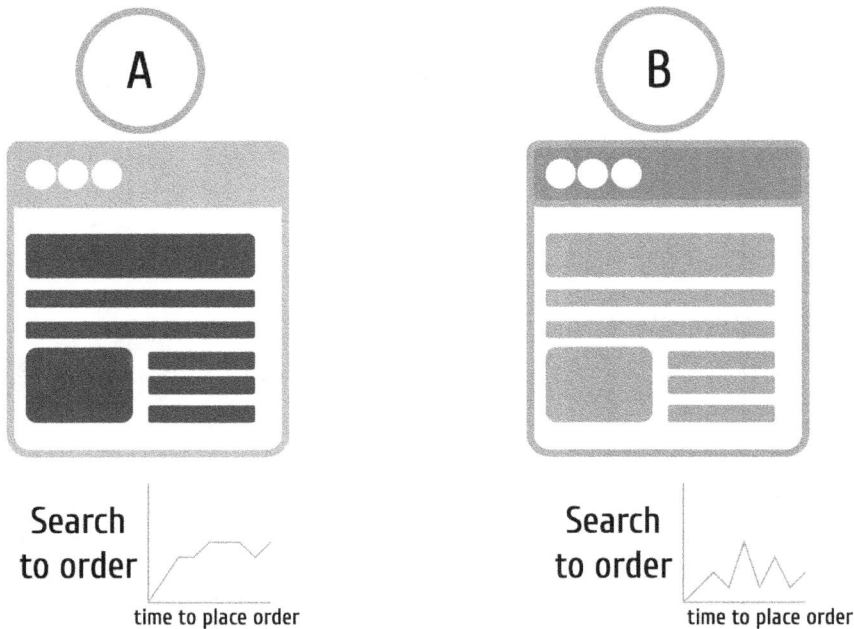

Figure 8.11 – A/B testing or split testing

In A/B testing, the old version of a feature and the new version of the same feature are available for different sets of customers, either the full user base or a selected user base. The data captured from user interactions is compared to understand and determine the effectiveness, such as the new version's user experience over the old version's. As shown in the preceding diagram, user experience – measured by the time taken to place an order when using a **Search to order** transaction, is captured across old (**A**) and new (**B**) before releasing **B** to a broader set of customers.

To get the true value of release on demand, architects need to pay attention to the architecture and design of systems and services. In the next section, we will review some of the key design elements.

Architecting for release on demand

Architecting for operations is an incredibly important factor for release on demand. Architects have to ensure the architecture of the system or service is designed to support the following points:

- **Progressively releasable**: Incrementally deployed services and system components are progressively releasable based on one or more release strategies.

- **Testable in production**: The system and services are testable in production to avoid unexpected faults post-release.

- **Full stack telemetry**: Adequate monitoring and analysis are in place during progressive release to ensure facts are collected and made available for decision making.

- **Running costs monitoring**: Ensure the implementation of cost monitoring to validate the utilization of the infrastructure to avoid a cost overrun.

- **Recoverability**: In case of failure during the window of safety, deployments are easily recoverable without hindering entire users.

So far, we have covered different aspects of CD and CI, automated deployment, and release on demand. A critical facet of software delivery is securing services and systems released to production. In Agile software development, security needs a radical rethink. The next section further explores securing systems by design.

Securing systems by design

Security is a significantly important pillar applicable across all phases of CD. In this section, we will review the security aspects of CD in detail.

DevOps and Agile software development often trade off security for speed and agility, steering off course from the significance of delivering secure applications. Enterprises adequately protecting assets by implementing built-in security controls with a security-aware mindset is paramount in precluding data, revenue loss, and brand reputation damage.

Historically, cybersecurity departments held the responsibility and authority to enforce and govern software development security concerns to safeguard an organization's intellectual property. However, the traditional stage-gate-based modus operandi of cybersecurity departments is far too complex, manual, and time-consuming. Therefore, traditional security approaches induce significant time delays and struggle to cope with the faster development and high-frequency release cycles of Agile software development. Cybersecurity departments thus need a radical transformation to reposition them from flow stopper to enabler of agility.

In Agile software development, security needs to be an integral part of the continuous flow of work with a shared responsibility to deliver secure applications faster to meet customer needs. To enable faster delivery, security engineers with a lean-Agile mindset need to adopt a shift-left pattern to apply security controls with the right set of modern tools and techniques. The shift-left pattern introduces security testing such as penetration testing and vulnerability tests earlier in the life cycle of the software development process. Detecting defects early reduces the time needed for code-test-deploy cycles.

In Agile, security is continuous and spans across architecture, development, the CD pipeline, and operations. Security controls are identified, developed, and tested by the team themselves, including fixing any deviations.

Several characteristics of security-aware teams are explained here:

- **Security activities as backlog**: In Agile, there is no free lunch; all work must be captured in the backlog, including security. Transparently managing security as part of the backlog helps to plan and execute security activities as part of the flow. Security validation errors, feedback, and risks have to be pushed to the backlog for prioritization.

- **Code security**: Cybersecurity must educate development teams on secure coding practices to consider security as a first-class citizen. Teams consciously upskill on new security developments by referring to industry practices such as the **Open Web Application Security Project** (**OWASP**) to understand various security threats. Static and dynamic code analyzers and code review tools automatically enforce standards and best practices to improve the efficiency, speed, consistency, and effectiveness of code security. **Software Composition Analysis** (**SCA**) can identify potential vulnerabilities of open source libraries.

- **Architecture security**: Using a lightweight and continuous approach for threat modeling by identifying threat boundaries is a significantly important activity for architects to identify the right security architecture that protects systems from internal and external attacks. Reducing the attack surface, securing APIs, encrypting message exchange, promoting security by design practices, and using appropriate compliance tools are mechanisms to help architect secure applications.

- **Infrastructure security**: Automating security policies as part of infrastructure as code integrated with the CD pipeline eliminates potential human errors in the infrastructure security configurations. Adopting network segmentation with the principle of least privileged access, integrating centralized identity and access management, consistent processes for automating security patches for vulnerabilities, and continually monitoring the production environment for security threats are great practices for operating secure systems. Well established and integrated process loops for fixing code based on learning from attacks enable organizations to be adaptive.

- **DevSecOps practices**: A mindset of shared responsibility across all team members is critical in developing secure applications. Automated and continuous scanning to detect vulnerabilities in code, infrastructure, and libraries as part of the delivery pipeline helps to find potential vulnerabilities at the earliest opportunity. In cases where automated testing is not possible, security experts must be embedded in the team to ensure there are no considerable delays in the delivery flow. Rightfully including constraints in the **Definition of Done (DoD)** and acceptance criteria specifically for security scenarios helps avoid critical segments of code going to production without sufficient security controls.

Architects play a critical role in ensuring adequate security considerations and good practices are followed by constantly engaging with the team. Trading off security and speeding up feature development to satisfy immediate customer needs can have far-reaching consequences on customer confidence.

Summary

As we learned in the previous chapter, the second segment of technical agility is DevOps and automation – a combination of embracing DevOps culture and implementing CD. DevOps is a culture that brings development and operational teams together without the wall of confusion. DevSecOps is an evolution of DevOps that instills security thinking into every step of Agile software delivery.

In this chapter, we saw that CD is a delivery practice to automatically move new features, improvements, and fixes to production by reducing the lead time and improving predictability. Consistently practicing CD dramatically improves the throughput of delivery and stability of systems and services. CD practice consists of CI, automated deployment, and release on demand. CI is a natural extension of development that helps to build, test, and deploy applications to one or more environments.

Next, we understood that automated test packs consolidate unit, functional, non-functional, and integration testing as part of the CI pipeline. Automated deployment pushes successfully tested packages to production in a dark mode using blue-green deployments without activating new features for customers. Such dark deployments utilize feature flags to validate features in production before releasing them to customers. Release on demand releases features progressively to customers using practices such as canary releases and A/B testing. Architecting for CD needs special considerations to enable an incremental build, automatability, testability, and deployability. Lastly, we saw that security is a substantially important activity in the CD practice that spans across all stages of pipelines. Security in DevSecOps and CD is a continuous process all team members are responsible for.

In this chapter, we have covered the second segment of technical agility – DevOps and automation. In the next chapter, we will continue to explore technical agility by focusing on the built-in quality aspect.

Further reading

- **History of DevOps**: https://devops.com/the-origins-of-devops-whats-in-a-name/

- **ACM computing survey**: https://arxiv.org/pdf/1909.05409.pdf

- **State of DevOps 2019 report**: https://services.google.com/fh/files/misc/state-of-devops-2019.pdf

- **Site reliability engineers**: https://landing.google.com/sre/

- **Continuous delivery**: https://martinfowler.com/bliki/ContinuousDelivery.html

- **Continuous delivery**: https://continuousdelivery.com

- **An empirical study of architecting for continuous delivery and deployment** https://www.researchgate.net/publication/327536527

- **Towards architecting for continuous delivery**: https://www.researchgate.net/publication/280776459

9
Architecting for Quality with Quality Attributes

"Architecture should speak of its time and place, but yearn for timelessness."

– Frank Gehry, architected the iconic Dancing House in Prague

The previous chapter explained the DevOps and automation segment of technical agility by examining the importance of fostering DevOps culture and architecting for continuous delivery. This chapter will focus on the third segment of technical agility – **built-in quality**.

Mission-critical applications must be anti-fragile and consistently deliver services without degradation of quality. Failures can have severe consequences for the business, such as revenue loss, customer dissatisfaction, and even, in some cases, health and safety issues. Therefore, a methodological approach for building quality into the software development process is crucial. To achieve this, we need to learn a mechanism of connecting measurable business outcomes such as revenue loss to quality attributes such as scalability. Built-in quality also needs to consider a balanced approach between risk and cost to eliminate potential wastage as quality costs money. To practice this, we need to foster an Agile approach by incrementally defining, negotiating, measuring, and improving quality. Architects play a significant role throughout the quality life cycle by embracing a systems thinking mindset.

This chapter starts by covering the importance of quality in software systems and explains the evolvability of quality, the need for systems thinking, and the concept of built-in quality. Then, we will review various quality attribute models and different ways to document quality attributes consistently. Later, we will examine different quality life cycle stages and related activities that ensure quality is not compromised in the delivery stages. Throughout this chapter, we will discuss the architect's role in ensuring the quality of solutions at scale.

In this chapter, we're going to cover the following main topics:

- Understanding software quality
- Adopting a quality attribute model
- Documenting quality attributes
- Using the quality life cycle in Agile software development
- Discovering and refining quality attributes
- Modeling and simulation
- Applying architecture trade-offs
- Developing quality attributes
- Assessing system quality

This chapter focuses on the **Built In Quality** focal point of the **Agile Architect's Lens**, which is illustrated in the following figure:

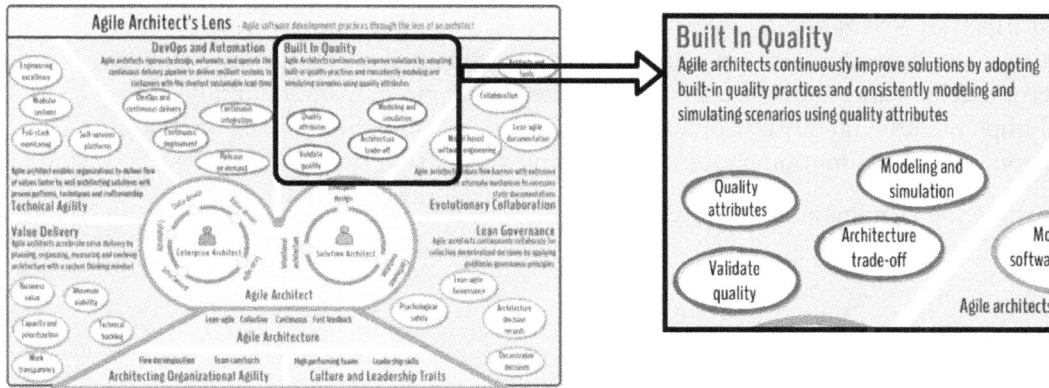

Figure 9.1 – Built In Quality – focal point

We will start this chapter by exploring the basics of software quality and its importance when building highly reliable and anti-fragile systems.

Understanding software quality

System quality is a measure of how well a system behaves under certain operating conditions and how well it evolves to handle future demands satisfactorily. One of the most significant responsibilities of Agile architects is to incrementally and explicitly capture quality needs and connect them to architecture decisions and backlogs. Many terminologies, such as quality characteristics, quality attributes, quality concerns, architecture concerns, technical requirements, **Architecturally Significant Requirements (ASRs)**, and **Non-Functional Requirements (NFRs)**, are interchangeably used for describing the quality requirements of software systems. This book uses quality attributes and ASR to refer to quality needs that constrain or influence architecture and design decisions.

Inaccuracy in detecting and implementing quality attributes can have adverse repercussions on the endurance of systems, which further leads to higher maintenance and operating costs. Poor-quality solutions result in substantially higher cycle times due to late detection, reworks, and retest of quality issues. As an example, **Innovate** was an ambitious and expensive program run by McDonald's, the global fast-food chain. With a 1 billion USD budget, Innovate started, in 1999, to build software to automate the entire fast food outlet. The idea was to collect information from all food outlets and present a boardroom view to its leadership so that operations could be monitored and improved in real time. Unfortunately, Innovate was canceled in 2002 and wrote off 170 million USD. Research studies observed that one of the key reasons for its failure was underestimating the technical complexities and quality needs beyond the stated user requirements. Therefore, focused attention on quality attributes is paramount.

IEEE 1061-1998 Standard for a Software Quality Metrics Methodology defines software quality as the ability of software to demonstrate behaviors such as reliability, interoperability, and so on. There is no one-size-fits-all mechanism for architecting and implementing quality attributes as they are directly influenced by the operating environment, consumption patterns, and business context.

The three forces influencing quality are shown in the following diagram:

Environment

Quality

Usage   Business context

Figure 9.2 – Forces influencing quality

For example, an architecture mechanism to improve performance may differ for an application deployed on mobile versus an application deployed on a cloud environment. Solutions vary with consumption patterns, such as a solution accessed once in a day versus a solution consumed 100 times in a second. Lastly, the performance mechanism for a solution used in a banking context varies from a tracking solution used in an Uber-style taxi service.

To get a clear understanding of these parameters, architects need to use the systems thinking concept, further explained in the next section.

Improving quality needs systems thinking

Architects building complex web-scale and reliable systems need a broader vision and comprehensive knowledge of the business and IT ecosystem. Localized designs and optimizations of specific components may not deliver the overall desired value to the business. Restricted thinking and a narrow solution mindset inhibit quality over time.

Systems thinking originated from the field of system dynamics and was and was introduced by MIT professor Jay Forrester in 1956. Systems thinking is an approach that helps solve problems with the system as a whole with a clear understanding of the system's components, boundaries, interactions, and environment.

The key parameters of systems thinking are captured in the following diagram:

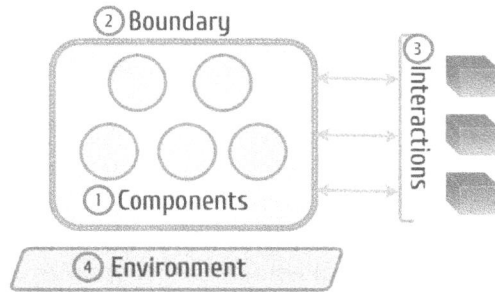

Figure 9.3 – Parameters of systems thinking

While conventional approaches focus on system components in isolation, systems thinking focuses on an expanded surface area even beyond the whole system by adding neighboring systems and their potential influences through interactions or sharing the environment.

The paper *Overview of Systems Thinking* by Daniel Aronson provides an excellent metaphor to explain systems thinking. He uses human action to reduce crop damage by insects as an example. When insects eat and destroy a crop, the farmer's immediate response would be to apply a pesticide to kill all insects effectively. On the surface, this may look like a perfect solution. However, there could be better solutions if the farmer takes a broader view of the issue before fixing it. Applying pesticides can surely kill insects of type A, but it could result in the growth of another type of insect, B, beyond control. Initially, insect A used to eat insect B. In this case, killing insect A is not optimal as it will not solve longer-term insect issues. Integrated pest management is a better solution, in which to kill insect A, the farmer adds more of its predators.

The philosophy of systems thinking is the belief that system components in isolation behave differently than when the whole system operated together. Noisy neighbors can negatively impact the operating quality of a component.

In Agile software development, systems thinking is highly relevant as development teams always work with incremental changes, potentially leading to isolated solutions. These solutions designed in silos may not fit well with the overall solution vision and may fail to operate at scale. The architect's responsibility is to provide the right directions based on a holistic, connected system view.

In addition to systems thinking, it is also important to realize that quality is not a one-time activity. It needs to be evolved with emerging business demands. In the next section, we will discuss more on the evolvability aspect of quality.

Incrementally applying quality with a build-to-adapt strategy

While quality is important in delivering reliable software solutions, excessive focus upfront on quality may not guarantee the solution's promises to customers infinitely. Build to last and build to adapt are two common strategies for baking quality into solutions during development. Build to last includes activities such as estimating and implementing 5 years' capacity, building fail-proof systems, designing by future-proofing, and so on, whereas build to adapt is using evolvable architecture, on-demand elastic scaling, anti-frAgile systems, and so on. The fundamental premise of Agile architecture is no **Big Upfront Design** (**BUFD**), which is equally applicable when building quality.

The Choluteca bridge is living proof that build to adapt is better than build to last. The following is an illustration of the current state of the Choluteca bridge:

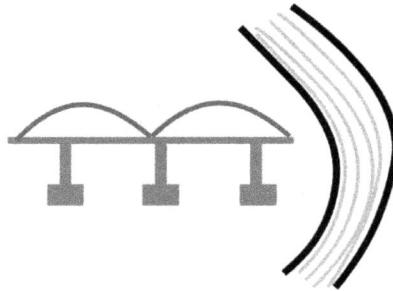

Figure 9.4 – Illustration of the Choluteca bridge

The Choluteca bridge is a 484-meter bridge built in 1998 in Honduras, a place prone to natural disasters. The goal was to build the strongest bridge that can withstand any natural disaster. Later that year, Hurricane Mitch devastated many places in Honduras, including nearby bridges and infrastructures. The Choluteca bridge – the high-quality architectural marvel – stood its test without any damage. However, sadly, the Choluteca river no longer flows under the bridge; it found a new path next to the bridge, making the bridge irrelevant.

Build to adapt is more appropriate for Agile software development where requirements may change the design even at the very last stage of delivery. Build to adapt is not just upscaling quality but also limiting. For example, in the unprecedented times of the global COVID-19 pandemic, many organizations reduced their scale of business operations, which means the cost of IT operations also had to be brought down to lower levels acceptable to present trading conditions.

Incremental development of quality also eliminates high upfront investment needs. For example, in the context of Snow in the Desert, the tracking microservice of the **Automatic Vehicle Tracking System (AVTS)** project does not need very high availability during the **Minimum Viable Product (MVP)** phase. But once the concept is proven and ready to scale across multiple markets, high availability is extremely important.

The continuous evolution of quality demands continuous quality assurance as well. An effective approach is to shift focus from post-quality assurance to applying quality at the source. The next section will go deeper into the built-in quality aspect of Agile software development.

Building in quality to eliminate waste

Built-In Quality (BIQ), or **Jidoka**, is one of the pillars of Lean manufacturing originating from the **Toyota Production System (TPS)**. BIQ is a practice of building quality at every step of the process instead of in post-development quality assurance routines. BIQ reduces wastage by reducing recall, rework, and retest cycle times post-development.

BIQ is a Lean-Agile practice for continuously improving value. Effective adoption of BIQ needs a radical shift in mindset to a quality-driven culture by moving away from the build and test cycles to continually improving quality through rapid incremental testing and feedback cycles. BIQ is the shared responsibility of the team with a sense of purpose to achieve the ultimate goal of delivering business value. It focuses more on prevention by building quality at the source than testing against an infinite number of KPIs. BIQ enables high-quality solutions and improves customer satisfaction and speed of delivery.

BIQ promotes a quality-first approach by adopting a sense of quality in everything, including designing solutions using the right architecture, patterns, and techniques, adopting good coding practices, and including the right quality assurance routines in the development and continuous delivery pipeline.

The three levels of BIQ maturity are shown in the following diagram:

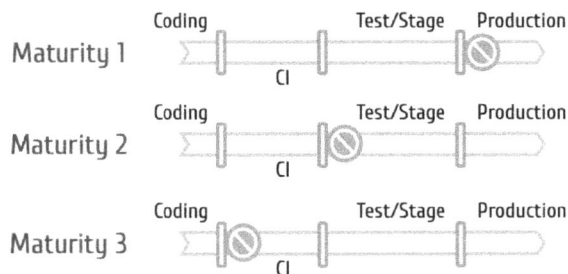

Figure 9.5 – The three maturity levels of BIQ

As depicted in the diagram, BIQ restricts defective work products flowing through the continuous delivery pipeline. The maturity is measured based on at what point faults are stopped from flowing to the next stage. For example, in **Maturity 3**, highly matured organizations practicing BIQ ensure defects are avoided by adopting well-established and streamlined preventive measures at the time of development, thereby forestalling faults entering the CI pipeline.

For accurately building quality, consistent definition and adoption of quality attributes are critical. The next section will examine different standard quality attribute models.

Adopting a quality attribute model

Quality requirements are classified as business quality such as time to market, architecture quality or ASR, and system quality. A typical software quality attributes model defines a taxonomy for the definition of system qualities. There are many quality attributes models available in the industry.

A few popular quality attributes models are described as follows:

- **IASA** defines *quality attributes as non-functional characteristics of a component or a system* and classifies them into four categories – usage, development, operations, and security. There are subcategories under each one of them.

- **FURPS** is another commonly referenced quality model, which stands for **Functionality, Usability, Reliability, Performance, and Supportability**. SAFe uses FURPS for explaining NFRs.

- The **IEEE Standard 1061** quality attribute model classifies quality attributes into efficiency, functionality, maintainability, portability, reliability, and usability.

- **ISO 25010** is one of the modern quality attribute models defined in 2007, built on top of ISO 9126 evolved from McCall and Boehm.

The paper *A Review of Software Quality Models for the Evaluation of Software Products* defined several quality models created over the years from McCall in 1977 to MIDAS in 2013. The research observes that ISO 25010 is the most comprehensive quality attributes model for software-intensive systems.

The following diagram shows the quality attributes wheel based on the ISO 25010 quality attributes model:

Figure 9.6 – Quality attributes wheel based on ISO 25010

At Snow in the Desert, the ISO 25010 quality attributes model is adopted consistently across all teams and ARTs to ensure everyone, including business and IT stakeholders, uses a single frame of reference for quality attributes. Beyond definition, for effective communication and management of quality attributes, structured documentation is also critical. In the next section, we will explore approaches for documenting quality attributes.

Documenting quality attributes

Quality starts from an accurate, structured, and consistent definition of quality attributes. Quality attributes are long-living representations useful for continuous improvements even after systems are deployed in production. Therefore, quality attributes have to be systematically maintained with periodic reviews.

Mannion and Keepence recommend using the **SMART** (**Specific, Measurable, Attainable, Realizable, Traceable**) model to capture measurable quality attributes with a high degree of accuracy without ambiguity. In this section, we discuss various SMART approaches for documenting quality attributes.

Using the scaled Agile approach for specifying quality attributes

Scaled Agile Framework (**SAFe**) uses NFRs as a terminology instead of quality attributes. The framework recommends defining NFRs within a *bounded context, as independent as possible, negotiable, and testable*. SAFe proposes a model as shown in the following diagram for describing NFRs:

Name	Scale	Meter	Baseline	Constraint	Target
Performance Response time	Time in seconds between tracking request send and response received	Average time observed by the monitoring system	NONE	3	5
Functionality Accuracy	Location send by the system versus actual location in meters	Based on sample data compared with physical data from driver	NONE	100	500

Figure 9.7 – Documenting NFRs based on the SAFe approach

As captured in the diagram, every NFR is associated with a **name**, optionally sub-names, a **scale**, and a **meter**. Once NFR is defined, values are captured against a **target**, **constraint**, and **baseline**. When developing greenfield systems, it is hard to define a baseline as there is no current system reference point. The constraint is the bare minimum value, and the target is the desired state.

Using a quality attribute scenario

A **Quality Attribute Scenario** (**QAS**) is a universal approach for documenting quality attributes as scenarios. The premise of QAS is a system or a component of a system running on an environment reacting measurably to an actor who initiates an action.

The following diagram captures the meta-model of the QAS along with an example:

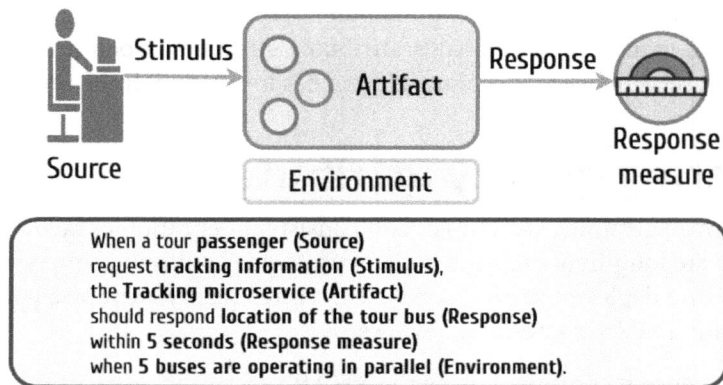

Stimulus
Artifact
Response
Source
Environment
Response measure
When a tour passenger (Source)
request tracking information (Stimulus),
the Tracking microservice (Artifact)
should respond location of the tour bus (Response)
within 5 seconds (Response measure)
when 5 buses are operating in parallel (Environment).

Figure 9.8 – QAS meta-model and example

A scenario from Snow in the Desert's **Automatic Vehicle Tracking System (AVTS)** is captured as an example in the preceding diagram, mapping **Source**, **Stimulus**, **Artifact**, **Environment**, **Response**, and **Response measure**. A mini-QAS is a simplified version that eliminates **Environment** and **Artifact**. QAS is extremely important for architects to get a clear understanding of the quality requirements. Structuring quality attributes with the QAS approach helps to foster better, more consistent communication with stakeholders.

Now that we have clarity on the quality attribute model and approaches for documenting QASes, we will now step through different the quality life cycle stages.

Using the quality life cycle in Agile software development

Delivering high-quality solutions in Agile software development needs continuous focused attention on quality. The following diagram defines the five life cycle stages of delivering quality:

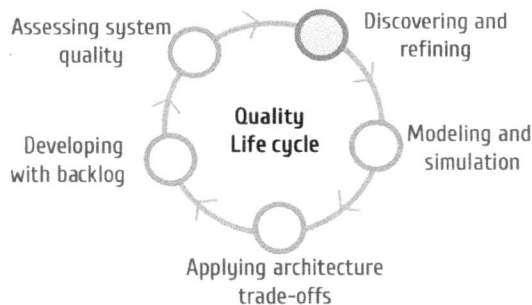

Figure 9.9 – Five stages of the quality life cycle in Agile software development

As captured in the diagram, the five life cycle stages for ensuring the system's quality needs are as follows:

- **Discovering and refining**: Discover quality requirements to satisfy current and future needs of the customer.

- **Modeling and simulation**: Develop models and perform simulations to detect challenges as early as possible.

- **Applying architecture trade-offs**: Apply architecture trade-offs to ensure no overdoing of quality engineering with objective decisions.

- **Developing quality with backlog**: Implement architecture mechanisms as backlog items to avoid dropping significant quality-related tasks.
- **Assessing system quality**: Adopt incremental quality assurance throughout the continuous delivery pipeline.

The quality life cycle is a continuous activity starting with discovering and refining quality needs parallel to business feature identification and elaboration. The first iteration during the inception phase is critical. It helps to understand most of the cross-cutting concerns, such as the business criticality of the application, usage patterns, user distribution, deployment constraints, and so on. Subsequent incremental iterations of the quality life cycle are smaller and more straightforward, generally performed for every sprint for upcoming features.

These five quality life cycle stages are further explained in the subsequent sections, starting with discovering and refining quality attributes.

Discovering and refining quality attributes

Unlike business requirements, it is implausible to expect to receive accurate quality needs and expectations from the business stakeholders. Hence, focused attention led by architects to discover quality requirements is an absolute necessity. The discovery of quality attributes broadly consists of identifying stakeholders, conducting QAWs, and refining QASes. The following sections cover these in detail, starting with capturing the stakeholder matrix.

Aligning stakeholders to quality attributes

The stakeholder matrix is an immensely useful tool for understanding stakeholder needs, performing trade-off analysis, determining the impact of architecture decisions, and mitigating potential risks. It is also incredibly powerful for architects to understand, communicate, and negotiate quality attributes with relevant stakeholders. The stakeholder matrix is generally captured once at the beginning and revisited in every iteration as required.

There are several models for stakeholder analysis. The following diagram shows Mendelow's power-interest grid model:

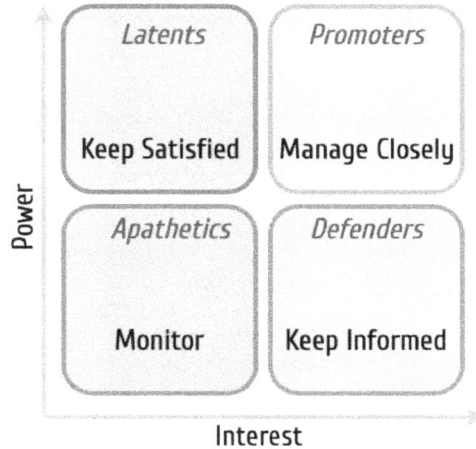

Figure 9.10 – Mendelow's power-interest grid model for stakeholder analysis

As shown in the diagram, each stakeholder is mapped into one of the four quadrants. **Imperial College London's influence-interest grid** uses a similar approach in which power is replaced by influence. **Mitchell's model** is a bit more comprehensive, based on power, urgency, and legitimacy, which will be discussed later in this chapter.

Once stakeholders are identified for a set of features, the next step is to conduct a quality attribute workshop, which is explained in the next section.

Conducting a quality attribute workshop

A **Quality Attribute Workshop** (**QAW**) is a mechanism to discover QASes by engaging all relevant stakeholders identified in the stakeholder matrix. The input for a QAW is a quality attribute model, and the output is a set of prioritized QASes mapped to stakeholders with business impacts. Instead of a full QAW, a mini-QAW is recommended in Agile software development to save time.

The mini-QAW during the inception phase may consume a few hours. In contrast, subsequent mini-QAWs at every iteration for upcoming features are generally conducted in a smaller, informal setup, no longer than 30 minutes.

A typical mini-QAW schedule during the inception phase is shown in the following diagram:

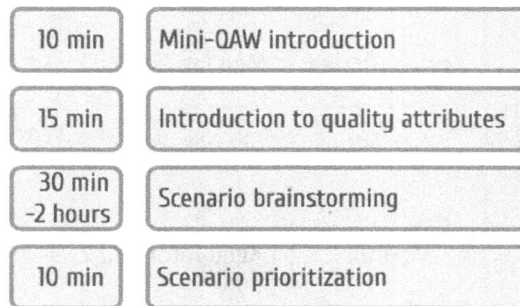

10 min	Mini-QAW introduction
15 min	Introduction to quality attributes
30 min -2 hours	Scenario brainstorming
10 min	Scenario prioritization

Figure 9.11 – Mini-QAW agenda

The following steps explain the agenda items shown in the preceding diagram:

- **Mini-QAW introduction**: This sets the stage by defining the purpose and activities during the workshop. If stakeholders are new to this solution context, architects also introduce the solution vision and intended architecture.

- **Introduction to quality attributes**: This is to make the attendees familiar with the quality attributes to develop appropriate QASes. It is essential to explain quality attributes using the business language mentioned in *Chapter 6, Delivering Value with New Ways of Working,* such as risk, cost, customer satisfaction, and competitive advantage.

- **Scenario brainstorming**: In this segment, architects choose to walk through either features or quality attributes. Stakeholders around the table brainstorm and capture QAS relevant to them based on their instinct and experience using sticky notes and attach them to the quality wheel, shown in *Figure 9.6*, next to the respective quality attributes. Stakeholders are encouraged to use informal and plain English for capturing QASes. Architects explain scenarios during brainstorming by correlating use cases from current systems and similar industry-popular systems and scenarios.

- **Scenario prioritization**: There are different approaches for prioritization, such as MoSCoW, **Quality Function Deployment** (**QFD**), the **dot** method, and so on. The dot method is the most straightforward and is quite popular with QAWs. Stakeholders add an allocated number of dots to quality attributes and QASes based on their priorities. Prioritization helps architects later in the trade-off analysis. Often, stakeholders also capture the business impact or risk of not meeting the QAS.

- Once scenarios are captured, architects work offline to refine and validate those scenarios before playing them back to stakeholders, which is explained next.

Refining QASes and playback

Refinement is an offline activity after the mini-QAW. During the refinement activity, architects convert QASes into a proper structure and map the stakeholder, business impact, and priorities captured during the mini-QAW.

The following diagram captures the output of refinement:

QAS	STAKEHOLDER	IMPACT	PRIORITY
Passenger getslocation data within 5 seconds for tracking request	CUSTOMER, OPERATIONS	HIGH	●●●●
Passenger getslocation data with accuracy of 500 meters, for tracking request	CUSTOMER, OPERATIONS	MEDIUM	●●●

Figure 9.12 – QAS after refinement

The diagram captures sample scenarios from the context of Snow in the Desert's AVTS initiative. Refined QASes are replayed to all stakeholders as necessary to validate and baseline before designing the right solution. Solution identification starts with modeling solution options, which is covered in the next section.

Modeling and simulation

Modeling and **simulation** are effective mechanisms for validating solutions against a given set of parameters without investing time and effort into development. Baselined QASes are captured in the architecture decision backlog for appropriate solution decisions. As discussed in *Chapter 5, Agile Solution Architect – Designing Continuously Evolving Systems*, these architecture decision backlogs are scheduled based on the **Last Responsible Moment (LRM)**.

As part of processing the architecture decisions backlog, the team keeps elaborating on multiple solution options. A critical activity is to model these solution options using various architecture and design models, such as sequence diagrams, domain model diagrams, context diagrams, component diagrams, and so on. Many models, such as workload models, deployment models, network models, user access models, and transaction models, are convenient in simulating and validating quality attributes.

A **transaction model**, shown in the following diagram, is one of the most helpful models to simulate quality conditions:

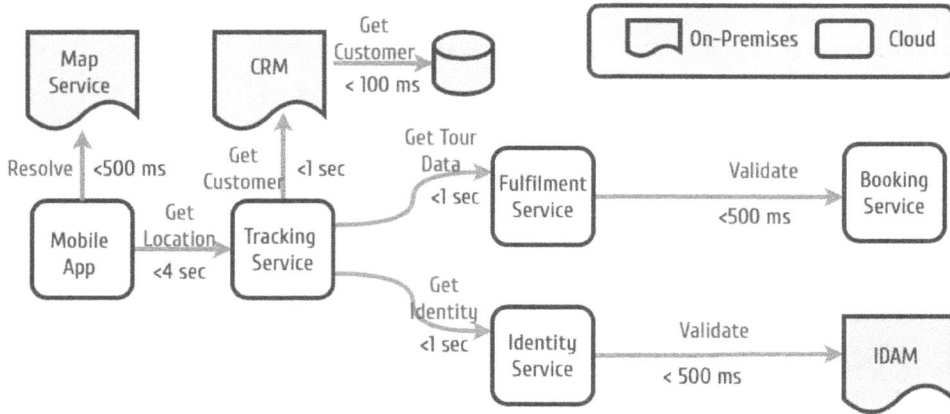

Figure 9.13 – Transaction model for the AVTS tracking service

The preceding diagram captures the *tracking* transaction in the context of Snow in the Desert's AVTS solution. Quality attributes such as performance, availability, and scalability are limited to the lowest common denominator in the transaction model across a chain of systems. As shown in the diagram, a transaction model captures how transactions are spanned across multiple components, subsystems, services, and systems. Transaction models help determine performance hotspots, latency challenges due to external calls, systems with legacy technologies, slow performers, systems with reduced levels of availability and scalability, and so on.

Another model that is equally useful is the **workload model**, shown in the following diagram:

CRITICAL TRANSACTIONS	USER	EXPECTED RESPONSE TIME	TRANSACTIONS PER HOUR	TYPE	%DISTRIBUTION
Tracking Service	Passenger, Driver, Operations	5 seconds	60000	Asynchronous Request-Response	99%
Registration Service	Passenger	10 seconds	50	Request-Response	<1%
Preference Service	Passenger	5 seconds	10	Request-Response	<1%
Others	Passenger, Admin	none	100	Request-Response	<1%
Others	System	none	50	Batch	<1%

Figure 9.14 – Workload model for the AVTS tracking system

The preceding diagram captures critical transactions of AVTS. As shown in the diagram, the workload model captures a mix of transactions, distributions, expected response times, user types, the nature of transactions, and so on. This model helps understand the system's behavior and acts as an input for building load testing scenarios.

It is quite common that an architecture decision may impact more than one quality attribute. Often, one adversely affects the other. In such cases, a trade-off analysis needs to be performed. The next section further explains the trade-off analysis process.

Applying architecture trade-offs

Architecture trade-off analysis is an extremely important activity for evaluating solutions against quality goals to make informed decisions. Implementing quality requires compromise as achieving everything is not possible. This trade-off is similar to what the *CAP theorem* states: *a distributed database system can only guarantee two qualities out of consistency, availability, and partition tolerance.* Some quality attributes may be cost-prohibitive, some may exponentially increase complexity, and some may have strong interdependencies with another quality attribute where improving one may bring the other down.

Vasa, a Swedish warship, is a classic example of building too much quality into a product without trade-offs. In the 1620s, Poland-Lithuania and Sweden were at war. The Swedish king wanted to show off their power and naval superiority by building a gun platform on a large, heavily armed ship that could also transport soldiers. The engineers built a vessel that no one had ever thought of before – a massive warship, 226 feet long, with 64 guns, 145 sailors, and 300 soldiers with two-gun decks. There were many firsts in Vasa – combining passengers and artillery, the number of guns, two-gun decks, and so on. Even though the ship was built as per the specifications, unfortunately, it sank on its maiden voyage as the bottom-heavy ship caught strong winds after sailing 1,300 meters from the harbor. The main reason for its failure was its instability as the ship's architect included way too many things without considering trade-offs. Even though the architect knew Vasa was a risky undertaking, they were not ready to go back to the king and share their challenges.

Architectural decisions conflicting with more than one quality attribute need trade-off analysis. Trade-offs are calculated based on parameters such as economic aspects, time to market, customer satisfaction, resource skills, and so on. Architects use stakeholder maps and business priorities captured as part of the mini-QAW for negotiating with relevant stakeholders.

There are a few standard methods for architecture trade-off analysis, which are discussed in the following sections.

Using the architecture trade-off analysis method

The **Architecture Trade-Off Analysis Method** (**ATAM**) is one of the oldest architecture analysis methods developed by the Software Engineering Institute. Using ATAM, architects can identify potential risks of a proposed solution and transparently communicate risks to relevant stakeholders.

ATAM evaluates architecture decisions against three parameters – trade-offs, sensitivity points, and risks. However, ATAM is complex, consumes time and effort, and is not visually appealing. In cost-critical scenarios, architects can also use the **Cost-Benefit Analysis Method** (**CBAM**) before concluding large-scale investment decisions.

Using the solution architecture review method

The **Solution Architecture Review Method** (**SARM**) is another substantially simple but more powerful method developed by Simon Field. It helps to identify the strengths and weaknesses of candidate solutions in terms of risk.

In SARM, QASes and ASRs are mapped to stakeholders using Mitchell's stakeholder model – based on power, urgency, and legitimacy. Once stakeholders are identified, ASRs are mapped to one or more stakeholders by capturing their interests as *Strong Interest*, *Interest*, or *No Interest*. SARM adopted a risk impact approach for prioritizing QASes. The risk impact approach is based on five parameters – *Catastrophic*, *Major*, *Moderate*, *Minor*, and *Negligible* – to indicate the impact of not meeting the QAS.

An evaluation of a sample scenario using SARM is given in the following diagram:

QUALITY ATTRIBUTE	SUB CATEGORY	ASR / QAS	STAKEHOLDER 1 OPERATIONS	STAKEHOLDER 2 IT	PRIORITY	OPTION 1 IN-HOUSE	OPTION 2 THIRD PARTY
RELIABILITY	MATURITY	Degree of reliability	STRONG INTEREST	INTEREST	CATASTROPHIC	LIKELY	POSSIBLE
MAINTAINABILITY	REUSABILITY	Degree to which existing platforms are reusable	INTEREST	STRONG INTEREST	MODERATE	ALMOST CERTAIN	RARE
COMPATIBILITY	CO-EXISTENCE	Ability to share a common environment	STRONG INTEREST	INTEREST	MODERATE	LIKELY	POSSIBLE
COMPATIBILITY	INTEROPERABILITY	Ability to integrate with other systems	INTEREST	STRONG INTEREST	MODERATE	ALMOST CERTAIN	LIKELY

Figure 9.15 – SARM for trade-off analysis

The diagram captures the solution decision for AVTS and its trade-offs. Two solution options for AVTS are in-house development or acquiring a third-party product, as explained in *Chapter 5*, *Agile Solution Architect – Designing Continuously Evolving Systems*. The diagram only captures a few quality attributes for simplicity. As indicated in the last two columns of the preceding diagram, the architect chooses the risk exposure for each solution option based on a risk likelihood model – *Rare*, *Unlikely*, *Possible*, *Likely*, and *Almost Certain*. Once the evaluation is done, SARM provides an opportunity to visualize the aggregate risk burden of each solution option together with the impact on stakeholders, as shown in the following diagram:

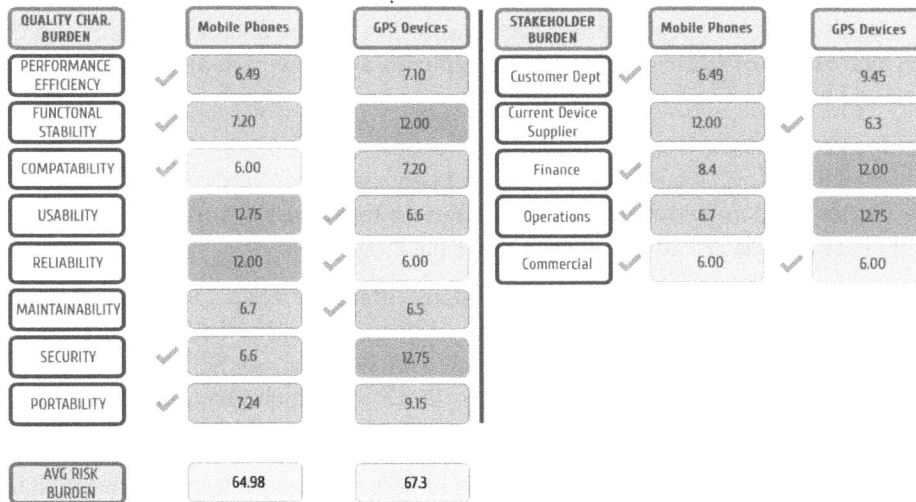

QUALITY CHAR. BURDEN	Mobile Phones	GPS Devices	STAKEHOLDER BURDEN	Mobile Phones	GPS Devices
PERFORMANCE EFFICIENCY	6.49	7.10	Customer Dept	6.49	9.45
FUNCTONAL STABILITY	7.20	12.00	Current Device Supplier	12.00	6.3
COMPATABILITY	6.00	7.20	Finance	8.4	12.00
USABILITY	12.75	6.6	Operations	6.7	12.75
RELIABILITY	12.00	6.00	Commercial	6.00	6.00
MAINTAINABILITY	6.7	6.5			
SECURITY	6.6	12.75			
PORTABILITY	7.24	9.15			
AVG RISK BURDEN	64.98	67.3			

Figure 9.16 – SARM risk trade-off and stakeholder impact model in the context of AVTS

The preceding trade-off diagram uses the decision point between GPS devices or mobile phones as tracking devices discussed in *Chapter 5*, *Agile Solution Architect – Designing Continuously Evolving Systems*. As captured in the preceding diagram, option 1, **Mobile Phones,** has the lowest risk exposure score, **64.98**, and hence it is better than option 2, **GPS Devices**, which has a risk exposure score of **67.3**. Similarly, option 1 satisfies most of the customers except the **Current Device Supplier**. SARM's visual model provides an excellent mechanism to perform trade-offs and make an informed selection of solutions from the candidate options. SARM can also be used for cost-benefit analysis when required.

Once the decisions are formulated, development teams must allocate sufficient time for developing quality attributes. Adding backlog items for quality-related development is decisive for development teams to plan and execute without dropping quality. The next section covers the recommended practices for including quality in the development backlog.

Developing quality attributes

Once the architecture decisions are made, the next step is to make that decision ready for the teams to implement. Pushing quality attributes to teams for development and compliance is a big challenge and is one of the most debated points in the Agile software development community.

There is no one-size-fits-all solution for this challenge; however, a combination of the following three approaches helps in overcoming this challenge:

- **Technical backlog items**: Some of the quality attributes need specific work from the team, such as developing a replication solution for the database or building the high availability of microservices. Such scenarios are converted and captured as user stories.

- **Backlog constraints**: In most cases, once the base solution is available, such as high availability or a shared data cache for faster customer access, several other user stories need to comply with the solution. In those cases, constraints such as acceptance criteria are added against those backlog items.

- **The Definition of Done (DoD)**: Similar to acceptance criteria, DoD captures quality attributes as a checklist item to ensure teams comply with quality attributes. Using DoD is particularly useful when each user story cannot be tested for a particular quality attribute. In such cases, DoDs are attached to a sprint or a release to ensure compliance.

  In most cases, quality attributes start as backlog items, then once implemented, they are attached as constraints to subsequent user stories or are added as part of DoD for sprints and releases. Capacity allocation, mentioned in *Chapter 6, Delivering Value with New Ways of Working*, is an important aspect when working with quality attributes to ensure spend enough time on developing quality. Adopting built-in quality needs to ensure quality right at the source.

The next section explores system quality assessment.

Assessing system quality

Stop-the-line is a technique used in manufacturing introduced by Taiichi Ohno of TPS, where a big red button is pushed whenever someone sees a critical quality issue in the production line. The idea was products with quality issues moved to production have much more severe consequences than stopping the line and fixing problems as and when they occur. Early detection and rectification of defects minimizes wastage. Hence, early and continuous quality assurance is required for the efficient delivery of quality solutions with optimal cost of development.

Brian Marick's Agile testing quadrants are a useful tool to understand various aspects of Agile testing. An adapted version of this was developed by Janet Gregory and Lisa Crispin is shown on the left side of the following figure:

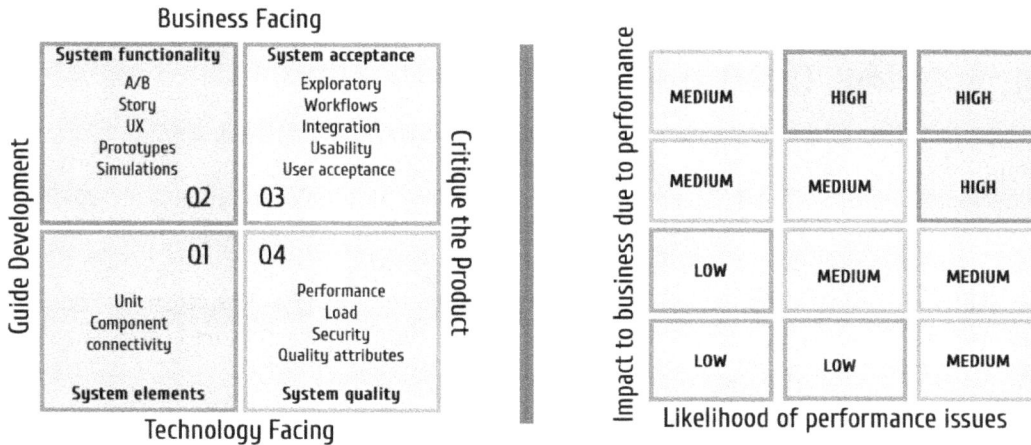

Figure 9.17 – Agile testing quadrants and risk-based performance testing model

As shown in the figure, architecture quality testing is in **Q4**, in which the best outcome is possible when the product is reasonably shaped up for production release. Often, quality tests are effort, time, and resource-consuming activities. If all tests are not possible, a risk-based automated performance testing approach by batches proposed by Tim Hinds and Steve Weisfeldt is useful for prioritizing a large number of performance testing scenarios. This model is shown on the right side of the preceding diagram.

Continuous and automated validations are possible with integrated quality checks at the time of development. Also, automating quality assurance routines with constant delivery pipelines eliminates human error.

Integrating quality checks in development

Integrating code profiling tools with IDEs helps us to understand code performance, such as the number of database calls, the response time of SQL, data access hotspots, the performance of messaging, overheads of external calls, CPU profiling, memory leaks, the response time of methods, throughput, thread profiling, and so on. Using performance unit testing frameworks as IDE plugins can detect throughput per second, the method execution time for various scenarios, and so on.

Automated quality assurance tests, such as performance, security, resource profiling, network profiling, and so on, have to be integrated with the continuous delivery pipelines with appropriate monitoring mechanisms to detect quality issues as early as possible. Fitness functions are useful techniques for validating architecture and quality attributes.

Using fitness functions

A **fitness function** is a concept borrowed from evolutionary computing and was introduced by ThoughtWorks. An architecture fitness function is a mechanism to define what is expected from an architecture capability. ThoughtWorks explains architecture fitness functions as *test functions that provide an objective integrity assessment of architectural characteristics*. It is an objective function that asserts how close implementation is aligned with the intent and objective. **Fitness Function-Driven Development** (**FFD**) is a useful technique for ensuring built-in architecture compliance.

Automated fitness functions are not just enough for covering all architectural aspects. **Monitoring-Driven Development** (**MDD**) is increasingly gaining popularity. MDD uses monitoring of various parameters in different environments for capturing system health continuously.

Evaluating using the Well-Architected Framework

AWS first introduced the Well-Architected Framework to analyze workloads' quality against five pillars: **operational excellence**, **performance efficiency**, **security**, **cost optimization**, and **reliability**. It was later adopted by other cloud providers, such as Microsoft and Google, as well.

The AWS Well-Architected Tool provides an excellent mechanism to review workloads against a consistent yardstick. It promotes best practices and patterns for cloud-native deployment, understands and manages risks appropriately, and identifies areas to invest with objective measures. Snow in the Desert adopted the Well-Architected Framework and consistently and regularly maintained dashboards to show the state of systems against five pillars.

The following diagram shows a sample dashboard coming out of the Well-Architected Framework:

Figure 9.18 – The Well-Architected Framework assessment outcome

Consistently maintaining the well-architected score in a dashboard, as shown in the preceding diagram, for all systems enables development teams to visualize the impact on quality after every release.

Summary

In this chapter, we have learned about the importance of quality in software systems. There is no one-size-fits-all solution for quality as it is heavily influenced by the environment, business context, and consumption patterns. Systems thinking, continuous evolution, and built-in quality are three key aspects architects need to be well aware of when building long-lasting, resilient systems.

We have learned about different quality attribute models and QASes for consistent use of taxonomies and documentation. We have studied the importance of quality at every stage of development by examining the quality life cycle stages, commencing with discovery and refinement, which occur parallel to business features discovery and refinement. The QAW is one of the most critical activities for architects to discover and prioritizes quality attributes by engaging all stakeholders identified using a stakeholder matrix. Architects further refine QASes gathered in the workshop and play them back to stakeholders for baselining. Baselined QASes are captured in the architecture decision backlog for making decisions.

Architects use modeling and simulation to validate solutions before development to detect potential early issues. Architects also use architecture trade-off assessments for making objective data-driven decisions. Backlog items and backlog constraints are created for development teams, guided by architects, to have focused attention on implementing quality attributes. Finally, architects foster continuous quality assessment techniques, such as IDE-based quality assurance, fitness functions, and the Well-Architected Framework, which are employed to ensure faults are identified at the earliest possibility and therefore minimize wastage.

In summary, this chapter covered how architects play a significant anchor role throughout the quality life cycle to ensure the solutions delivered meet the desired quality standards. Now that we have seen various aspects of technical agility, in the next chapter, we will be moving on to the next topic and exploring how to manage Lean architecture documentation in Agile software development.

Further reading

- **Overview of systems thinking**: `http://www.thinking.net/Systems_Thinking/OverviewSTarticle.pdf`

- **Four Pillars for Improving the Quality of Safety-Critical Software-Reliant Systems**: `https://resources.sei.cmu.edu/asset_files/WhitePaper/2013_019_001_47803.pdf`

- **Software Architecture & Quality Attributes**: `https://sites.google.com/site/misresearch000/home/software-architecture-quality-attributes`

- **ATAM**: *Method for Architecture Evaluation*: `https://resources.sei.cmu.edu/asset_files/TechnicalReport/2000_005_001_13706.pdf`

- **Quality Attribute Workshops**: `https://resources.sei.cmu.edu/asset_files/TechnicalReport/2001_005_001_13862.pdf`

- **SARM**: `https://sarm.org.uk/documentation/`

- **A Review of Software Quality Models for the Evaluation of Software Products**: `https://www.researchgate.net/publication/269417429_A_Review_of_Software_Quality_Models_for_the_Evaluation_of_Software_Products`

- **AWS The Well-Architected Framework**: `https://docs.aws.amazon.com/wellarchitected/latest/framework/welcome.html`

- **Google Architecture Framework**: `https://cloud.google.com/architecture/framework/design-considerations`

- **Agile testing quadrants**: `https://labs.sogeti.com/guiding-development-Agile-testing-quadrants/`

10
Lean Documentation through Collaboration

For the most part, things never get built the way they were drawn.

– Maya Lin, designer of 'A Fold in the Field', a sculpture park in Auckland

The previous chapter covered technical agility and different pillars contributing to technical agility, such as patterns and techniques, DevOps and continuous delivery, and built-in quality. This chapter advances to documentation, one of the most debated topics in Agile, and reviews various documentation aspects with a specific focus on delivering architecture artifacts.

Amateur Agile teams treat documentation as a no value-added activity. While excessive documentation creates flow impediments, little documentation can have severe consequences in terms of the sustainability of systems' delivery flow and reliability. How much documentation is sufficient is a repeatedly debated point in Agile software development. This chapter will get deeper into the documentation topic to clarify common sources of confusion. The traditional approach of documentation produces needless and non-consumable documentation, which adds cost to the company. Embracing alternative techniques to minimize documentation and adopting Lean-Agile thinking helps to realize value in the documentation. Building adequate architectural documentation with an incremental approach adds significant value to overall Agile delivery.

In this chapter, we're going to cover the following main topics:

- Persisting knowledge with documentation
- Using Lean-Agile ways of documentation
- Documenting software architecture
- Applying model-based software engineering

This chapter focuses on the **Evolutionary Collaboration** focal point of **the Agile Architect's Lens**:

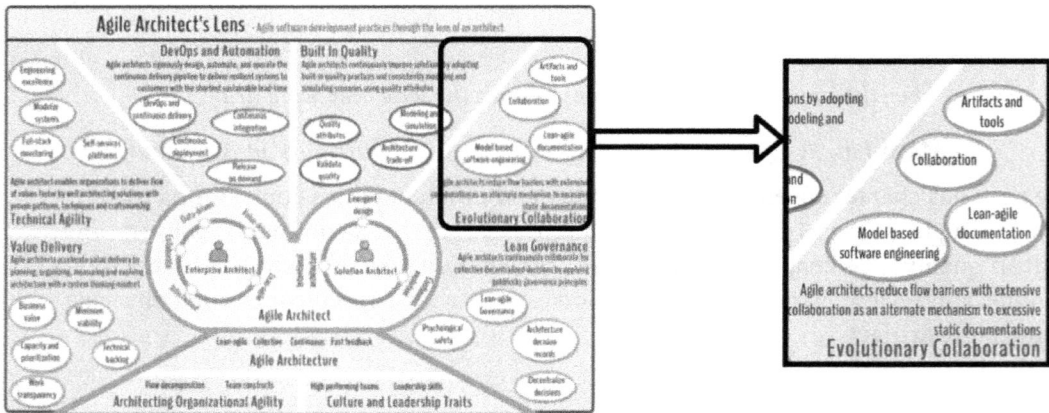

Figure 10.1 – The evolutionary collaboration – Focal point

This chapter explores the importance of documentation in Agile projects and how traditional documentation approaches hinder the velocity of Agile software development. We will explore alternative mechanisms for documentation and then analyze how to use the Lean-Agile methods to document unavoidable scenarios. Later in this chapter, we will also highlight different architectural artifacts and structures, and reinforce them with a few examples from Snow in the Desert.

Persisting knowledge with documentation

Working software over comprehensive documentation is one of the Agile manifesto principles that leads to delusions such that Agile projects do not need to be documented at all. Agile development methodologies encourage a relentless focus on delivering value, which often causes the elimination of architecture documentation as it is regarded as a no-value activity.

A closer look at the articulation of the Agile manifesto values indicates that signatories have recognized the value of documentation, but working software has more value. *Robert C Martin* observed that *documentation is not anti-Agile. If it were, then there would not be any Agile books, articles, and blogs.*

Jim Highsmith, one of the signatories of the Agile manifesto, warned the developer community that Agile and documentation are not oxymorons. Adequate documentation is essential for the success of Agile software development. Irrespective of the development methodology, since most enterprise software remains in service for a long period, documentation is an important means of knowledge retention.

Alistair Cockburn, another signatory, recommends that documentation needs to be light but sufficient. The Crystal methodologies introduced by *Alistair Cockburn* suggest that documentation needs to be created, but that individual projects ought to decide what they need.

Large-Scale Scrum (**LeSS**) observes that the only software documentation satisfying engineering design criteria is the source code. However, LeSS also acknowledges that source code alone is not sufficient for all purposes. Often, source code with a million lines of code is not suitable for understanding the overall big picture architecture. LeSS also recommends using the **System Architecture Document** (**SAD**) for documenting the system architecture.

The best analogy for explaining the importance of documentation is in-memory and persistent database stores. In an in-memory data grid, fault tolerance can be achieved with data replication across multiple nodes. This is similar to avoiding a single point of knowledge loss by spreading knowledge across many people through collaboration. Snapshots of in-memory data have to be dumped into file storage for recovery, which is identical to considering knowledge as source code. In the event of failure, recovery from file storage takes time, similar to retrieving knowledge from source code. If a database is used as the backend for the in-memory data grid, structured data can be stored and retrieved quickly when required from the database, while most of the data is served from in-memory. In this case, the database is analogous to the documentation that teams produce.

Persisting knowledge is appropriate in many scenarios. We will explore some of these scenarios in the next section.

Understanding the reasons for documentation

Documentation in software development is not a complete no-go as it can lead to catastrophic effects in the long term. A closer review of the purposes of documentation for its consumers helps in understanding its value. The reasons for documentation may entail one of the following scenarios:

- Retention of knowledge gained over time for onboarding and educating new team members, reviewing and understanding historical events, sharing learnings, and acquiring information on past decisions and rationale.

- A mechanism for communication by handing over artifacts from one team to another, such as IT to business, product management to architecture, architecture to engineering, and engineering to operations.

- Government and regulatory bodies often require tons of documentation with a certain level of detail for mandatory certifications and compliance. Often, regulators and auditors ask for traceability and proof in the form of documentation.

- Rule books and playbooks for engineers to understand, develop, and support critical software systems. These documents also serve as the knowledge base for on-call support engineers to enable them to respond quickly to customer queries.

- In the case of frameworks, libraries, and reusable assets, documentation is essential for consumers to understand purpose and usage patterns. Poor documentation stalls adoption and hinders productivity.

- User manuals, help documentation, and release notes are very important for multi-year development products.

- Specifications and legal contracts with suppliers and remote teams, such as service level agreements, are essential for large enterprises managing mission-critical software systems.

Persisting knowledge, therefore, is significantly important and cannot be entirely circumvented. However, embracing new ways of documenting is necessary as traditional approaches no longer support the faster pace of Agile software development. We will go into more depth on this subject in the next section.

Why the traditional way of documenting is no good

Traditional software development methods are stage-gate driven in which documentation is submitted for approval in order to move on to the next phase. In such scenarios, documentation is treated as a mechanism for constructing internal contracts between non-trusted teams. These contract documents, such as design documents, are thrown over the wall to the next team for proceeding further. Often, localized progressive decisions post-handover are not updated in the document, leading to an obsolete state.

The following diagram is an illustration of the stage-gated process:

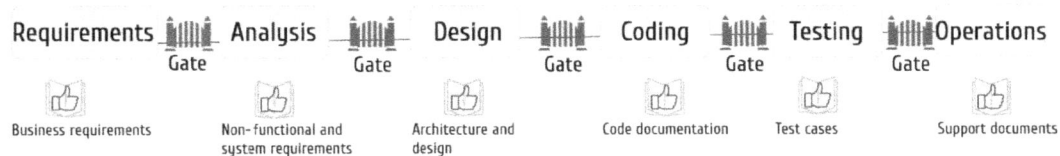

Figure 10.2 – Stage gate-based approvals based on documentation

As shown in the preceding diagram, each gate expects certain documents for approval. The article *Exploratory Study of Architectural Practices and Challenges in Using Agile Software Development Approaches*, by *Muhammad Ali Babar*, observes that there are advantages and disadvantages to formally documenting software architectures. As per the survey conducted by the author, most participants viewed formal documentation as a 30-40% overhead with no value addition.

There are many reasons why the traditional way of formal documentation is considered a non-value-added activity. A few of the critical challenges are as follows:

- **Waste of effort**: In Agile software development, some systems such as e-commerce websites perform incremental releases multiple times a day, where architecture changes are considerably frequent. In such cases, traditional documentation approaches cannot match the pace of the change cycles and, hence, slow down the continuous delivery flow.

- **No one reads**: Traditional documentation approaches are mostly produced for the sake of stage-gate approvals, and not used thereafter. Besides, large-scale documentation with hundreds of pages is not consumable at a practical level.

- **Over budget**: The production of excessive and unnecessary documentation, often measured by the number of pages, requires substantial manual effort for authoring and reviews with additional costs of storage and maintenance. Often, these static documents with no search capabilities also significantly increase the consumption costs.

- **Change tolerant**: Once software delivery commitments are met, development teams often pay no attention to documents already produced. As a result, these documents hardly match the reality of designs and code. The consequences of defective content are far more dangerous as this leads to disconnected understanding among stakeholders.

- **Siloed documentation**: Traditionally, documents are created and maintained in silos, such as architects managing architecture documentation, and engineers managing detailed design documents. Capability-based documentation leads to scattered information, often with restricted access.

The traditional approach of software documentation is mostly a flow stopper in Agile software development. However, on the contrary, poor documentation causes the evaporation of architecture and design knowledge. As a result of knowledge evaporation, developers devote more effort to fixing issues and, more critically, their fixes may break other parts of the system due to a lack of knowledge.

Embracing alternative ways to avoid documentation and adopt Lean-Agile approaches helps to improve the flow in Agile software development projects. We will elaborate on this in the next section.

Using Lean-Agile ways of documentation

To effectively improve Agile software development flow, a critical examination of documentation is required to understand what can be eliminated, what needs to be persisted, how simplification can be achieved, and the appropriate time for delivery. Transparent communication and collaboration improve shared knowledge among team members, thereby reducing the need for documentation and freeing up valuable resources to focus on core delivery activities. When documentation is essential, adopting a purpose-driven approach with the right tools is beneficial in Agile development.

Evolutionary collaboration over documentation

In Agile software development, a handoff is regarded as one of the flow impediments. Therefore, documentation as a mechanism for handing off work across teams is also inappropriate; instead, leverage opportunities for interaction between team members.

Evolutionary collaboration fosters continuous face-to-face interactions without boundaries and constraints as a primary mechanism for developing a shared understanding across all team members. Teams nurture a culture of free-flowing communication in their workspace without the need to have pre-set meetings. Agile organizations promoting evolutionary collaboration invest in building informal collaboration spaces with massive digital and physical whiteboards, large rolls of paper all around with plenty of sticky notes and markers, big plotters, video recording, and playback facilities. They also encourage a verbal culture, fewer meetings and ceremonies, the acceptance of hand drawings as artifacts, and so on.

Digital workspace and collaboration tools play a significant role in evolutionary collaboration. Large-format, laser-powered phosphor displays enable individuals and teams to visualize, interact, naturally engage, and finally transform outcomes digitally for long-term persistence without any extra effort. Digital wiki-style knowledge management systems enable continuous engagement with collaborative authoring, faster feedback cycles, seamless content aggregation, and transparent communication. Real-time conversational tools allow channel-driven communications to push relevant content to targeted groups selectively instead of disconnected traditional email communications.

Human beings are visual learners. We process information based on what we see. Visuals stimulate us and stay longer in memory than verbal messages. *Scott Ambler* observed that displaying models publicly on boards around the team space, or even internal websites, promotes better communication. *Alistair Cockburn* urged the use of information radiators, which are hand-drawn, printed, or electronic displays, in a highly visible location where stakeholders can see them as they pass by. Teams effectively collaborate using large whiteboards to create architecture design models as part of solution design workshops. LeSS observes a linear correlation between effectiveness and the amount of whiteboard space as it encourages creative flow and participation. LeSS also recommends keeping the diagrams on the wall throughout the development cycle as it inspires new ideas and facilitates quick design adjustments with just-in-time stand up conversations.

Some of the techniques beyond collaboration, which are helpful in eliminating documentation, are stated here:

- **Documentation as code**: Moving traditional documents to code is a better way to extract value. Reusable code instead of blueprints and reference architectures, patterns as code, architecture validations as code, policies as code, infrastructure as code, and so on are much more consumable and help to save money and time.

- **Generate documentation**: Reconstructing required documentation with a code-first approach, such as generating API documentation and models from code, not only saves time but also helps to document the truth.

- **Focus on models**: Collaboratively engaging relevant stakeholders using model-based designs is an excellent alternative to extensive textual documentation. The consistent use of models affords a shared understanding of the design among stakeholders.

- **Using visual aids**: Short videos, animations, and scribes are excellent mechanisms for the repeated sharing of messages, such as for educating and onboarding engineers.

The alternative mechanisms discussed are excellent opportunities to reduce documentation efforts. However, be mindful that it is not possible to eliminate documentation entirely. We will discuss how to effectively compile documentation in the next section.

Just barely enough documentation when required

Documentation is a long-lasting digital memory. Agile software development tries to eliminate excessive documentation with collaboration. While collaboration is an excellent communication mechanism that brings a quick shared understanding, it cannot be a like-for-like replacement for documentation. Adequate documentation for information that has long-term relevance is required in order for the enterprise to achieve knowledge retention and contractual and regulatory obligations.

While there are no rigid guidelines, documentation in Agile software development must be **incremental, represent the truth, and be responsive to change**. Besides, good Lean-Agile documentation is driven by four key elements, as shown in the following diagram:

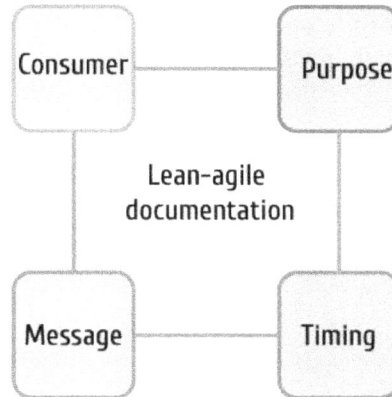

Figure 10.3 – Four key elements of Lean-Agile documentation

As shown in the preceding diagram, documentation in Agile needs a **Consumer**, a **Purpose**, a concise **Message**, and appropriate **Timing**. The following points explain these in detail:

- **Consumer**: A document has to be written for a consumer; if there is no consumer, no document is required. In some cases, multiple viewpoints of the same content are produced to help specific consumers.

- **Purpose**: Every document needs a purpose. Good documentation is when the contents produced are finely tuned to the purpose of the consumer. Always look for alternative and simplified mechanisms to satisfy the purpose with minimal effort and cost.

- **Message**: Understand what message needs to be conveyed to satisfy the purpose from the consumer perspective. It is not about hundreds of pages; it is all about the ease of consumption. The content of the document has to be high quality, accurate, legible, and concise.

- **Timing**: At what point consumers require a piece of information is also an important consideration. Documentation produced as late as possible potentially carries more credible information.

A careful analysis is recommended to decide whether the documentation is required or not and determine what form or shape is better suited in a specific scenario. The following diagram helps in this context:

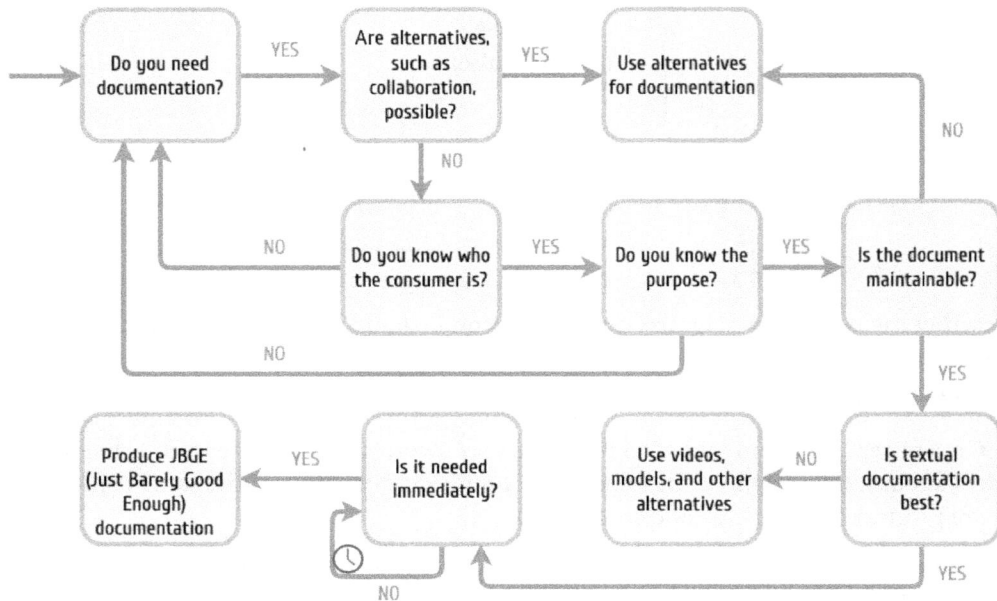

Figure 10.4 – Workflow to determine the need for documentation

The diagram is adapted from **Disciplined Agile (DA)** to and DocOps. The workflow helps to determine whether the documentation is needed, as well as identify the target consumer, the purpose, at what point the documentation is required, and the correct tools.

Delaying documentation as much as possible helps architects and development teams to understand consumers' expectations and accordingly use the right content, format, and tools when producing documentation. Using documentation as a checklist item part of the **Definition of Done (DoD)** ensures that sufficient and mandatory documentation is appropriately generated as part of the development.

How much documentation is sufficient?

There is always the question of how much documentation is enough? The most straightforward answer is how much documentation is required by the consumer to meet their purpose, and which is impossible to achieve with other techniques. It is rare to see projects with no documentation. Large projects and critical projects require more solid documentation. *The Cockburn Scale*, introduced by *Alistair Cockburn*, is potentially useful for understanding the relative scale of adequate documentation. This scale categorizes projects based on criticality and size. The criticality is defined on a scale of 4, as shown on the vertical axis of the following diagram:

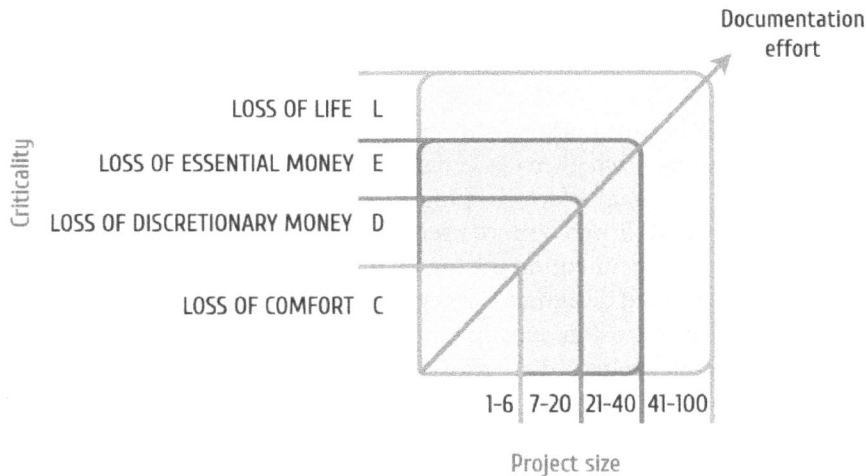

Figure 10.5 – Documentation effort linked to criticality and project size

The project size is measured in four categories based on the number of people working, as shown in the horizontal axis. The effort needed for physical documentation goes up as the combination of complexity and size goes up, as shown in the preceding diagram. Alternative mechanisms, such as consistent collaboration, are hard to maintain for the projects in the top-right corner. Much more stringent documentation is needed for projects that impact on people's lives.

When developing documentation in Agile software projects, adhering to a set of principles accelerates the continuous flow. The following section examines some of the critical principles.

Adhering to the principles of documentation

Purpose-oriented documentation helps customers to gain significant value from documents. Good documentation needs to align with the following principles:

- **Product centric**: Documenting all the necessary information in one place enables better clarity in communication. Product-centric documentation follows a single repository for all types of documents for a product, such as business features, architecture, design, and test cases. As a general rule of thumb, you should separate project execution-related documents, such as team actions in retro and plans, from real product-related documents, such as architecture. The latter has a long shelf life, whereas the former is short lived. Integrating consumers and businesses into product documentation aids in transparent communication, collaboration, and faster feedback.

- **Document what is maintainable**: As discussed in *Chapter 4, Agile Enterprise Architect – Connecting Strategy to Code*, create documentation only if it is maintainable with minimal effort. Keeping simplicity is paramount. For example, a strategy on a postcard is much more useful than a hundred-page strategy document. Focus on documenting the big picture, such as architecture, instead of details such as detailed design. It is risky if documents are obsolete as they are significant causes of confusion. Documenting as late as possible, and generating documents from code, helps reduce efforts to maintain documents accurately.

- **Don't repeat yourself**: Adopt a good documentation tool that can aggregate content from multiple sources instead of repeating information. Information curation from the right sources helps to ensure that documentation is not duplicated and accurate. Duplication increases maintenance efforts and is likely to go out of sync. This approach also helps to use fit-for-purpose tools for various aspects, including stories, modeling, and textual documentation.

- **Create consumable artifacts**: Instead of static documentation, adopt interactive collaboration tools with extensive search capabilities for enabling evolvable documentation. Extensive use of visuals and models help consumers to acquire knowledge with ease. Avoid setting up information boundaries unless it is critical to do so.

- **Ensure good communication**: Documentation by itself is not useful if not communicated or accessible to relevant stakeholders. Designing a good communication structure around documentation enables a seamless flow of information.

In *Chapter 2, Agile Architecture–The Foundation of Agile Delivery*, we discussed architecture as a continuum. Along the same lines, architecture documentation also has to be progressive. It is hard to classify architecture documentation as high level and low level. The following diagram captures the stages of architecture documentation in an Agile software development project:

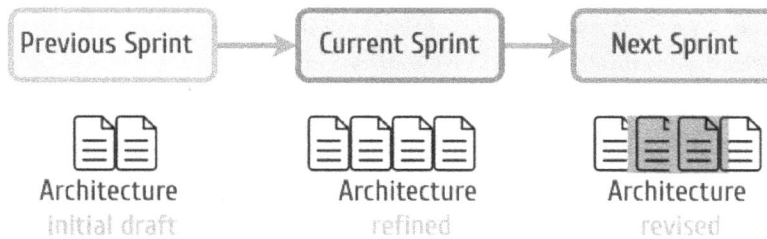

Figure 10.6 – Stages of architecture documentation in Agile

As shown in the diagram, the recommended approach is to call architecture documentation the **initial draft** before the current sprint, **refined** in the current sprint, and **revised** after the current sprint. Refactoring documentation is an essential activity for supporting the evolution of documents.

In the next section, we will deep dive into architecture-specific documentation aspects.

Documenting software architecture

Architecture documentation refers to recording guidelines, practices, architecture intents, decisions, rule books, reference architectures, and patterns using various tools such as wiki-style documentation tools, modeling tools, and Agile team tools. This section elaborates on this idea.

Different proposals for documenting architecture

Architecture documentation evolved over the years from rigid template-based documentation to an essential information-based approach. While many of those legacy ideas proposed by traditional frameworks such as **Rational Unified Process** (**RUP**) are still valid in Agile software development, the ways of documenting and the depth of documenting are different.

View-centric approaches were popular in the early 2000s. The essence of architectural viewpoints is that different stakeholders want to see different architectural views based on their interests. The different view frameworks are as follows:

- The **4 +1 view** proposed by *Philippe Kruchten*, later adopted in the RUP, has four architecture viewpoints: **logical**, **process**, **development**, and **physical**.

- The **Siemens Four Views** model is another one that has **conceptual**, **module**, **code**, and **execution** views.

- Philips Research, the electronics giant, created the **CAFCR model** of architecture, which calls for five views: **customer**, **application**, **functional**, **conceptual**, and **realization**.

- In the year 2000, **IEEE 1471-2000** proposed a flexible view approach instead of prescribing a fixed set of views to solve stakeholder-specific concerns.

- In 2002, researchers at the *Carnegie Mellon Software Engineering Institute* proposed a framework for documenting stakeholder-specific views called *Views and Beyond*, which promoted the use of relevant views similar to IEEE 1471-2000.

In the Agile software development context, *George Fairbanks* introduced the **Architecture Haiku** as an easy-to-build design description assuming that architects can document their proposals in an A4 size paper. The major sections of an Architecture Haiku are depicted in the following diagram:

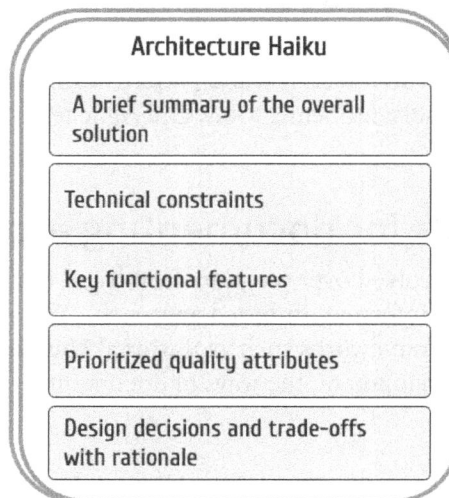

Figure 10.7 – Architecture Haiku

Haikus are often created following collaboration sessions done by teams using mechanisms such as whiteboard drawings. Haiku is a straightforward approach to capturing the architecture summary most concisely. However, in reality, Haiku itself is not sufficient. For example, a single way of representing architecture is not effective; different viewpoints are still necessary for better communication with various stakeholders.

Design decisions need to be well documented, and each decision record may need one-pagers to represent architecture options and trade-offs.

Scaled Agile Framework (SAFe) recommends having a **solution intent** as a central repository to store all documents that convey the **what** and **how** aspects of a solution. It is a critical knowledge base that captures minimal and sufficient information, current and future stages, specifications, designs, and test cases. It enables teams to collaborate on solutions, including vendors. Some of the key principles proposed by **SAFe** are favoring models over documentation, collaboration, a single repository, keeping documentation high level, and keeping it simple.

LeSS uses the **Software Architecture Document (SAD)** as a means to document solutions. LeSS emphasis using physical collaboration spaces and physical whiteboards for documenting SAD. Teams continuously update the physical diagrams through collaborative discussions during development. These artifacts are transformed digitally post-development.

Disciplined Agile (DA) uses the term **travel light** and proposes having just enough models and just enough documentation. Similar to LeSS, DA also highlights the importance of using physical display walls and whiteboards.

Documentation ecosystem at Snow in the Desert

The effectiveness of the documentation approach heavily depends on tools. Managing everything in a single tool may not be feasible nor correct. Snow in the Desert uses the principles of documentation to develop minimal but sufficient product-centric documentation.

The following diagram captures the structure of architecture documentation followed at Snow in the Desert:

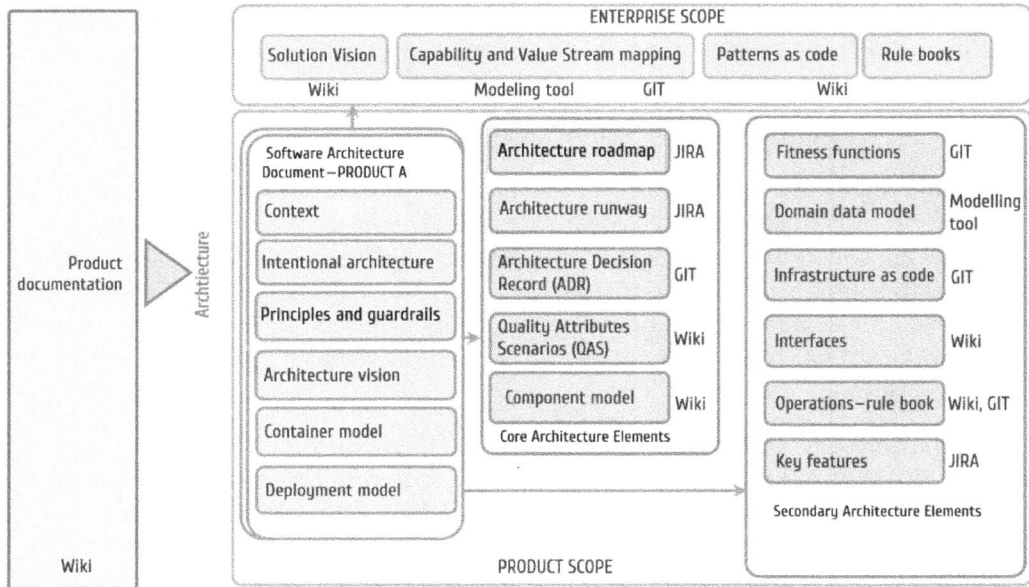

Figure 10.8 – Architecture documentation landscape at Snow in the Desert

As shown in the diagram, Snow in the Desert follows a product-centric approach, and architecture documentation is just one element of the overall product documentation. The **Software Architecture Document,** as shown in the diagram, captures the overall architecture, whereas detailed elements may be captured on different wiki pages linked to the main architecture document. Previous chapters have covered most of the building blocks mentioned in the diagram. **Architecture Decision Records** (**ADR**) will be explored in the next chapter.

The following diagram illustrates a tool-centric view based on the approach adopted by Snow in the Desert:

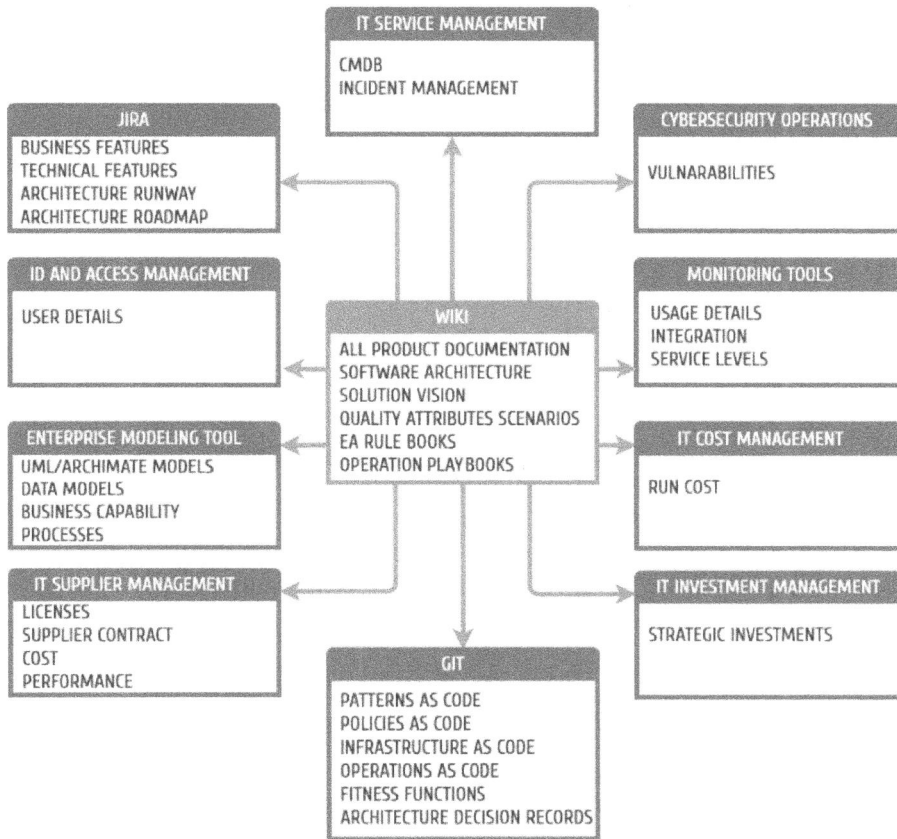

Figure 10.9 – Architecture documentation ecosystem of tools with Snow in the Desert

The diagram shows a fully integrated model for enterprise and solution architects to document various architectural elements. It helps answer many questions instantaneously and therefore saves a considerable amount of time by avoiding repeatedly discovering information. It is extremely important to keep all the tools in the documentation ecosystem connected and synchronized.

Models are an integral part of architecture documentation, and to the extent possible, models should be used instead of free text. Models offer better communication, are easy to understand, can help to spot defects, and reduce maintenance overheads. We will go deeper into different aspects of modeling in the next section.

Applying Model-Based Software Engineering

Model-based software engineering has acquired massive popularity over the years. We have touched on the modeling aspects in *Chapter 9, Architecting for Quality with Quality Attributes*, and examined a couple of models, including the transaction and workload models. We have also explored **Domain-Driven Design** (DDD) in *Chapter 7, Technical Agility with Patterns and Techniques*. Modeling is one of the most critical activities for architects as it helps eliminate wastage by detecting defects early in the flow. All scaling Agile frameworks mention the importance of modeling before development to determine the design's potential shortcomings.

The International Council on Systems Engineering (INCOSE) defines **Model-Based Systems Engineering** (MBSE) as a formalized application of modeling to support system requirements, design, analysis, verification, and validation activities, beginning with the conceptual design phase and continuing throughout development and later life cycle phases. MBSE helps increase productivity, allows better communication, and easy impact assessment of changes, promotes and faster flow. MBSE moves software design from document-centric to model-centric. It supports system flows and architecture, system requirements analysis, and system process flows. A model analyzes a system's design to identify faults prior to implementation.

The model-centric approach brings consistency with a standard language. MBSE may use SysML, UML, ArchiMate, or other notation languages. However, traditional modeling languages such as UML are often low-level constructs, not expressive, require a certain level of technical knowledge, are not engaging, and are disconnected from the natural whiteboard drawings. Many architects, therefore, prefer block diagrams on PowerPoint and other drawing tools for developing architecture diagrams. Even though those PowerPoint diagrams are aesthetically sound, it is hard to find consistency even within the organization's boundaries.

Simon Brown introduced the **C4 model** for representing architecture and design in an expressive way. It helps architects to communicate their ideas effectively and efficiently. C4 is not a notation language, like UML and ArchiMate. C4 uses the concept of abstraction first, in which higher-level models are abstract, and lower-level models show more concrete information.

The following diagram shows the C4 metamodel:

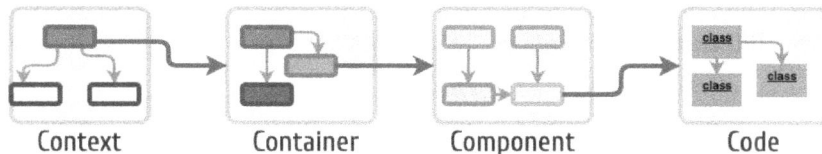

Figure 10.10 – C4 metamodel for architecture documentation

As shown in the diagram, architecture models start with a **Context** diagram, and then drill down to **Container**, **Component**, and **Code**. The lowest-level diagram is code represented as UML class diagrams often generated with a code-first approach.

In the context of Snow in the Desert's AVTS scenario, the following diagram discussed in *Chapter 5*, *Agile Solution Architect – Designing Continuously Evolving Systems*, captures the context-level view:

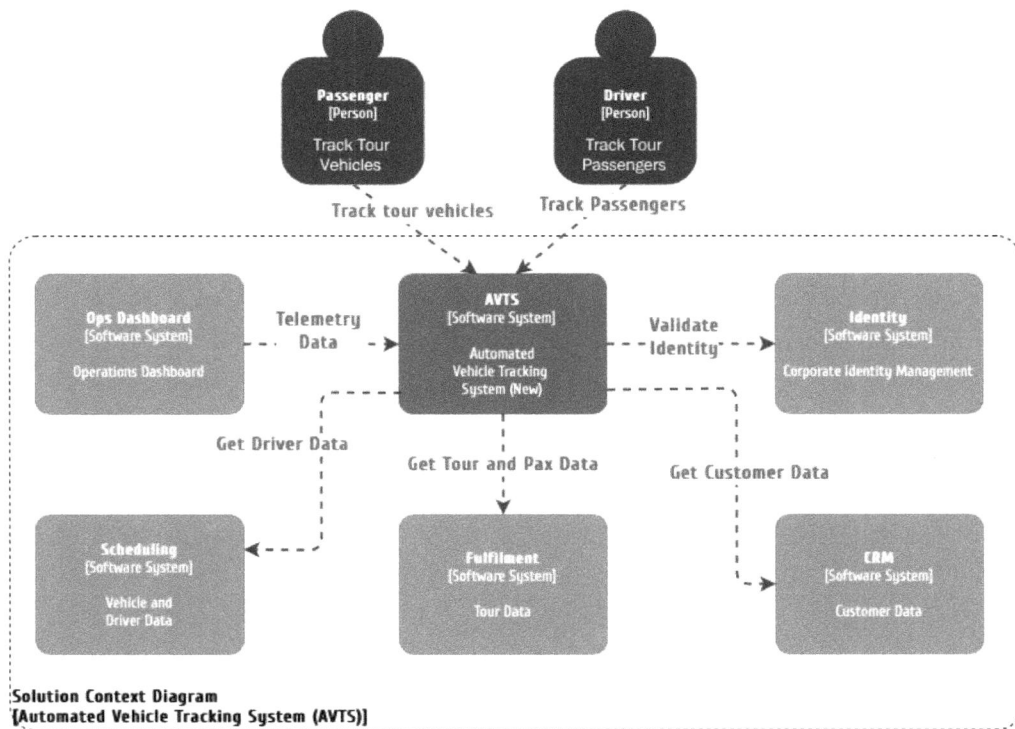

Figure 10.11 – AVTS context diagram using C4

The following diagram, the container model, discussed in *Chapter 5, Agile Solution Architect – Designing Continuously Evolving Systems*, captures the next level of context and explains the different components of the system:

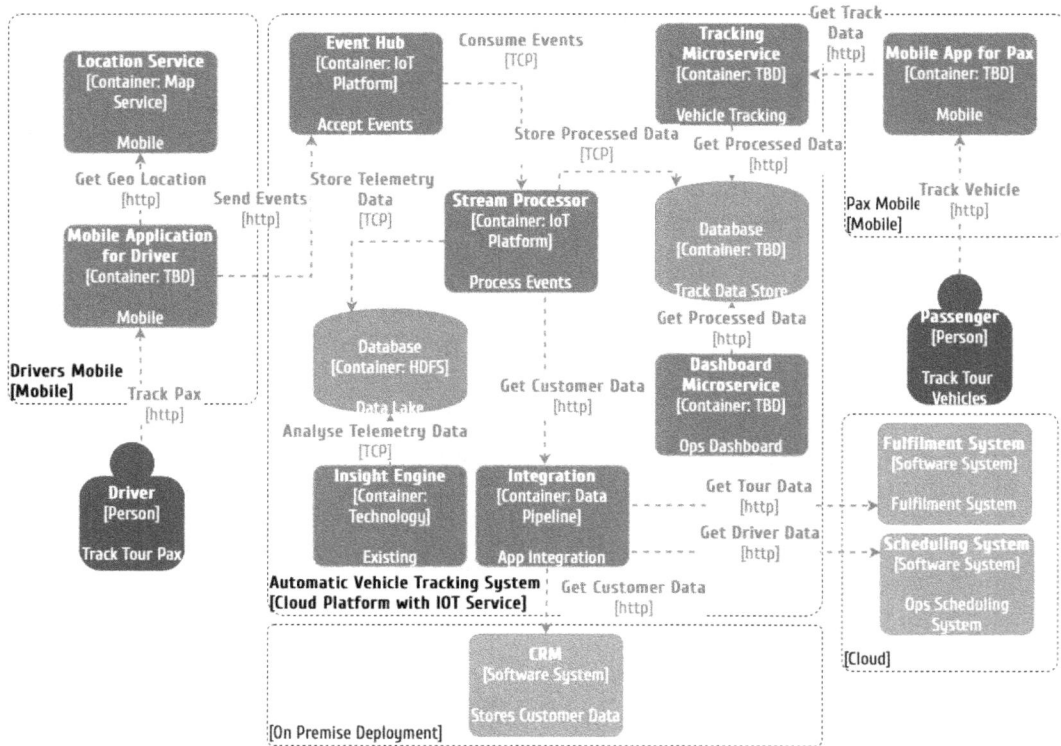

Figure 10.12 – AVTS container diagram using C4

The component model discussed in *Chapter 7, Technical Agility with Patterns and Techniques*, as shown in the following diagram, captures the third level, such as the details of a particular service or component:

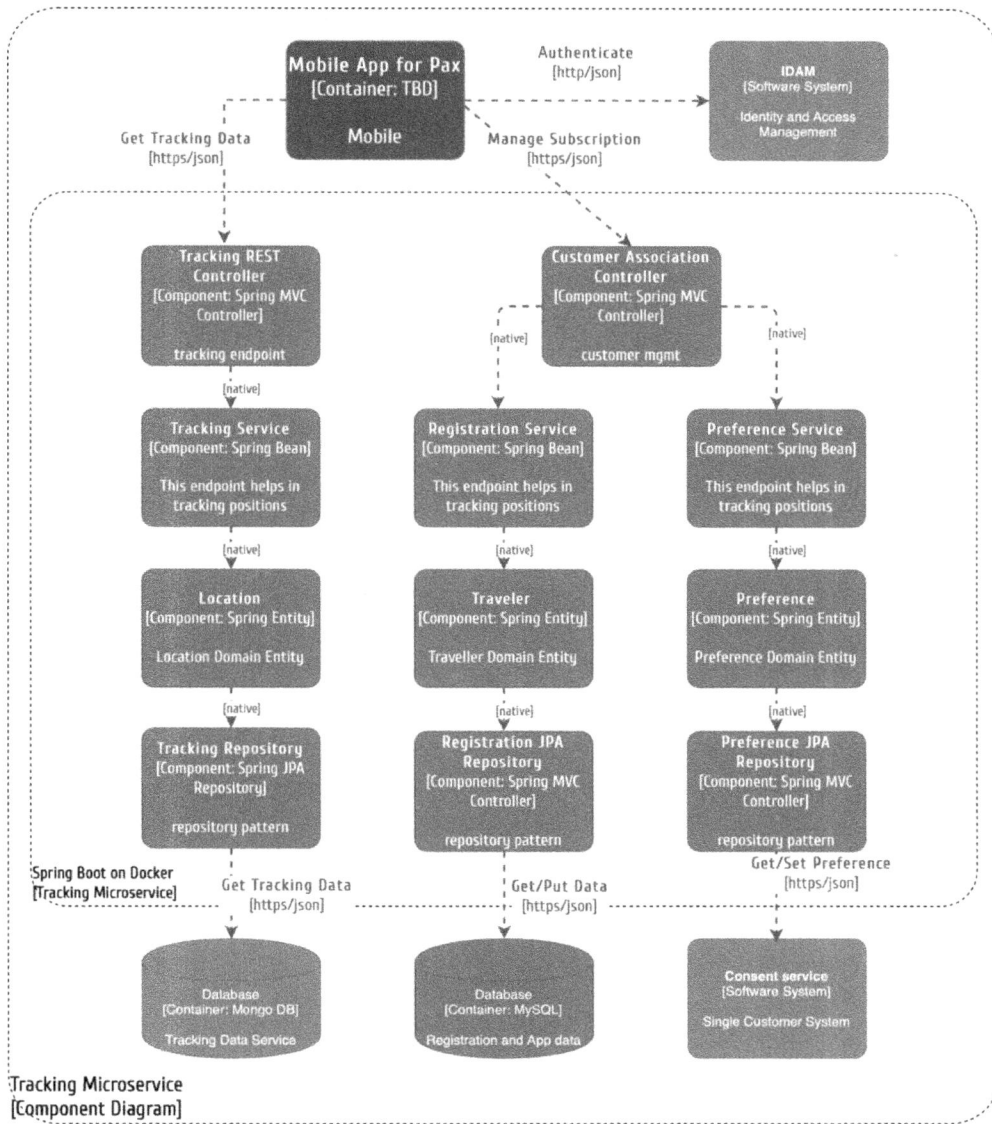

Figure 10.13 – AVTS component diagram using C4

There are many other supplementary diagrams possible with C4, including a system landscape diagram, a deployment diagram, and a communication diagram. The deployment model from the AVTS scenario is depicted in the following diagram:

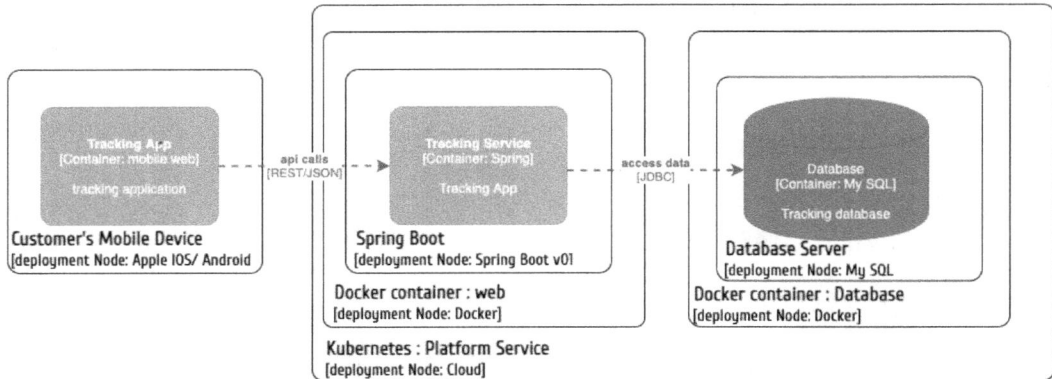

Figure 10.14 – AVTS deployment diagram using C4

As explained with a series of AVTS architecture diagrams, C4 helps in zooming in on more details and provides different architecture views targeting different stakeholders with expressive diagrams.

Summary

In this chapter, we have learned the importance of documentation in developing software systems following Agile development methodologies. We have learned the reasons for documentation, why documentation is considered a flow barrier, and some of the challenges associated with traditional documentation methods, such as exceeded budget, no one reading the documentation, the fact that it is a waste of effort, change tolerance, and documentation silos.

Furthermore, we have explored the Lean-Agile documentation and the adoption of evolutionary collaboration as an alternative approach for documentation. We have learned about alternative methods, such as documentation as code, generating documentation, focusing on models, and the use of visual aids. Additionally, this chapter examined the adequacy of documentation and principles for good documentation, such as purpose-driven documentation, consumer-driven documentation, delivering just barely enough documentation, and the timely delivery of documentation. We have also explored how much documentation is enough using the Cockburn scale. In addition, we have reviewed various elements of architecture documentation and explored an integrated tools ecosystem that is pertinent to good documentation. Models are vital for architecture documentation, and in this context, we have examined MBSE and, specifically, the C4 model. Later in this chapter, we also learned about different architectural artifacts with examples from Snow in the Desert.

In this chapter, we have covered an architect's view of documentation in Agile software development. The next chapter will explore how to achieve architecture alignment with Lean-Agile governance and safety nets.

Further reading

- **Lean Documentation: Strategies for Agile Software Development**: `http://agilemodeling.com/essays/agileDocumentation.htm`

- **An Exploratory Study of Architectural Practices and Challenges in Using Agile Software Development Approaches**: `https://core.ac.uk/download/pdf/59342997.pdf`

- **Comparing the SEI's Views and Beyond Approach for Documenting Software Architectures with ANSI-IEEE 1471-2000**: `https://resources.sei.cmu.edu/asset_files/TechnicalNote/2005_004_001_14498.pdf`

- **C4 Model**: `https://c4model.com`

11
Architect as an Enabler in Lean-Agile Governance

Creativity is allowing yourself to make mistakes. Art is knowing which ones to keep.

– Scott Adams, the artist and creator of the Dilbert comic

In the previous chapter, we discovered alternative approaches to minimize documentation efforts and covered the adoption of Lean methods for essential documentation. Lean documentation helps to accelerate the sustainable, continuous delivery of values. Governance is perceived as another flow impediment in Agile software development. In this chapter, we will investigate and learn how to adopt architecture alignment without affecting the speed of delivery.

Architecture governance is traditionally a lengthy, low-frequency, board-based discussion based on extensive documentation. Agile software development promotes interactions and engagements over documented evidence-based reviews. With the help of a revolutionary rethink, establishing a fit-for-purpose, Lean-Agile governance can accelerate and facilitate business agility and innovation. By fostering autonomy and empowerment with a continuous sharing context, teams can self-govern architecture decisions, eliminating block times. Adhering to a Lean-Agile mindset and striking the right balance between alignment and autonomy is significantly important in Agile software development. Transparency, openness, and honesty are the foundations of good governance. It also requires an egoless culture where teams must feel physiologically safe when making critical software decisions.

In this chapter, we're going to cover the following main topics:

- Understanding architecture governance

- Bringing agility with Lean-Agile governance

- Balancing autonomy with agility

- Documenting architecture decisions

- Ensuring psychological safety

- Measuring the quality of architecture decisions

This chapter focuses on the **Lean Governance** focal point of **the Agile Architect's Lens**:

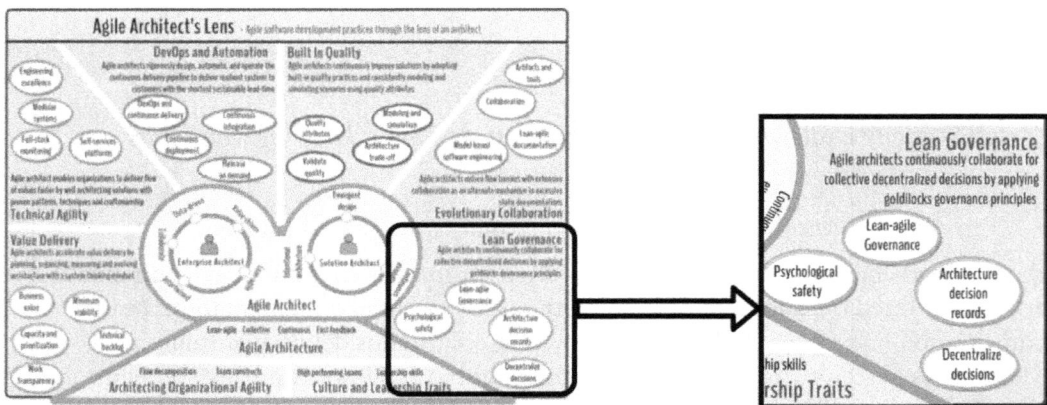

Figure 11.1 – The evolutionary collaboration – Focal point

This chapter will study the pitfalls of traditional governance and explore the principles and benefits of the Lean-Agile approach for governance. We will also explore working without command and control and the ways of balancing autonomy and alignment with the right levels of governance. Later in this chapter, we will explore the people aspect of governance, architecture decision documentation, and measurement.

Understanding architecture governance

Governance is the act of ensuring that corporate assets and interests are protected while achieving business objectives and outcomes. Governance aligns strategic funds with the long-term goals, as well as short-term improvements, of the enterprise. Governance is broadly classified into three levels, as shown in the following diagram:

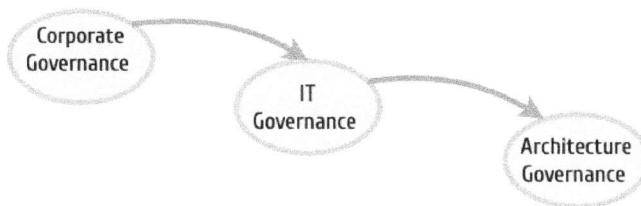

Figure 11.2 – Governance hierarchy

The three levels of governance are explained as follows:

- The **corporate governance** structure of the enterprise, with policies, principles, practices, and monitoring alignment to such corporate regulations.

- **IT governance** controls and guides IT investments and processes related to resources and assets through a set of strategies, practices, processes, and policies.

- **Architecture governance** ensures that the evolution of the solution landscape consistently aligns with the strategic objectives of the enterprise.

Disciplined Agile (DA) defines governance as the leadership, organization structure, and streamlined process to enable IT to work as a partner in sustaining and extending its ability to produce meaningful value for its customers. **The Open Group Architecture Framework (TOGAF)** defines architecture governance as the practice and orientation by which enterprise architectures and other architectures are managed and controlled at an enterprise-wide level.

Architectural governance aims to ensure that technology changes are introduced to the business appropriately and sustainably. There are two types of architecture governance – **strategic alignment** and **solution expectation**, as shown in the following diagram:

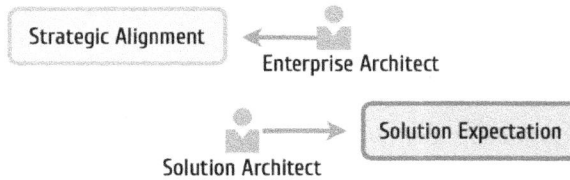

Figure 11.3 – Aspects of architecture governance

Strategic alignment ensures that the system landscape is continuously evolving to meet changing customer needs while aligning and protecting IT strategies. Strategic alignment effectively balances business strategies and IT strategies. The enterprise architect is primarily responsible for governing the architecture's strategic alignment. The solution expectation is whether the solution delivered is aligned to the purpose, guardrails, intentional architecture, and quality requirements. A solution alignment continuously balances sustainability and speed of delivery. The solutions architect is primarily responsible for governing the solutions architecture alignment. Solution architects may further extend the solution alignment to cover technical design aspects such as using best practices, the correct design patterns, and the quality of the solution.

Architecture governance exists in software delivery for valid reasons, such as reducing the risk of failure, the possibility of solutions, and not meeting customer requirements. For example, if the enterprise vision is to move all applications to the cloud, every deployment needs to step toward that goal. If decisions are not aligned and governed, it leads to the organization never reaching its intent.

Some of the advantages of good architecture governance practices are summarized as follows:

- Investments are made in the right initiatives that drive toward fulfilling the vision of the enterprise.

- Aligning to regulatory compliance so as to avoid brand damage and revenue leakage.

- Identifying architecture risks earlier in the delivery cycle before the cost of reworking becomes unmanageable.

- Achieving economies of scale by reusing assets, technologies, and patterns.

- Optimizing the system's landscape to achieve operational efficiency.

- Improving quality and sustainability by balancing the solution architecture and designs against the speed of delivery.

Governance is not required when working with small teams. Complex software delivery involving many teams working in parallel require appropriate levels of governance. Effective governance can entail cost savings, customer satisfaction, and reduced lead times. However, traditional architecture governance needs critical introspection.

Challenges with traditional governance

Traditional governance is problematic and follows a stop-and-check, stage-gated, rigid process. These governance checkpoints often use extensive documentation to validate adherence to strategic objectives and architecture intent.

Stage-gated governance is illustrated in the following diagram:

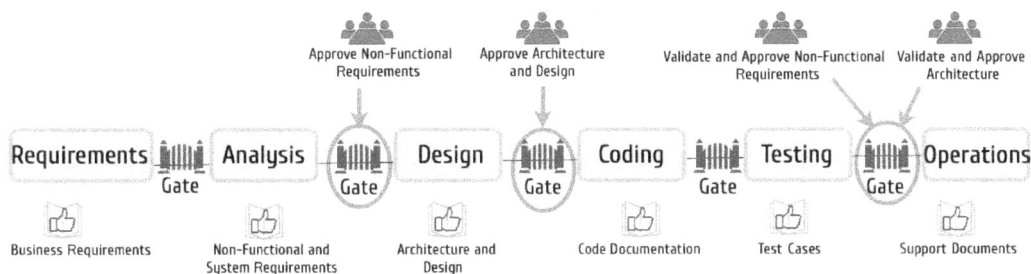

Figure 11.4 – Stage-gated governance process

As discussed in *Chapter 3, Agile Architects – the Linchpin to Success*, traditional architecture governance follows a bureaucratic process heavily disconnected from the ground reality, illustrated using metaphors such as *architecture police*, *ivory tower*, and *architecture astronauts*.

Some of the key challenges associated with traditional architecture governance are as follows:

- Traditional governance reflects the ivory tower model by following elaborately documented policies, patterns, and guidelines. Such guidelines are not nimble and may not reflect current ground-level realities, often being obsolete and unconsumable. Even in many cases, such policies may not exist or may not be agreed upon or communicated across all relevant stakeholders. Many organizations traditionally follow a single-standard-based approach to fit everything into one standard, which limits innovation.

- The architecture forums use command and control often, and are hierarchical in nature. Each decision may also have to go through many bodies before securing development approval. These governance forums take place in a highly formal environment similar to a board meeting. Many stakeholders use these meetings as a place to show their power and are therefore also ego-driven.

- Decision makers are not fully connected, and therefore decisions are taken without sufficient context. Traditional architecture decision makers use a standard-first approach, and zero tolerance in governance results in the importance of business values and the larger customer context being overlooked. Many architecture boards use subjective and inconsistent decisions that are not framework or data-driven.

- Decisions, once taken, are not transparently communicated to impacted teams. Therefore, teams have to operate without understanding the rationale behind decisions and what value they add. In such cases, it is a lost opportunity for teams to acquire any learnings from these decisions.

- Architecture decisions are taken once for the overall architecture, very early or late in the delivery cycle. Many subsequent decisions are not tracked and aligned. Exceptions approved by the board are often not connected to backlog and funding. In frequent board meetings and back-and-forth conversations, delays follow as teams have to wait for board meetings to take place.

While governance is an important activity, it is often seen as an anti-pattern in Agile software development. *Allen Holub* observed that one of the reasons why many Agile projects fail is due to wrong governance practices that are not based on value. The report *Governance for Agile Delivery* observes that the Agile community recognizes that multiple layers of governance do not necessarily improve the quality of technical solutions, speed up delivery, or reduce risk. At the same time, organizations running large programs across many Agile teams need adequate governance.

In the next section, we will see how to implement governance in Agile software development.

Bringing agility with Lean-Agile governance

Irrespective of the software development methodology, protecting enterprise investments with strategic and solution alignments is pivotal. While architecture governance is inevitable, it requires a dramatic overhaul to align with Agile software development values and principles. One of the manifesto principles for Agile software development is valuing individuals and interactions over processes and tools. Successful implementation of architecture governance in Agile software development requires a radical rethink by placing individuals and interactions at the center.

The impact of unexpectedly turning the steering wheel of a speed car cruising at *200 km/hour* is enormously dangerous compared to a cruising speed of *20 km/hour*. Similarly, when project development is occurring at a high speed, it needs razor-sharp focus and great discipline. Poor discipline can have significant consequences, such as improper investments, unsustainable flow, and non-reliable systems. Good discipline comes with unwavering support from leadership, knowledge, and maturity.

In Agile software development, light-touch governance is still required no matter how well the teams perform and are disciplined. Architecture governance in Agile needs to observe the following principles:

- Incremental and iterative non-invasive inflow inspections instead of stage-gated stop and check validations

- Empowerment and autonomy granted to teams instead of top-down command and control

- Collaborative and collective decision making over large-scale documentation and endless analysis

- The exchange of continuous engagement and feedback over disengaged architecture decision boards

Agile software development practices are already embedded with adequate non-invasive inflow checkpoints, such as early and frequent testing, incremental releases, ceremonies for feedback, and inspect and adapt sessions. In the next section, we will go into more detail on Lean-Agile governance.

Embracing Lean-Agile governance

Lean-Agile governance shifts the focus to people, responsive, pragmatic, incremental, and value-driven alignment. Anchored around shared understanding, Lean-Agile architecture governance promotes the continuous communication of context, purpose, rationale, and the business values of guidelines and strategies to the team. Lean governance focuses on maximizing business value by protecting strategic investments and solution alignments without impacting delivery at pace.

Lean-Agile governance offers teams an opportunity to perform self-governance by enabling empowerment and autonomy without losing visibility and control. In Lean-Agile governance, architects and other governors need to be closer to the team, sharing the same understanding of the context to validate trade-offs objectively by considering business value. This closer engagement reduces the friction between architects and teams.

Agile Path (www.Agile-path.com) introduced the concept of **Event-Driven Governance (EDG)**. EDG is intended to be lightweight, Lean, virtual, and trust-based. In EDG, the governance process is dormant unless specific governance events trigger it. In the EDG approach, governance is primarily based on community-driven and self-governance, wherein a virtual governance body comes into existence only when necessary and has been activated.

All scaling frameworks emphasize the Lean governance practices for large enterprises embracing Agile software development at scale. The foundation of DA is based on self-organization with effective governance. It observes that regular coordination meetings, lightweight milestone reviews, and so on help achieve architecture alignments. Besides, the architecture owner part of the team is ultimately accountable for architecture decisions. One of the **Scaled Agile Framework (SAFe)** principles is decentralized decision making, and this stipulates that delivering value in the shortest sustainable lead time requires decentralized decision making. Escalated decision-making requests inhibit flow delays and result in suboptimal designs due to a lack of context awareness.

Lean-Agile governance is based on a set of principles that we will discuss in the next section.

Principles of Lean-Agile governance

The successful adoption of Lean-Agile governance necessitates a substantially different mindset and culture. The Lean-Agile architecture governance requires thinking Lean and embracing agility at every step of alignment. Effective and efficient governance promotes a set of principles well aligned with Agile software development practices and philosophies.

The different steps of Lean-Agile governance are described in the following diagram:

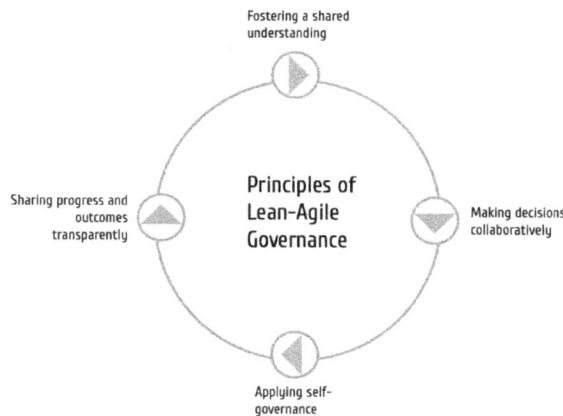

Figure 11.5 – Principles of Lean-Agile governance

These four principles of Lean-Agile governance are explained in the following sections.

Fostering a shared understanding

Well-informed teams better align architecture decisions with organizational goals and architecture intentions, eventually improving productivity, and reducing block times and rework costs. Therefore, all critical stakeholders linked to delivery flow must be continuously aligned on the context and organization's knowledge required for efficient and effective decision making.

Architects play a crucial role by acting as a champion of communication. The architects also share the essential aspects of business and IT strategic alignment, business purpose, operational business constraints, assumptions, economic frameworks, and guardrails for teams to make decisions autonomously. Well analyzed localized decisions improve flow speed without compromising sustainability.

Making decisions collaboratively

As discussed in *Chapter 2, Agile Architecture – The Foundation of Agile Delivery*, collective intelligence, collective ownership, and collective knowledge enable teams to develop a strong belief in their ability to make the right decisions. Architects are part of the team, and engage and collaborate throughout the delivery cycle to motivate and facilitate members to make the right decisions and guide them when they go off guardrails.

As discussed in *Chapter 5, Agile Solution Architect – Designing Continuously Evolving Systems*, solution design workshops are great avenues for making decisions collaboratively. Collaboration minimizes the effort required to manage post-decision-making governance. Collaboration is not limited to architecture decisions, also defining the governance process and enterprise architecture assets that are useful as regards decision making, such as architecture patterns.

Applying self-governance

Lean-Agile governance promotes self-governance. Self-governance is a practice in which teams perform all the necessary governance functions within the team, without intervention from external entities or presenting decisions to boards for approvals. While architects are part of the team, they also need to establish the required framework and processes for team members to self-validate their decisions. Self-governance includes community discussions, determining when to escalate, measuring the quality of architecture decisions, requesting feedback, assessing solutions, and so on.

The well-architected framework discussed in *Chapter 8, DevOps and Continuous Delivery to Accelerate Flow*, is an example of a self-assessing framework. Later in this chapter, we will showcase a measuring framework for architecture decisions. Architects also enable decentralized decision making by educating and fostering the technical excellence required for architecture decision making. They also offer self-service roadmaps and technology radars to facilitate the adoption of new technologies.

Sharing progress and outcomes transparently

Agile architecture governance promotes feedback instead of rigid board-level discussions and stringent processes. **Definition of Done (DoD)** is one mechanism for ensuring that critical architecture alignments are not discarded.

The use of information monitors, inspect and adapt ceremonies, sprint demos, retrospectives, and architect sync sessions are good opportunities for architects to receive feedback and help identify and mitigate risks. Using continuous monitoring through automation for real data collection helps validate the status of architecturally significant requirements instead of relying on documentation.

Benefits of Lean governance

Some of the key benefits of Lean-Agile governance, adapted from the *2017 Agile Governance Survey*, are as follows:

- **Lightweight**: Lean-Agile governance is incremental and aligned with the flow with minimal disruptions, documentation, and wait times.

- **Improves productivity**: There are no unnecessary back-and-forth conversations as the context is shared, allowing you to spend more time on value delivery development work instead of reaching a consensus.

- **Increases quality**: Decisions taken by people who have the context and knowledge with appropriate support results in better quality decisions for delivering sustainable quality solutions. As you build it, you own it; the team who takes the decision also owns the decisions.

- **Promotes IT investments wisely**: Lean business cases and approvals are part of the Lean-Agile governance to ensure that large investments are adequately assessed and gauged before pushing to teams for implementation.

- **Improves team morale**: Lean-Agile governance is people-oriented, and offers autonomy and empowerment for teams to make collective decisions. These behaviors make them self-motivated as their ideas and decisions are valued. High morale teams deliver better quality solutions faster.

- **Focuses on business value**: Moving from a standard-first approach to a business value-first approach helps deliver valuable software that meets customer needs.

- **Foster innovation**: Teams are empowered so that they can fearlessly make decisions with full awareness of the context and guardrails of their decisions. Empowerment with context promotes an innovative culture.

We have seen the concept of Lean-Agile governance, which strongly promotes autonomy and empowerment. However, the question still remains: how much autonomy?

Balancing autonomy with agility

Lean-Agile governance uses collaborative and decentralized decision making by enabling autonomy and empowerment without losing alignment. However, the question remains as to the limit of autonomy and decentralization to strike the right balance between protecting investments and continuous flow.

Luna, Kruchten, and *Moura,* in their paper, *State of the Art of Agile Governance: A Systematic Review,* observed that the level of governance must always be adapted according to the organizational context. The level of governance required to achieve business agility must be balanced and adjusted when needed, considering each organization's particular conditions and timing. Just enough for one organization may be too much for another one. As the authors observed, there is no one-size-fits-all governance.

As demonstrated in the following diagram, governance can be at both extremes:

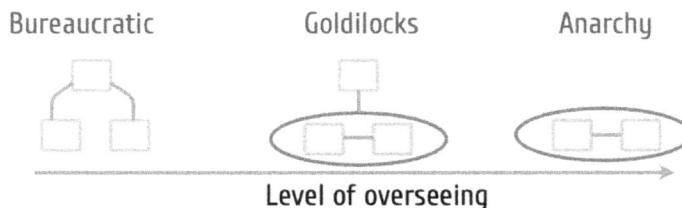

Figure 11.6 – Different levels of architecture governance

No governance leads to anarchy, a state of chaos with no accountability. In anarchy, either single individuals randomly take decisions without consensus or teams reach agreement too easily. The latter is more frequent and is called groupthink. Groupthink is a psychological phenomenon where groups make decisions without critically evaluating alternative ideas or viewpoints because of their cohesiveness and harmony. Group members make decisions prematurely to minimize conflict. On the other hand, extreme governance can have a serious impact not only on the flow, but also on employee morale.

Just enough governance is still necessary for protecting investments and to achieve economies of scale. Goldilocks governance, from the Goldilocks principle, originated with the children's story entitled *The Three Bears*, in which a young girl named Goldilocks consistently looks for a bowl of porridge at just the right temperature. Goldilocks governance means just enough governance based on the organization's needs, the maturity of the people, the culture, the blast radius caused by an incorrect decision, the business impact, architecture styles used, and so on.

Unfortunately, even Goldilocks governance varies between organizations, and therefore organizations need to be judgmental in setting up governance. **Silverline** in the following diagram is an indication of the path of adoption of governance:

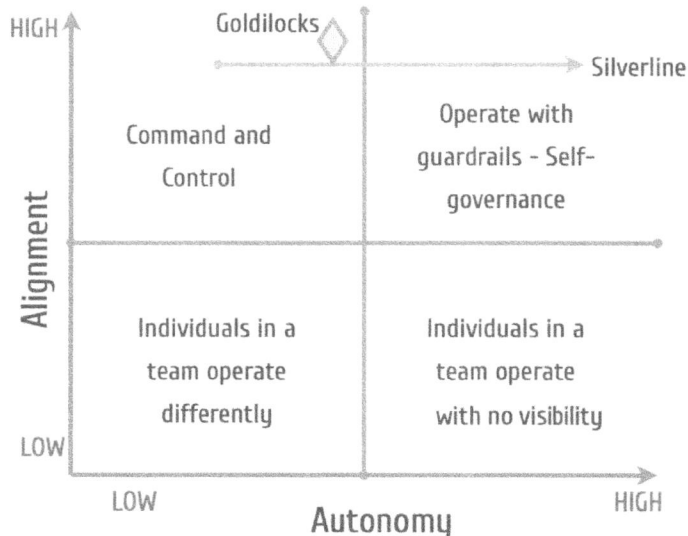

Figure 11.7 – Silverline of governance

The preceding diagram illustrates the positioning of autonomy and alignment. There are four quadrants, and the line shows typical waterfall organizations transforming to Agile.

Organizations need to traverse the Silverline governance path incrementally when implementing architecture governance based on the organization's current context. At the beginning of this journey, Lean governance principles are applied, people aspects are taken into account, and just enough empowerment and autonomy are afforded to teams. The mindset and culture changes are augmented with high-frequency governance forums, where people with context regularly meet to ensure that incorrect decisions are reverted at the earliest possible juncture.

The challenge is still determining what decisions can be decentralized and what decisions cannot be. The next section throws some light on this topic.

Determining the level of decentralization of decisions

Jeff Bezos's letter to *Amazon* shareholders released in 2015 proposes an approach to demarcate what decisions can be decentralized. It observes that some decisions have a severe consequence and are not irreversible, such as one-way doors, or Type 1 decisions. He furthermore observed that senior experts must make these decisions based on data meticulously through deliberation and consultation without rushing. On the other hand, most of the decisions are changeable or reversible without a significant impact. Amazon called these two-way doors, or Type 2 decisions. Amazon's recommendation is that Type 2 decisions have to be taken quickly by high intelligence individuals or small groups. Governance collapses when Type 2 decisions are taken through a long and complicated consensus process where the value and cost of wrong decisions outweigh the process itself. These extended processes are a flow stopper in Agile and also bring down innovation.

In summary, different strategies are required for reversible decisions and irreversible decisions. Jeff also commented that most of the decisions could be made with 70% of the information. Waiting for more information slows down the decision process.

Reversible decisions are candidates for decentralization, whereas irreversible decisions need to be centralized and taken by senior leadership, architecture review boards, or even single individuals such as a CTO. Individuals at senior leadership take intuition and gut-based decisions based on experience at times beyond financial data.

Spotify follows a similar concept to determine whether teams can take decisions or should escalate to higher-level boards. At Spotify, teams create **Architecture Decision Records (ADR)** and **Requests for Comments (RFC)** online or in a meeting for review and feedback. An adapted process is summarized in the following diagram:

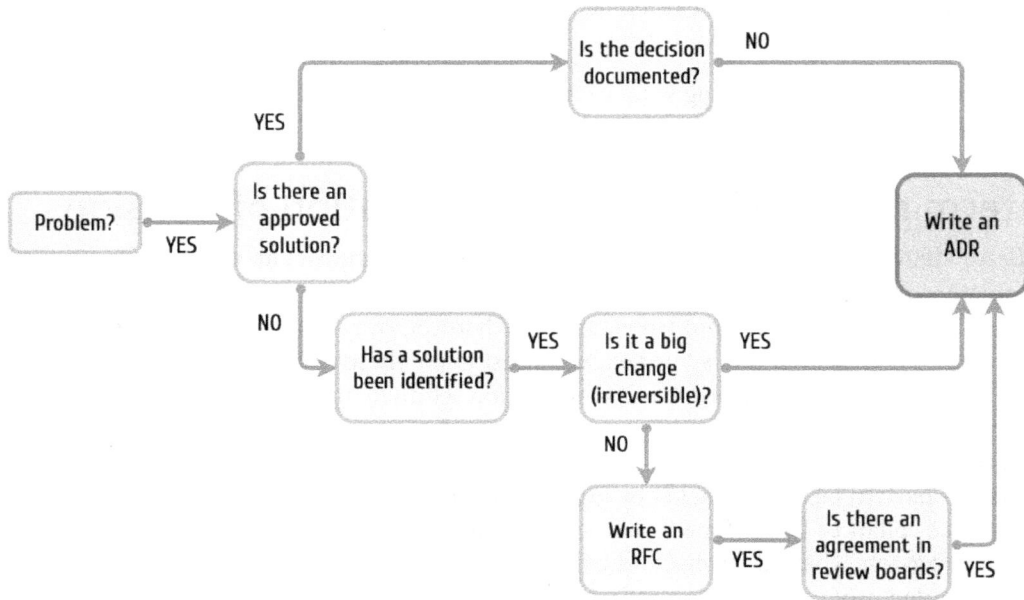

Figure 11.8 – Spotify's RFC approach

In all cases, decisions are documented in an ADR. We will discuss more on ADRs later in this chapter.

SAFe recommends a similar mechanism to determine what can be decentralized and what cannot. SAFe suggests three parameters to decide whether decisions can be taken locally – the frequency of decision, time criticality, and economies of scale, as shown in the following diagram:

DECISION	FREQUENCY Y=2, N =0	TIME CRITICAL Y=2, N=0	ECONOMIES OF SCALE Y=0, N=2	TOTAL	DECISION TYPE UPTO 3 CENTRALIZE
Identify technology stack for microservices	2	1	2	5	DECENTRALIZE
Identify IoT platform for AVTS	0	1	0	1	CENTRALIZE

Figure 11.9 – SAFe approach for identifying decentralized decisions

Both centralized and decentralized decision making has its advantages and disadvantages. *Donald G Reinertsen* observed that centralized decision making brings economies of scale and reduces duplication, resulting in highly optimized systems. However, centralized decision making slows down the process due to bandwidth and the distortion of information with senior architects and executives.

On the other hand, decentralization results in high-quality decisions taken by people with higher intellectual knowledge having in-depth contextual information.

Implementing Goldilocks governance – a case study

Snow in the Desert uses a mix of centralized and decentralized decisions. Decentralized decisions are taken based on the SAFe approach by the teams in collaboration with architects. The following diagram illustrates the governance structure with Snow in the Desert:

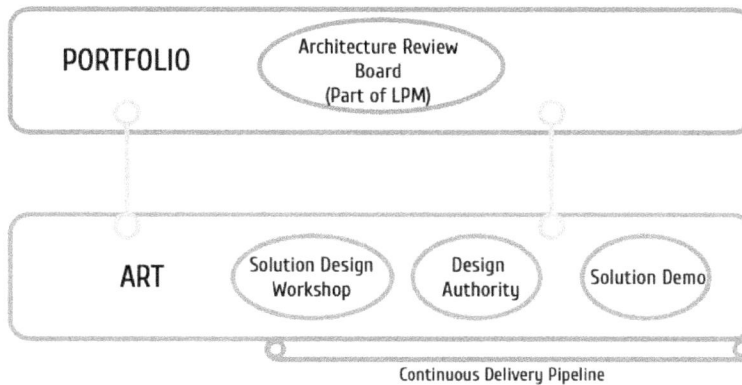

Figure 11.10 – Architecture governance with Snow in the Desert

Snow in the Desert mainly uses a **prove-before-standardize** approach for adopting new technical standards, which fosters innovation to a great extent. With this approach, new tools are discovered, tested, and learned by teams locally, unless those are high investment decisions. They transparently share this information for the awareness of the organization. Teams acquire, build, and operate such tools and technologies with the smallest blast radius until value is delivered. They then use cross-pollination and knowledge sharing to promote the success of new technologies and tools, which are then set as enterprise standards.

In large enterprises, arriving at consensus is one of the biggest challenges in decision making. The four types of decision making defined by *Michael Tardiff* are shown in the following diagram:

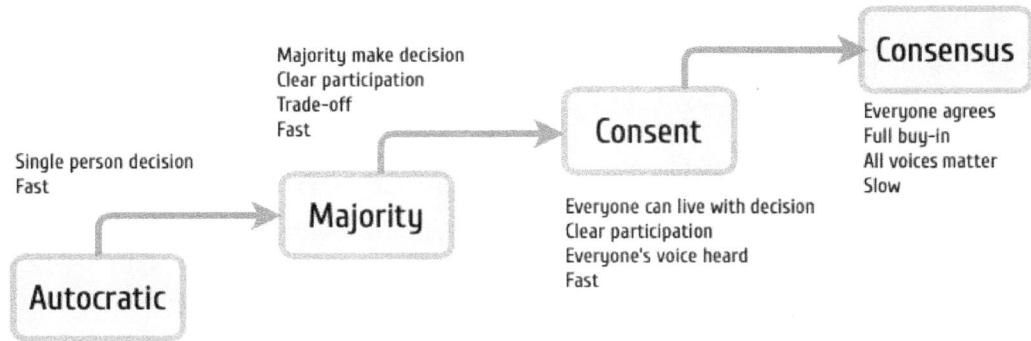

Figure 11.11 – Michael Tardiff's decision model

With Snow in the Desert, architecture decision-making forums follow a consent-based approach for faster, inclusive decision making. Whether architecture decisions are centralized or decentralized, it is incredibly valuable to document and share decisions as assets consistently across the organization.

Documenting architecture decisions

Decisions form an excellent knowledge base, and this enhances predictability in decision making. In a decentralized decision-making environment, sharing decisions consistently helps teams to understand the rationale and thought process behind decisions. It enables teams to make similar arguments and take similar approaches when dealing with comparable situations.

Transparently publishing architecture decisions also enables non-invasive controls and oversight. Enterprise architects and other senior leaders can review published decisions and provide their feedback at the earliest possible juncture before the team progresses far into the implementation cycle.

In the previous chapter, we have covered architecture documentation, where ADR is one of the critical elements of architecture documentation. ADR helps architects and teams to document architecture in a consistent, structured form.

Compared to traditional architecture decision making, in Lean-Agile governance, there are no big architecture decisions. Large architecture decisions hide many details. In Agile software development, architecture is broken down into a decision backlog that consists of several fine-grained decisions, as explained in *Chapter 5, Agile Solution Architect – Designing Continuously Evolving Systems*. Every architecture decision backlog item is transposed in an ADR.

Lean-Agile governance advocates a lightweight version of ADR. The following diagram shows the ADR structure proposed by *ThoughtWorks*:

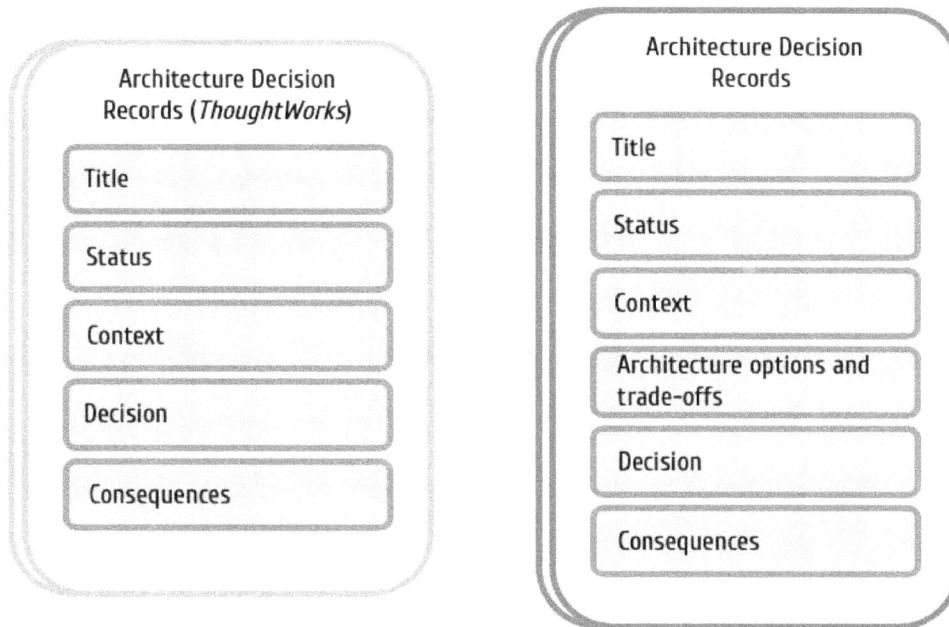

Figure 11.12 – Lightweight architecture decision records

Snow in the Desert uses an adapted version of lightweight ADR proposed by ThoughtWorks for documenting architecture decisions. These ADRs are version-controlled documents stored in Git. The structure of ADR used with Snow in the Desert is shown on the right-hand side of the preceding diagram.

One of the critical aspects of Lean-Agile governance is its people-oriented nature. We will elaborate on the people aspect of governance in the next section.

Ensuring psychological safety

The principles of Agile software development are centered around a set of intrinsically motivated individuals. Lean-Agile governance, therefore, needs to enable these individuals with empowerment. However, enterprises that empower people with the right intentions often fail to see the mindset of individuals.

Flicking quickly to decentralized decision making can also put development teams and individuals under tremendous pressure. Enterprises need to offer a sense of comfort and psychological safety to their people for freely making decisions without fear or chronic anxiety. Decisions made under pressure can go wrong, and this could impact the employees' mental health and well-being.

For critical decisions, if individuals are not confident and are under pressure when making decisions, enterprises need to provide adequate platforms to transfer the accountability of decisions. In such cases, they seek explicit help from architecture governance forums to endorse decisions. These are significantly important safety nets, especially during the initial period of the organization's Agile transformation. Teams feel psychologically safe with safety nets, which can be slowly taken off as teams become acquainted with decision making.

In a psychologically safe environment, individuals fearlessly make decisions. In the case of individuals making mistakes, enterprises must offer a supportive environment where individuals can accept their errors without fear. Such environments aggressively foster an innovative culture.

Organizations implementing Lean-Agile governance by promoting self-governance and empowerment need a significant culture shift. Leaders need to ensure that capable individuals are present in teams making decisions and that they are trusted. Rejecting decisions demoralizes and devalues employees. Avoid such terminology in architecture review boards and ADRs. Architecture and design review boards must enable and guide individuals with sufficient information pointers to direct them on the right path through constant engagement and interactions. Individuals should not feel that they are being overlooked.

Trust and respect are significant in Lean-Agile governance. Effective governance is possible only if the people govern solutions, and the people who make decisions share good relationship. Individuals must feel that they are valued, respected, and recognized for their work in order to foster an empowered culture. They must be encouraged to make decisions even if they make mistakes. Each individual in the team may be different; understanding this intimately and individually to support them with an open and honest culture is essential for long-term success.

The governing body often creates a governance charter, architecture principles, guardrails, patterns, and intentional architecture as a reference point for governance. People do not like someone to tell them how to work. However, they take ownership of what they contributed. Therefore, in a collaborative environment, the individuals who do the work need to be included in setting up the governance process and related artifacts. They have the right to say how they want to be governed. That will bring more acceptance and help them to be open and diligent.

A healthy tension between teams is always good as long as they work toward fixing issues. Nominating a capable individual in the team with authority helps to moderate effectively in case of conflicts. Additionally, educating individuals on appropriate skills and competencies helps level the playing field, which eventually saves time. Architects must nurture team members through on-the-job training, mentoring, and constant guidance.

DA emphasizes the people's aspect of governance quite clearly. Some of the important aspects are as follows:

- Work with a Lean governance mindset.
- Exhibit servant leadership and lead by example.
- Motivate and enable people to do the right thing.
- Communicate clearly, honestly, and in a timely manner.
- Transparently communicate the governance process and structure.
- Learn continuously beyond the review artifacts.
- Verify artifacts and mitigate risks.
- Consider the long and short term.

As we've learned so far, Lean-Agile governance is anchored around people and self-governance. A quick self-assessment framework helps individuals to understand the trade-offs of the decisions made. We will explain this further in the next section.

Measuring the quality of architecture decisions

Architects at times end up taking suboptimal decisions with eventual integrity in mind due to unforeseen situations, as mentioned earlier. On the other hand, in many cases, all of the required information may not come through, even at Define acronym. In certain other situations, decisions need to be taken based on compromises and trade-offs. In all decision-making scenarios, it is important to understand the balancing of various parameters used in the decision-making process to understand shortcomings.

The simplest approach to measuring the quality of architecture decisions is based on five core parameters: business value, lead time, sustainability, risk, and cost. Many architects do not include cost in architecture decisions and regard it as a delivery parameter. However, in Agile architecture, the cost of quality and the cost of architecture need to be considered as a key parameter for decision making. The following diagram shows the five parameters and simple questions to assess the proposed architecture decision against:

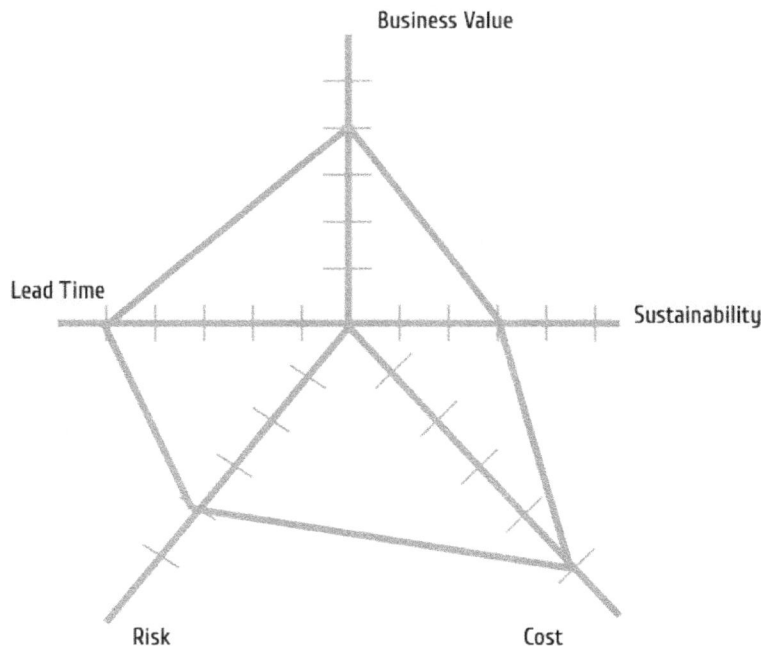

Figure 11.13 – Measuring the quality of decisions

We can use this diagram to evaluate, on a scale of 1 to 5, the risk, cost, sustainability, business value, and lead time and plot them on a spider chart to bring visibility and understand the decision's trade-offs.

Summary

This chapter centered on why many Agile development teams perceive governance as an anti-pattern and how to adjust governance to influence outcomes positively.

We have concentrated on architectural governance, which ensures that technology innovations are introduced to the business competently and sustainably by adequately aligning with the strategy and solution intent. The traditional ivory tower governance approach is disconnected, commands a control, out of context, and lacks communication.

We have examined Lean-Agile governance, which is incremental and iterative, and non-invasive in terms of the flow. In Lean-Agile governance, teams are empowered and autonomous with a high degree of collaboration for collective decisions. They exchange feedback continuously instead of waiting for board approvals. We then looked at the four principles of Lean-Agile governance: fostering a shared understanding; making decisions collaboratively; applying self-governance; and sharing progress and outcomes transparently. Lean-Agile governance is lightweight, improves productivity, increases quality, promotes IT investments wisely, develops team morale, focuses on business value, and fosters innovation.

The balancing of alignment and autonomy is critical in Lean-Agile governance, as they are mutually exclusive. We have reviewed Goldilocks governance as a middle ground and the Silverline path for the adoption of self-governance. Providing psychological safety for individuals and teams when making decisions is imperative. Lastly, we looked at an approach for documenting ADR and measuring the quality of architecture decisions against risk, cost, sustainability, business value, and lead time.

In summary, adopting the right levels of Lean-Agile governance can help accelerate value delivery with a continuous sustainable flow. In the next chapter, we will explore how to architect organizations for maximizing agility.

Further reading

- **Governance for Agile delivery**: `https://www.nao.org.uk/report/governance-for-Agile-delivery-4/`

- **State of the art of Agile governance**: *A systematic review*: `https://arxiv.org/pdf/1411.1922.pdf`

- **Software Architecture Decision-Making Practices and Challenges: An Industrial Case Study**: `https://arxiv.org/pdf/1610.09240.pdf`

Section 4: Personality Traits and Organizational Influence

In this section, we will review the personal transformations required for every architect. We will also examine the influence of teams and organizations on architects' operating models.

This section contains the following chapters:

- *Chapter 12, Architecting Organizational Agility*
- *Chapter 13, Culture and Leadership Traits*

12
Architecting Organizational Agility

We shape our buildings; thereafter they shape us.

- Winston Churchill

In the previous chapter, we covered the aspect of Agile architecture's governance and explored how architects work on alignment without authority. We have also discussed the critical artifact—**Architecture Decision Records** (**ADR**) for documenting architecture decisions. This chapter will examine the architect's critical role in architecting organizations for agility.

In our increasingly volatile world, business agility is the prime reason why many organizations embark on a rapid transformation journey. As a critical business enabler, IT supports business agility by streamlining its software delivery process. Embracing Agile software development at scale, fostering technical excellence, adopting DevOps and continuous delivery pipelines, and nurturing and practicing a Lean-Agile mindset are ways to enhance and make the software delivery process more nimble. However, reshaping an organization's structure to enable autonomy is inevitable in IT transformation. An effective delivery structure aligned to the continuous flow of value enables the autonomy of a team. For many organizations, this means breaking the traditional silos of capability walls. Agile architects play a significant role in redesigning the IT organization since they understand the holistic, big-picture view of software systems and their interdependencies.

In this chapter, we're going to cover the following main topics:

- Correlating business agility with IT agility
- Improving the flow of value in IT
- Organizing people around value
- Structuring teams within flow teams
- Implementing communities of practices
- Stepping into next-generation IT

This chapter focuses on the **Architecting Organization Agility** focal point of the **Agile Architect's Lens**:

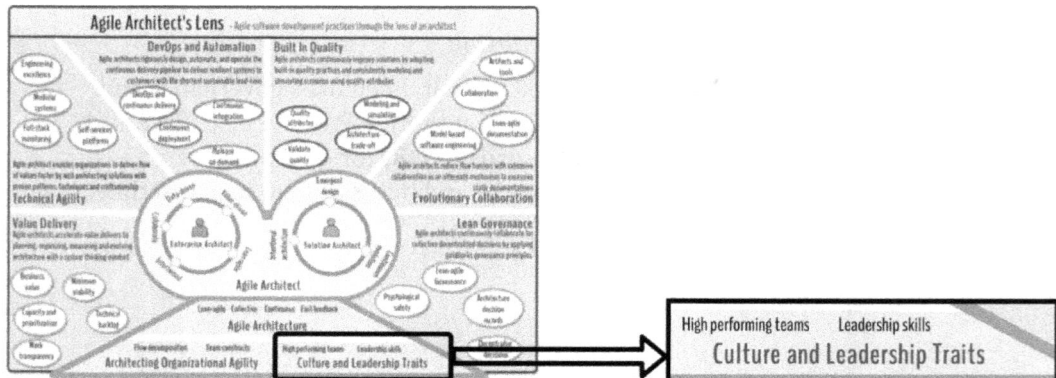

Figure 12.1 – Architecting Organization Agility – Focal point

This chapter will kick off by establishing the correlation between business agility and IT. We will then introduce the concept of the flow of value and examine opportunities for optimization. Then, we will go through the flow decomposition framework to understand how to design teams around the flow of value. Later in this chapter, we will also examine how to organize people within flow teams to function effectively and efficiently. We'll close this chapter by exploring the concept of next-generation IT.

Correlating business agility with IT agility

Agility is vital in today's increasingly **Volatile, Uncertain, Complex, and Ambiguous (VUCA)** world, where many organizations are challenged on their ability to react quickly and efficiently to contain damage and take advantage of new opportunities. As the world saw with the COVID-19 global pandemic, organizations cannot merely imagine, predict, or plan certain market vulnerabilities. By building agility, resilient organizations prepare for changes by staying nimble to adapt to changes in strategies with ease, to gain competitive advantage.

Business agility is the ability to sense market opportunities and respond rapidly through innovative solution delivery using short delivery cycles. This helps to increase revenue and customer satisfaction at pace with low cost of ownership. Organizations need to resize, reshape, and remodel their business for resuming business operations post-COVID-19. Many legacy organizations are embarking on a business transformation to bring business agility for their sustainability by sensing this urgency.

Organizations use agility at scale as the core transformation strategy to shape optimal operations, faster time to market, respond to planned and unplanned demands, and adapt to changes in priorities and strategies. Business agility is at its best when an organization can respond rapidly in a creative, adaptive, and resilient manner to cope with uncertain, complex, and ever-changing market dynamics. IT departments, as a critical enabler, need to be prepared to support the business rapidly, effectively, and efficiently to enable business agility.

The following diagram illustrates how business agility and IT agility are correlated:

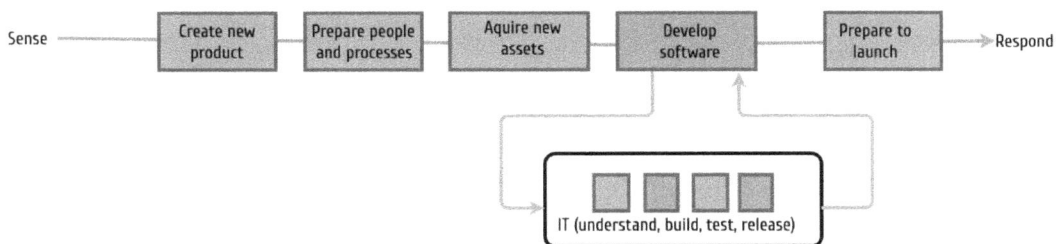

Figure 12.2 – Correlation between business and IT agility

Sensing market opportunities and responding to customers at the earliest opportunity with quality products and services suitable for their needs and wants is the primary strategic objective of any business. As shown in the diagram, most business enablers need software support. Developing software systems is critical for many business organizations to respond competently, and therefore most business transformations significantly impact IT departments and the ways of system developments.

Modern IT departments are prepared to adapt, lead, and respond to business demands continuously at a sustainable pace without compromising on quality. In such IT organizations, Agile software development practices play a significant role in incrementally delivering maximum possible business value with frequent release cycles.

The most critical aspect for any IT department is to optimize the flow of value or flow of work relentlessly, depicted as the **understand**, **build**, **test**, and **release** cycle in the preceding diagram. The next section will explore further the flow of value.

Understanding business and IT flow of value

As reviewed in *Chapter 2*, *Agile Architecture – The Foundation for Agile Delivery*, Lean manufacturing processes always try to minimize the lead time by optimizing the manufacturing flow. In Lean manufacturing, the flow kicks off when the customer places an order and delivers value when the customer receives the vehicle ordered. The value flows through several steps, which are candidates for optimization. **Value stream** and **value chain** are often the terms used for representing this flow of value. The **Scaled Agile Framework (SAFe)** calls this **value streams**.

The following diagram shows an example flow of value in the context of Snow in the Desert:

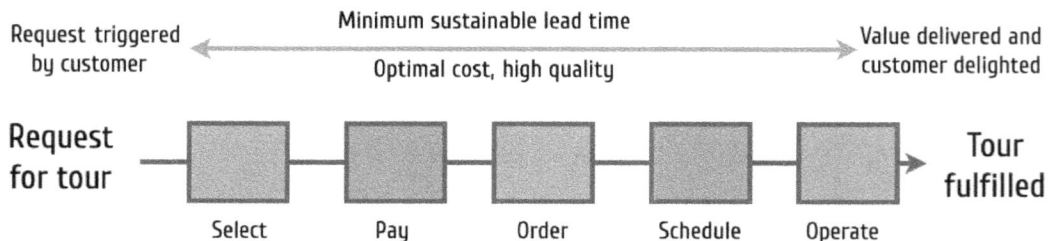

Figure 12.3 – An example flow of value in the context of Snow in the Desert

As shown in the diagram, the flow of value from a business perspective starts when a customer requests a tour through an online channel, a travel outlet, or a partner. The final value is delivered when the requested tour is fulfilled. The business's objective should be to optimize the flow steps so that customers receive value as early as necessary. Businesses do this by balancing sustainability, cost, and quality.

While the previous example was from a business perspective, IT as a business owns a certain flow of value, as shown in the following diagram:

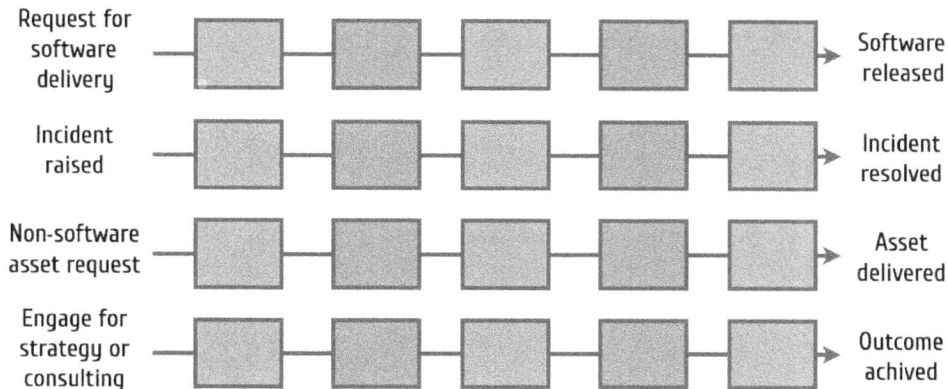

Figure 12.4 – Flow of value in IT as a business

The preceding diagram captures only an indicative list of IT flows in the context of Snow in the Desert. Different IT organizations may have other flows of value, such as cybersecurity services and so on. IT **flows of value** are defined as the value delivered by IT to its customers. The first flow represents the core IT software delivery flow, enabled with Agile software development practices having the *understand*, *build*, *test*, and *release* steps.

Other IT flows, such as incident management, asset delivery, strategy consulting, and so on, are non-software delivery flows. Even for such flows, IT adopts Lean-Agile principles and mindsets to ensure requests are served with minimal lead time, good quality, and optimal cost.

Several aspects need consideration for improving the flow of value. The next section explores more on flow improvements.

Improving the flow of value in IT

For non-software IT flows such as IT asset delivery, optimizing the flow involves reviewing steps in the flow and eliminating, optimizing, or automating necessary steps to increase delivery speed. The following diagram shows the IT asset delivery flow:

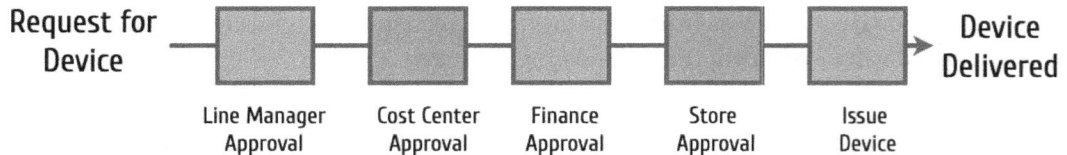

Figure 12.5 – IT asset delivery flow

As shown in the diagram, the flow commences when an employee requests a device. This request goes through several steps before the device is issued to the requester. At Snow in the Desert, line manager approval is based on whether the employee holds a device already, cost center approval is based on the budget left, finance approval is based on the monthly spend limit, and store approval is based on the right devices for the right purpose and their availability. A critical review of this reveals that all approvals except the store approval could be automated, and therefore the flow speed can be dramatically improved.

Software delivery flows can be accelerated by adopting Agile software delivery practices at scale, implementing DevOps and continuous delivery, building technical excellence, and fostering a culture of Lean-Agile mindset. The Lean-Agile practices eliminate waste in the flow steps by avoiding handoffs, implementing automation, and continuously collaborating. However, for the efficient adoption of Agile at scale, in addition to the four parameters, the organization's structure also needs to change. Previous chapters covered the first four aspects of flow improvements except the organizational structure.

The following diagram captures the five elements of optimizing the software delivery flow:

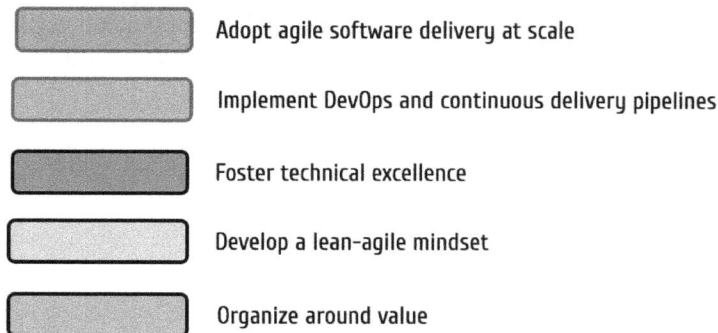

Figure 12.6 – Different aspects of IT software delivery flow improvements

Traditional IT organizations are often organized around lines of business or IT capabilities. These silos of team orientation lead to handoffs and wait times and add significant challenges to collaboration. The concept of organizing around value originated from Lean manufacturing. In Lean manufacturing, fixed capacity autonomous teams are allocated to various parts of the production line or flow of work. Work moves from one team to the other seamlessly. The rest of this chapter focuses on the Agile software delivery aspect and organizing around value.

Organizing people around value

In software, successful Agile teams are fully autonomous. Similar to manufacturing, teams need to be organized around the flow of value to achieve the required levels of autonomy. For large organizations, this restructuring is substantially complex. Frameworks such as **SAFe**, **Disciplined Agile** (**DA**), and **Large-Scale Scrum** (**LeSS**) suggest methods for restructuring teams aligned to flow.

The following diagram illustrates the concept of flow-centric teams:

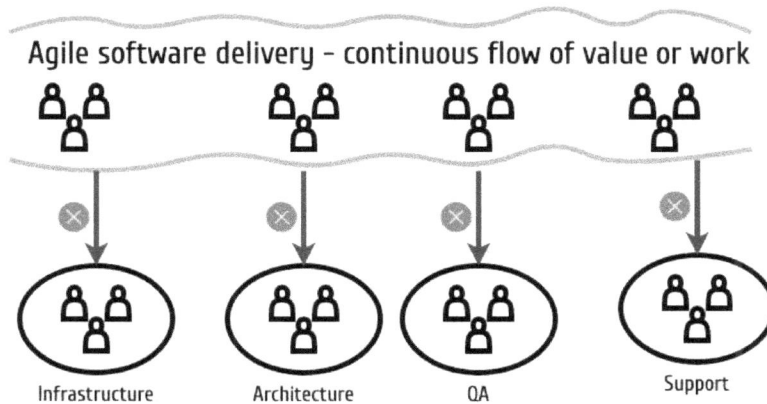

Figure 12.7 – Flow-centric autonomous teams avoid external dependencies

As shown in the diagram, the work continuously flows through a software delivery pipeline, where teams are organized around those flows. A fully autonomous team must be self-sufficient without needing to reach out to external teams.

Organizing around value is a way to improve flow by breaking the power barriers of traditional organizations. Therefore, aligning to the flow of work significantly impacts the organizational structure. The challenge for large enterprises is that adding all the required people to a single software delivery flow creates unmanageably massive teams. Therefore, a systematic method is required to organize people into manageable units without losing focus on delivery flow. The next section covers the solution to this problem in detail.

Examining the need for architecting the organization and teams

Software architecture and organization structure are inherently connected. A well-architected organization can produce a well-architected software system. *Melvin Conway* stated that *"an organization that designs a system would produce a design whose structure is a copy of its communication structure,"* known as **Conway's law**. In reality, even if it is not a copy, architectures are greatly influenced by the organizational structure. In the same lines, *Ruth Malan* paraphrased this as *"if the architecture of the system and the architecture of the organization are at odds, the architecture of the organization wins."* Since Conway's law points to a fact, such as the law of gravity, enterprises can only make it right by architecting organizations and team boundaries well.

In the IEEE article *The Architect's Role in Community Shepherding, Damian Tamburri, Rick Kazrnan,* and *Hamed Fahimi* observe that *"beyond architects designing software components and champion architecture qualities, they also have a responsibility in complicated organizational rewiring to avoid community smells."* The authors further explain community smells as architecture challenges due to disconnected and unorganized teams and subteams working in silos.

Snow in the Desert's legacy organization structure is a great example to prove Conway's law. The following diagram captures the essence of Snow in the Desert's pre-transformation state:

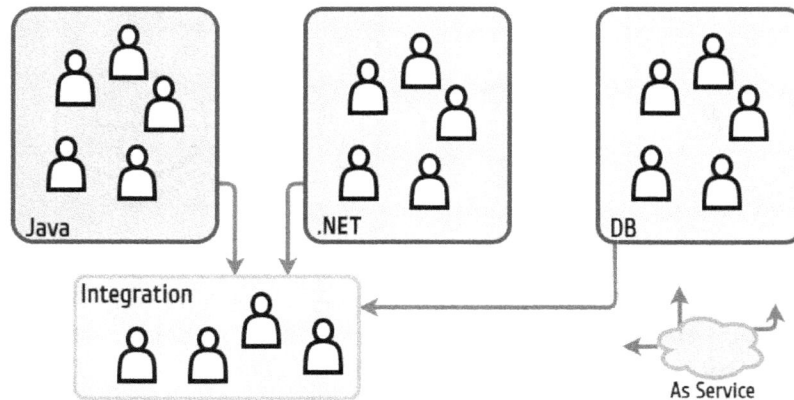

Figure 12.8 – Technology-centric organization – an illustration of Conway's law

The software engineers were organized based on the technology they are experienced with, such as Java, .NET, and databases. Others, such as architecture, testing, and so on, are organized by capabilities with loosely defined service levels. These service levels are defined based on gatekeeper rules such as *if requirements are signed off, artifacts will be handed off in N number of days*.

An example of community smell, in this case, is that each technology group uses its own solutions for integration, monitoring, and so on. Integration between these groups goes through an integration team. The integration team uses an **Enterprise Service Bus** (**ESB**) for all integrations. Therefore, the architecture principle of integration automatically turns out as cross-technology integration.

In line with *Ruth Malan's* observation, modern organizations use a *reverse Conway maneuver* or *inverse Conway maneuver* for designing organization structure. To succeed in this, enterprises need a desired architecture blueprint to drive organizational changes. Architects have a critical role in designing such blueprints as they are well placed to understand how systems are interconnected and their re-architecting possibilities.

The following diagram shows flow-aligned teams – **flow teams** – designed with the right boundaries, which are autonomous with limited dependencies on the outside:

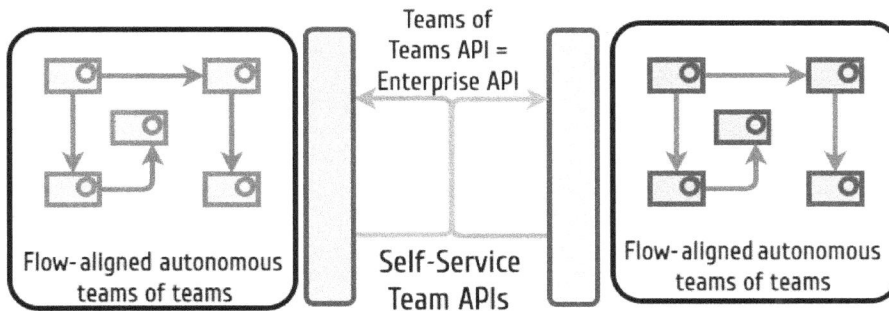

Figure 12.9 – Flow-aligned autonomous teams, inverse Conway's law

Once the structure is created, teams use self-service APIs for team-to-team interactions using a teams API. These team APIs eventually evolve into enterprise APIs. Autonomous teams can deliver value faster since external work dependencies are limited and therefore require minimal co-ordination efforts and wait time. If there are dependencies, they are well managed through clear inter-team boundaries.

The subsequent sections cover details on designing team boundaries aligning to the flow.

Identifying flow and organizing teams with flow decomposition

Flow decomposition is a method for designing flow-aligned teams or simply flow teams in large organizations. Flow decomposition starts with identifying different services offered by IT, as shown in *Figure 12.4*. The following diagram replicates the IT flows with the resource count against each flow in the context of Snow in the Desert:

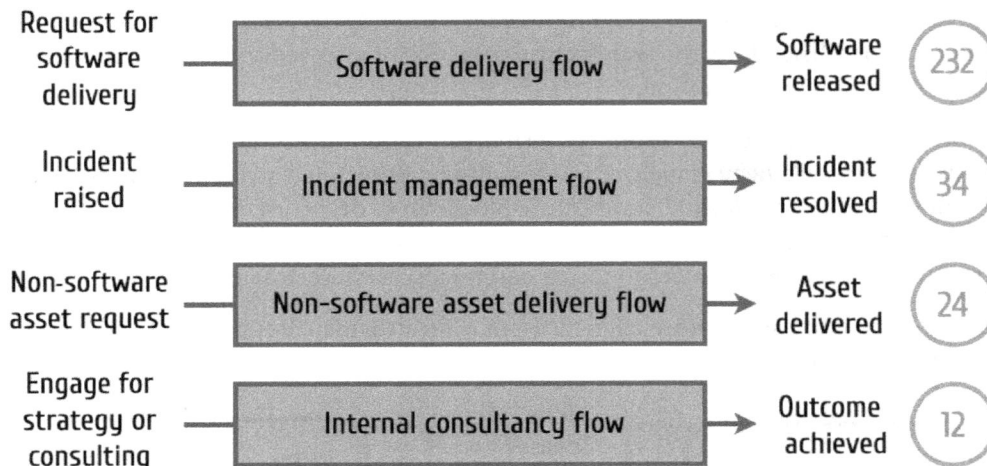

Request for software delivery	→	Software delivery flow	→	Software released	232
Incident raised	→	Incident management flow	→	Incident resolved	34
Non-software asset request	→	Non-software asset delivery flow	→	Asset delivered	24
Engage for strategy or consulting	→	Internal consultancy flow	→	Outcome achieved	12

Figure 12.10 – IT flows at Snow in the Desert with staff density

While the last three flows have a smaller number of people associated with them, most developers are allocated within the **software delivery flow**, which makes it too big a team to manage.

There is no perfect answer for the right number of people in a flow-aligned team. The anthropologist and evolutionary psychologist *Dunbar* proposed that an individual can maintain stable relationships with up to 150 people. In practice, any large team above 100 is difficult to manage. SAFe recommends having *50–125* as an effective team size. In the flow decomposition method, limit the resource count to 100. Since the **software delivery flow** has more than 100 resources, flow decomposition needs to be applied.

The flow of value mimics the business steps involved in delivering value. In an ideal state, businesses will also re-organize themselves against the flow of value. Often, in large enterprises, businesses are still structured around capabilities. Even in such cases, IT needs to re-organize based on flows, not against the lines of business.

Flow decomposition uses a five-step decomposition process, as shown in the following diagram:

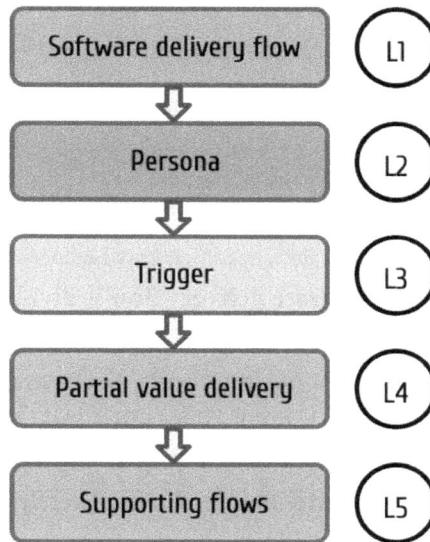

Figure 12.11 – Flow decomposition techniques

As shown in the diagram, the five levels of decomposition are marked as *L1 to L5*:

1. Our first step is to identify the software delivery flow within IT, which is already captured in *Figure 12.10*.

2. The next step of flow decomposition is to decompose based on **personas**. In the case of Snow in the Desert, the software delivery flow can be further decomposed into three other flows.

The following diagram illustrates this scenario:

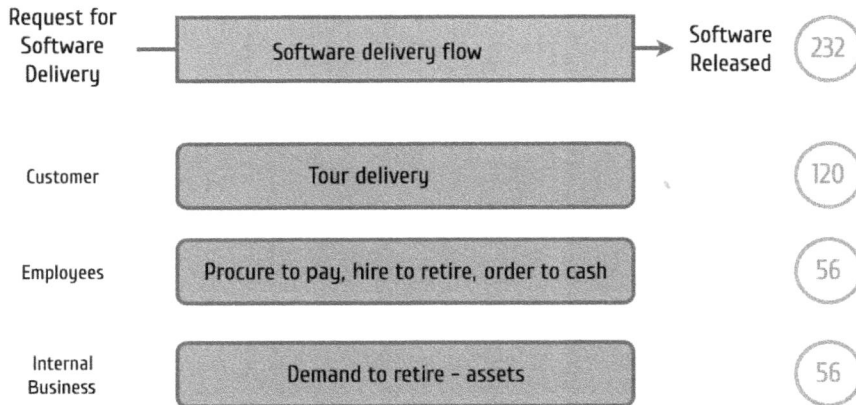

Figure 12.12 – Flow decomposition based on personas

The diagram shows that the **software delivery flow** is divided into three distinct flows based on three key personas – **Customer**, **Employees**, and **Internal Business**.

A flow could span across multiple business units. For example, the **Tour delivery** flow connects two business units – *Front Office* and *Field Operations*.

Like many other large enterprises, customer flow at Snow in the Desert still has the largest allocation of people, which exceeds 100. Therefore, it needs further decomposition.

3. Next is to look for **triggers** in the customer flow. The following diagram illustrates this with an example:

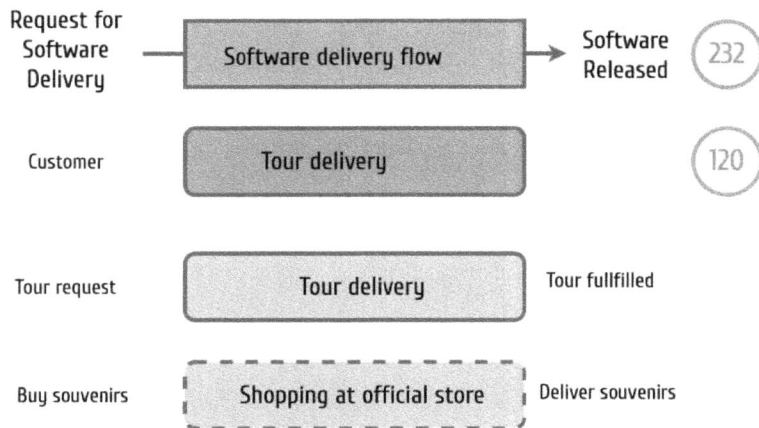

Figure 12.13 – Flow decomposition based on triggers

As shown in the preceding diagram, **Shopping at official store** is a fictitious scenario to explain the concept of decomposition based on triggers. Since all customer interactions are based on a single trigger at Snow in the Desert, this step is skipped. In the banking context, customers interacting for a loan service, account service, forex service, and so on are different triggers.

This approach may still lead to a higher number of people in each of the flows.

4. In such cases, the next step is to break down the flow into **intermediate bits of value**, as shown in the following diagram:

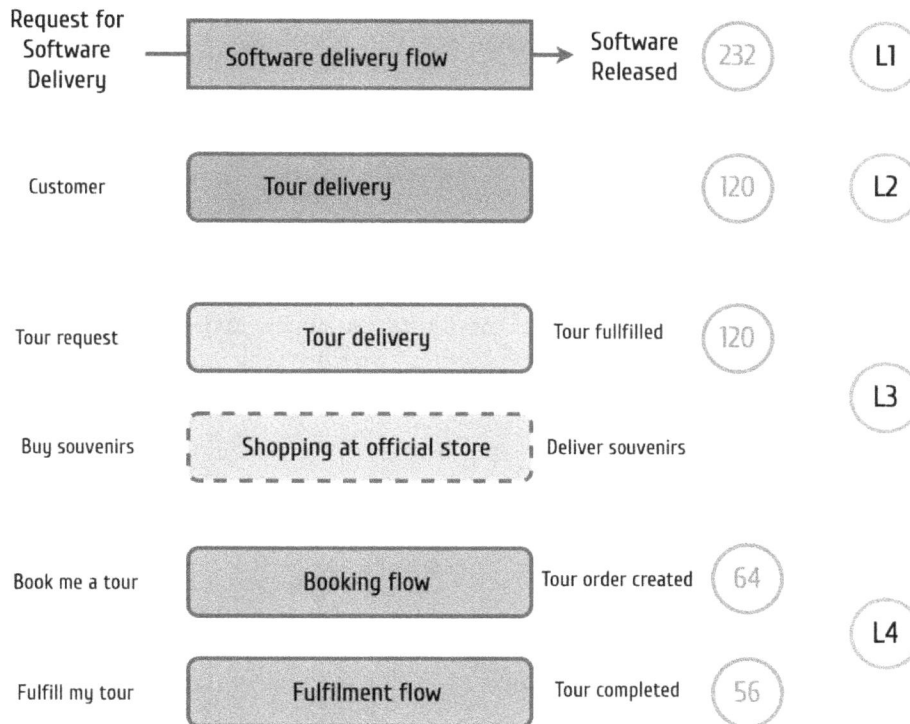

Figure 12.14 – Flow decomposition based on partial value delivery

As shown in the diagram, the **Tour delivery** flow is divided into **Booking flow** and **Fulfilment flow** as these two are often loosely connected and usually happen at different times. At the end of **Booking flow**, customers receive confirmation of their booking, which is considered a partial value.

5. If the team sizes are beyond 100, then use **supporting flows** to break down such
    flows. However, this needs extreme care as the coordination effort at that level could
    substantially hinder the flow speed. Since at Snow in the Desert all flows are under
    100, no further breakdown is required. There may be additional internal flows
    supporting business flows, such as **Product planning flow** to support **Booking
    Flow**, as shown in the following diagram:

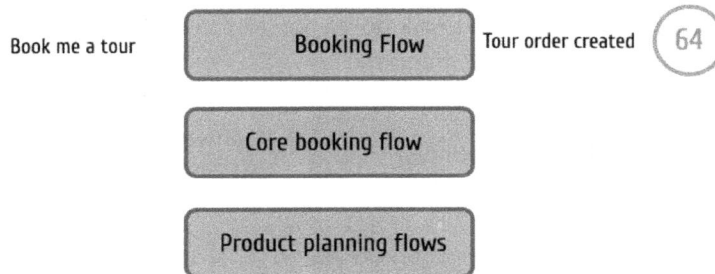

Figure 12.15 – Flow decomposition based on sub-flows

There may be an increase in the **coordination coefficient** at every level in this
decomposition process. The coordination coefficient is the additional effort and
complexity needed for coordinating two or more sub-flows using dedicated or virtual
resources. In this case, **Booking flow** and **Fulfilment flow** are very tightly coordinated.
The following diagram shows the relationship between the coordination coefficient and
the levels of decomposition:

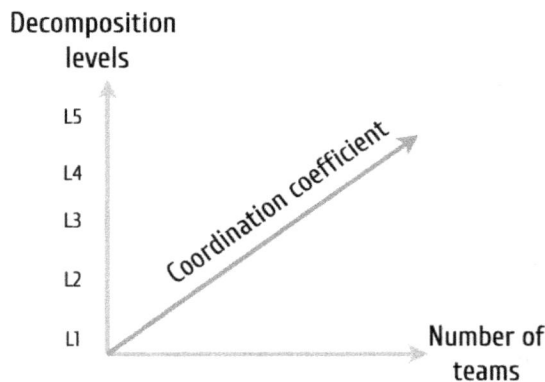

Figure 12.16 – Relationship between the coordination coefficient and levels of decomposition

As shown in the diagram, L5 needs significantly tighter coordination, which may slow
down delivery. So far, in the context of Snow in the Desert, there are four distinct flows
identified, each with less than 100 people. The next step is to assign systems to these flows
to build autonomous **flow teams**.

Assigning systems to flows

The next step in organizing teams around the flow of value is mapping systems to flows. This can be done by understanding which systems are needed to support the flow. The following diagram captures flows identified in the context of Snow in the Desert and how systems are mapped to those flows:

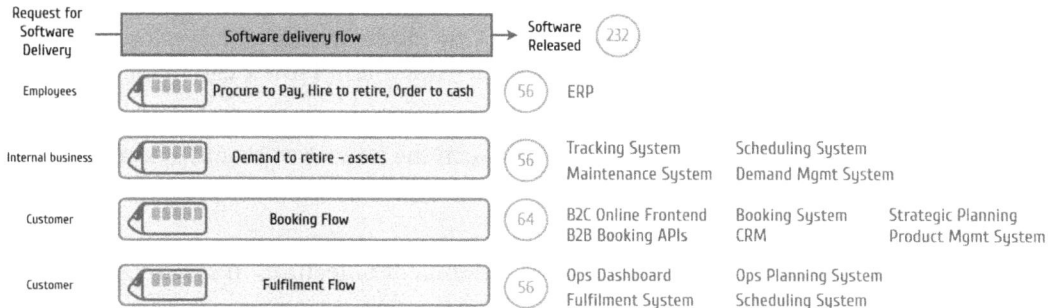

Figure 12.17 – Flow teams mapped to systems at Snow in the Desert

There are many more systems in Snow in the Desert, but allocating critical systems gives a good understanding of teams' autonomy. Most of the other systems are often satellite systems associated with these larger systems. In this process, interestingly, organizations may end up with a few systems left unallocated. A critical review is required to understand why those systems are left unallocated. Are they not used? Are there any flows missing? Or are they used across multiple flows?

One of the challenges that enterprises may face is that the same system is required across multiple flows. These are candidates for potential rearchitecting.

The following diagram captures some of the possible solutions:

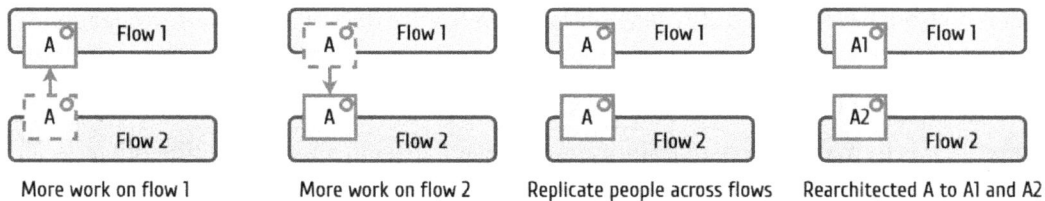

Figure 12.18 – System shared across multiple flow teams

As shown in the diagram, there are four possible solutions:

- The first step is to identify work dependency against flows. For example, in most cases, features in **Flow 1** change system **A** more than features in **Flow 2**. In this case, system ownership is better aligned to **Flow 1**.

- In the reverse scenario, system **A** ownership goes to **Flow 2**. If work density cannot be clearly determined, then system **A** ownership is spread across **Flow 1** and **Flow 2** where resources will be duplicated. However, this approach is not recommended unless the system fully supports distributed development. In such situations, ownership for continuous improvement is a challenge unless one of the flow teams takes the system's ultimate authority.

- An alternate option is to rearchitect system **A** by modularizing and developing microservices around flow boundaries so that **Flow 1** and **Flow 2** can manage their respective microservices.

- It is incredibly important to distinguish between the flow of work and system dependencies when determining system ownership. For example, system **A** and system **B** may be highly dependent.

- However, most of the time, to fulfill the development of features in system **A**, there is no update required in system **B**. In a work dependency scenario, even though the integration between **A** and **B** is less, changes in **A** mostly need changes in **B** to fulfill the end-to-end business needs. In such cases, it is apt to collocate both systems **A** and **B** under the same flow. Higher degrees of work dependency leads to delivery slowness as a result of overhead in coordination.

System-to-flow mapping is performed without constraints, assuming in the long run that teams are re-architecting those dependent systems to achieve loose coupling between flow teams. The enterprise has to be mindful that, in the near term, it may add additional coordination efforts.

A point to note is to avoid creating platform and technology flow teams. A platform team crossing more than 100 developers itself is an issue that needs to be investigated. Creating a platform flow team creates additional handoff points, which leads to waiting times in delivery. It also develops unhealthy tension between feature delivery and platform teams.

Giving autonomy to flow teams

The next step is to make flow teams autonomous by identifying and embedding other resources required to fulfill the flow team's needs. This includes architects, testing resources, infrastructure teams, platform teams, supplier team members, subject matter experts, PMOs, coaches, and so on. The intention must be to cut down all external work dependencies. However, there may be cases where specialist scarce resources, such as data scientists, information architects, and so on, cannot be spread across all flow teams. In such cases, they will be accessed as a service.

The following diagram shows the flow team and types of interactions with other teams:

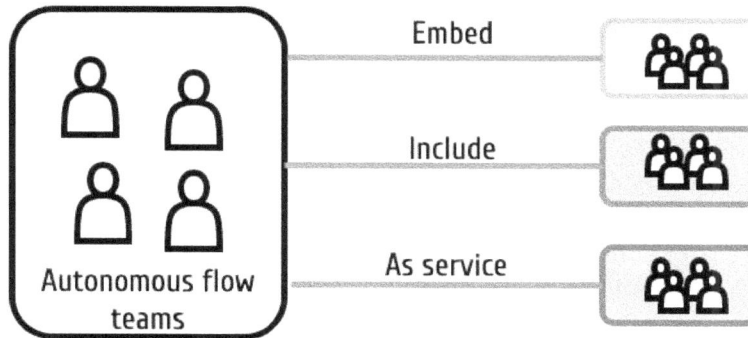

Figure 12.19 – Autonomous teams external team engagement model

As shown in the diagram, flow teams embed specialists into the team, such as **Site Reliability Engineers** (**SREs**), cybersecurity members, architects, and so on. Other specialists, such as data scientists, are included for a particular duration, such as a sprint – they come as dedicated resources, perform, and go back to their homes. If both options are not possible, such resources are accessed as services. Backlog items are pushed to other teams' backlogs when they are engaged as a service to ensure plans are integrated, visible, and committed. Often, many flow teams include a people person to look after and facilitate the team's needs.

SAFe calls flow teams **Agile Release Trains** (**ARTs**). Essentially, flow teams are teams of teams. If the flow teams are L3, L4, and L5 levels, a coordination layer needs to be considered with dedicated or virtual members. In SAFe, this coordination layer is termed **large solution trains**. The following diagram shows the interflow team coordination using virtual team members:

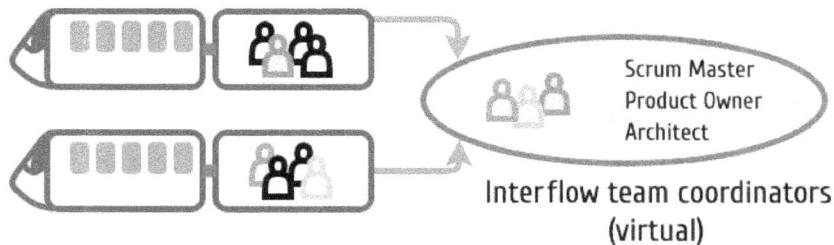

Figure 12.20 – Virtual inter-team coordination of flow teams

In the preceding diagram, coordination teams are virtual, with members coming from different flow teams. Flow teams are long-living teams committed and locked in for a fixed period. They may undergo optimization over a period of time. These teams are generally based on a fixed capacity, determined based on budget guardrails. Once established, subsequently, work continuously moves to teams. Flow teams work as a highly cohesive team with a shared goal to deliver value with a common cadence.

The following diagram shows the structure of IT as a mix of flow teams and non-flow teams:

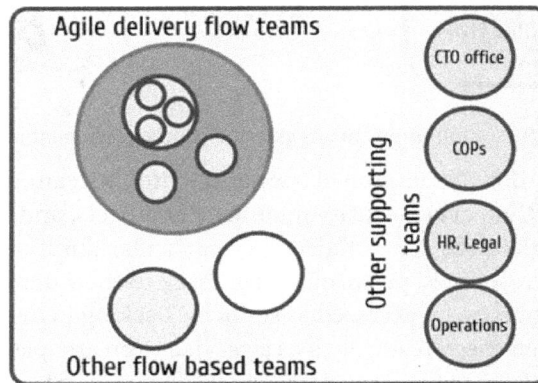

Figure 12.21 – Non-hierarchical IT teams

As shown in the diagram, there are several flow teams at multiple levels with almost no hierarchy, including Agile software delivery and other IT flow teams. Apart from flow teams, there could be CTO office, operations, HR, legal, and other flow support teams, such as the Agile Center of Excellence, cloud support teams, cybersecurity, and so on.

Comparing capability-centric versus flow-centric teams

The flow-based team approach fundamentally breaks the traditional siloed capability-based approach. This change right away impacts enterprise architects as they lose the capability view and are no longer in a great position to work on enterprise capability maturity. Without capability system maps, it is hard for enterprise architects to identify synergy, reusability, and rationalization opportunities. What if the same capability is implemented by two flow teams using different technologies or solutions?

The following diagram shows the capability-to-flow mapping, a useful mechanism for enterprise architects to bring back the capability model:

Capability 1 Capability 2 Capability 3

Flow 1

Flow 2

Flow 3

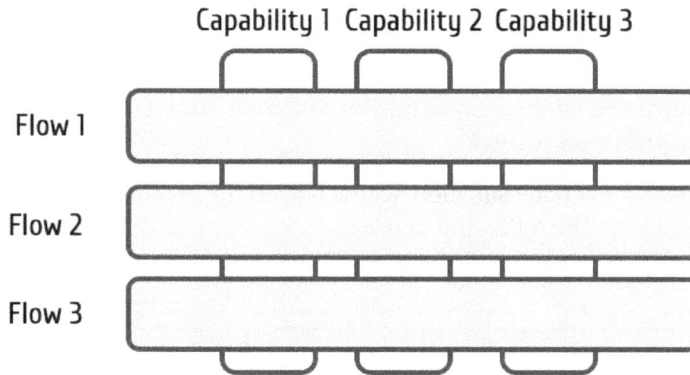

Figure 12.22 – Flow-to-capability mapping for balancing speed and sustainability

While flow teams relentlessly try to improve flow speed, the capability view gives enterprise architects the opportunity to operate on software systems' sustainability. Hence, the flow-to-capability mapping is very important for enterprises in decision making.

Validating the organization of teams

Once flow teams are designed, they need thorough validation before operationalizing. The following diagram captures the flow team validation approach:

Initiative/Work	ART 1	ART 2	ART 3	ART 4
Work item 1	X			
Work item 2			X	
Work item 3				X
Work item 4		X	X	
Work item 5	X			

Figure 12.23 – Flow team validation tool

As shown in the preceding diagram, the flow teams are mapped against several recently delivered initiatives or features. The objective is to mark which flow teams deliver those features in the new flow team construct. If the construct is perfect, a single work item is touched by no more than one flow team.

Structuring teams within flow teams

Flow teams are autonomous and are as big as 100 people. To operate effectively and efficiently, flow teams need an internal structure. Different strategies can be used for structuring teams within flow teams.

The primary principle is to create sub-flow teams within flow teams, but this time the flow is at the feature level. The following diagram captures three different possibilities of feature-based team structuring:

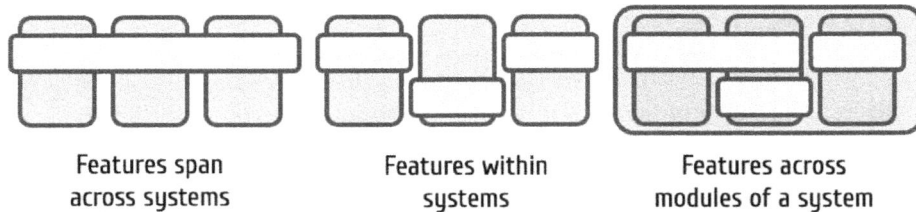

Features span   Features within   Features across
across systems   systems   modules of a system

Figure 12.24 – Compositions of feature-based teams

As shown in the diagram, a feature can span across multiple systems, be within a system, or span across modules of a system. In this context, a feature is a set of continuously evolving requirements. For example, in a tour booking scenario, payment, search, and booking are long-lasting and evolving features. A stable, fixed-capacity team is required to develop and maintain those features. A flow team may have a mix of teams such as feature teams spanning across systems, a feature team per system, or a feature team across modules of a system.

Apart from the feature team, there may be a **system team**, **enabler team**, and **facilitation team**, as shown in the following diagram:

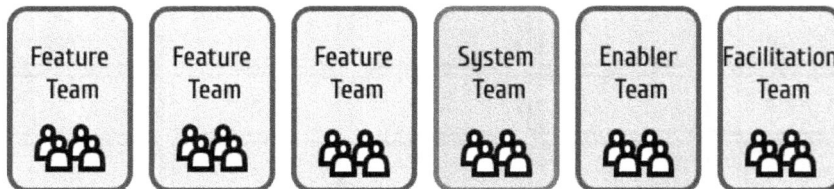

Figure 12.25 – Different types of teams

The four types of team constructs are explained as follows:

- **Feature team**: These are engineering teams developing business features, following one of the models explained earlier – feature teams across **systems**, within a system, or across modules of a system.

- **System team**: Responsible for continuous integration, deployment, and release management. They are also responsible for systems monitoring, production-readiness checks, performance, vulnerability assessments, and other system-related activities. They also look for proactive system improvements, such as capacity adjustments and so on. An outside infrastructure team can almost be eliminated by adopting self-service platforms such as the cloud.

- **Enabler team**: Enabler teams are for the quick enablement of feature teams. They undertake complex technical challenges such as the first-time implementation of **Continuous Integration and Continuous Delivery (CI/CD)** pipelines, the deployment of software such as a Kubernetes cluster, and so on. The main motive is to ensure that feature teams are not losing focus. Enabler teams are transient; they use the **build-operate-transfer (BOT)** approach.

- **Facilitation team**: This includes the PMO, finance, the people manager, supplier management, and other roles that are not related to engineering but are very much required for the flow team's success. They offload unwanted management and operational activities from engineering and system teams' roles. Facilitation teams are non-technical teams and therefore are not represented in many of the scaling Agile frameworks.

Following the two-pizza team principle coined by Amazon, every team needs to be constrained between *7 and 15* people to reduce communication overhead. A simple approach for team formation is through self-selection workshops documented in the book *Creating Great Teams: How Self-Selection Lets People Excel*. Examples of self-organizing principles are as follows:

- Do what is best for the company.

- Teams need to be made up of eight or nine people as an ideal state with a broad boundary of 7 to 15.

- Each team should have at least one person from each of the functional groups.

- At the end of the selection process, look at team composition and skills and make any necessary adjustments.

- After many rounds, fine-tuning happens with a single principle – what is best for the company?

Use **cognitive load** to determine how many people are required for a system, irrespective of the current workload. *John Sweller* defines cognitive load as the "*total amount of mental effort being used in the working memory.*"

Flow teams should avoid organizing teams around technical system infrastructures, such as an architectural layer, programming language, middleware, user interface, and so on, as this creates many dependencies, impedes the flow of new features, and leads to frAgile designs.

Also, avoid creating platform teams unless it is critically necessary. Platform teams induce handoffs, delays, and tension between feature teams and platform teams. The following diagram captures scenarios for organizing platform teams:

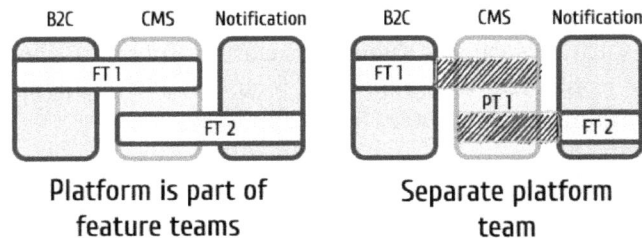

Figure 12.26 – Platform team composition

As shown in the diagram, platform teams can be part of the feature team or a dedicated team. **FT 1** and **FT 2** are feature teams that own respective parts of the CMS platform, as shown on the preceding diagram's left side. On the right side, the CMS is owned by a separate dedicated team, **PT 1**, which takes work from both **FT 1** and **FT 2**. The choice is mainly dependent on the platform's technical capability and the skills required by the developers. The recommendation is always to go with the former option.

The following diagram shows a full flow team:

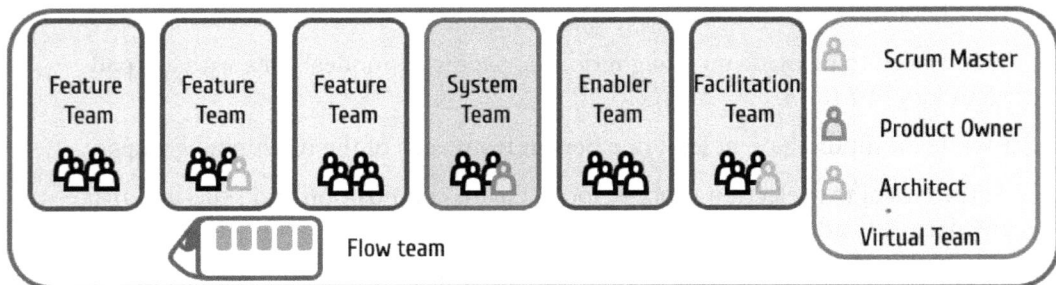

Figure 12.27 – Putting it all together, a flow-based team

Different flow teams may have different team topologies based on what is the best fit for them. In general, flow teams consist of feature teams, system teams, enabler teams, and facilitation teams, and a virtual leadership team consists of a Scrum Master, a Product Owner, and an architect. In SAFe, this is called **ART leadership**.

Engineering, architecture, and other capabilities are now fragmented across multiple flow teams. While this is best suited for autonomous delivery teams, the biggest challenge with this approach is fostering these resources' capabilities. We will examine the role of communities of practice in the next section.

Implementing communities of practice

Communities of Practice (**CoPs**) are cross-cutting learning and development groups responsible for upskilling and cross-skilling resources. For example, an architecture CoP may educate and grow its members to advance cloud, microservices, IoT, Agile architecture, and other newer technologies.

The following diagram shows the formation of a CoP:

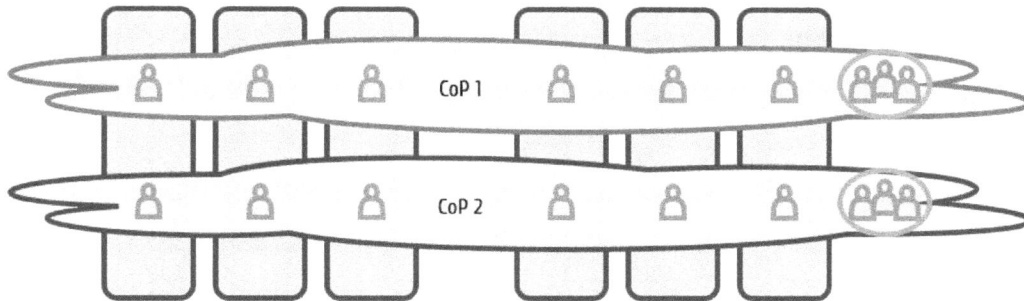

Figure 12.28 – CoPs

The vertical blocks in the preceding diagram indicate the flow teams. CoPs may have dedicated members and are responsible for coordinating, facilitating, documenting, and sharing cross-flow team learnings. In particular to architecture, these practice teams are responsible for building architecture patterns, principles, guidelines, reference models, and so on together with flow team members.

Stepping into next-generation IT

Flow teams are excellent vehicles to deliver business value faster without compromising on sustainability and quality. They work even better for achieving business agility when business organizations are also transformed as a flow-based organization by breaking the traditional business unit silos. This helps in optimizing the steps in the flow and delivering software faster together with a fully aligned community of business owners. In such cases, the business owners may become true product owners.

If the business is organized based on flows, flow-based IT teams can be better collocated and integrated into the business teams. Embedding teams with business brings higher motivation levels and a sense of purpose as both teams work on a shared vision and context. The CoPs connect these teams together for community-based growth. IT operations, other IT value streams, the CTO office, cybersecurity, and other essentials continue to stay within IT boundaries.

The following diagram captures the integrated business and IT model:

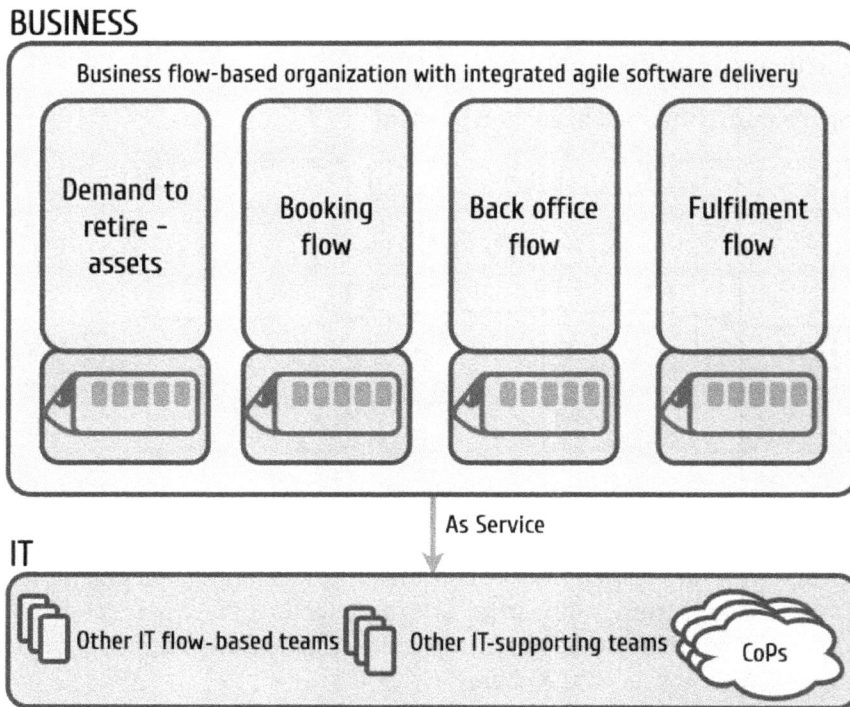

Figure 12.29 – Integrated IT and business model

As shown in the diagram, at Snow in the Desert, IT flow teams are fully embedded within the respective business teams to maximize cohesion and, therefore, deliver better and faster value.

Summary

This chapter focused on the importance of architecting organizations and the architect's role in creating a sustainable team structure fine-tuned for delivery.

In this chapter, we have learned about business agility and the correlation between business and IT. As an enabler, IT needs to be prepared to be nimble to support the business as new market signals emerge. There are five parameters important for enabling operational agility in IT – adopting Agile software delivery at scale, technical excellence, embracing DevOps and continuous delivery, a Lean-Agile mindset, and organizing around value. We have also examined the options of optimizing a flow by relentlessly eliminating waste, such as unnecessary manual approvals. Identifying the flow of value within IT and understanding the distribution of resources is essential for organizing people.

This chapter also introduced the concept of flow decomposition, an effective approach for creating manageable autonomous flow teams by limiting team size to around 100. The five flow decomposition levels are based on the software delivery flow, personas, triggers, partial value, and support flows.

We then explored how to organize empowered flow teams. There were four different team types that we identified and discussed – feature team, system team, enabler team, and facilitation team. We also looked at some of the challenges with flow teams, such as enterprise architects' challenges around capability, sharing systems between flow teams, creating platform teams, and growing individuals in a federated team environment. Finally, we closed the chapter by introducing the next-generation IT organization structure by embedding flow-based teams of IT and business together, making for a Leaner IT organization.

The next chapter explores architects' culture and mindset aspects, to succeed in Agile software development.

Further reading

- **Team topologies**: `https://itrevolution.com/wp-content/uploads/2020/01/Team-Topologies-at-Parts-Unlimited-The-Unicorn-Project.pdf`

13
Culture and Leadership Traits

"The secret of change is to focus all of your energy, not on fighting the old, but on building the new."

– Socrates

The previous chapter explored the critical role of architects in digital transformation by architecting organization for agility.

Beyond the usual ways of working and technical excellence, in Agile software development, culture, behavior, and mindset play a significant role. The organization and its leadership influence individuals and the team's culture. Adopting new leadership approaches and styles, investing in people and competency, communicating with clarity, and developing a psychologically safe environment are positive behaviors of a high-performing organization. High-performing teams exhibit high cohesion, high moral values, and bring more revenue to the organization in the long run. We will examine this by studying the behavior, mindset, and characteristics of a high-performing organization and teams.

Architects in Agile architects need a robust collaborative culture underpinned by personal qualities such as trust, transparency, and a positive growth mindset. They also require additional interpersonal skills such as conflict resolution, that naturally arise when high cognitive individuals collaborate. Providing the right environment and consistent motivation helps developing individuals' personal and interpersonal behaviors. In this chapter, we will also focus on personal qualities such as empathy, respectfulness, curiosity, care, and so on. We will also study collaboration, conflict resolution, developing cognitive intelligence, motivation, feedback, and other traits required for Agile architects beyond the conventional competencies such as influence, negotiation, and so on.

In this chapter, we're going to cover the following main topics:

- Understanding the need to change
- Examining culture in high-performance organizations
- Choosing the right leadership
- Personality traits of an architect
- Interpersonal traits of an architect

This chapter focuses on the **Culture and Leadership** focal point of **The Agile Architect's Lens**, as shown in the following screenshot:

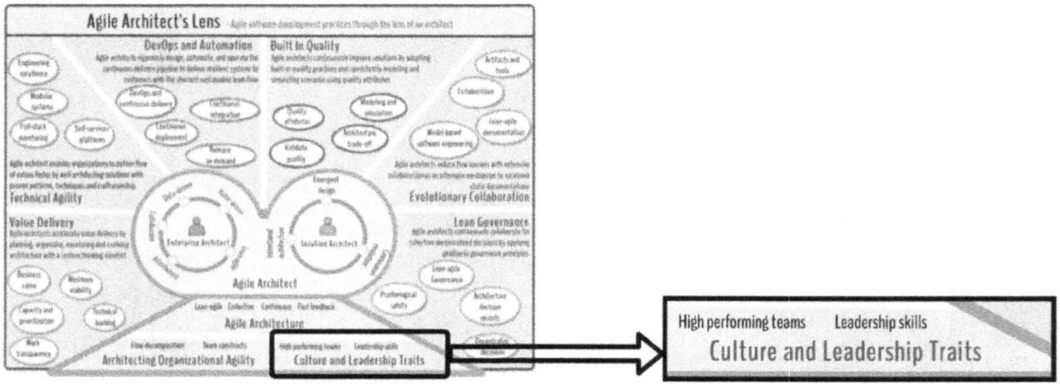

Figure 13.1 – Culture and leadership – focal point

This chapter begins by helping you understand why Agile architects must acquire new competencies in Agile software development. We will then examine the characteristics of high-performing organizations and teams and review the right set of approaches and leadership styles for Agile architects. Later, we'll discuss the top 10 personal qualities required for Agile architects within the relevant context. Finally, we'll look at the interpersonal qualities of Agile architects by sharing useful tools and models.

Understanding the need for change

One of the Agile software manifesto principles centered around people is *"Build projects around motivated individuals. Give them the environment and support they need, and trust them to get the job done"*. As we discussed in *Chapter 3, Agile Architect – Linchpin to Success,* a construction architect's journey to becoming a gardener and interior designer is beyond merely changing their ways of working. There needs to be an entirely radical transformation of culture, mindset, and social behavior.

The following diagram captures the change in the operating environment for architects in Agile software delivery:

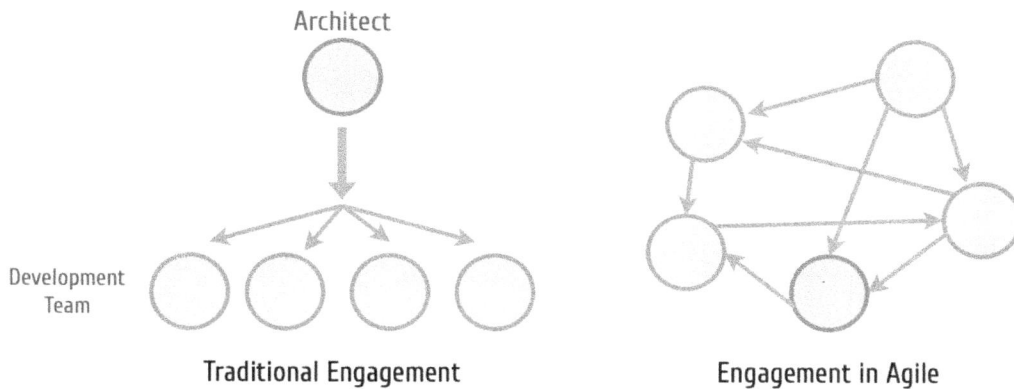

Figure 13.2 – Architect's engagement in traditional versus Agile environments

As shown in the preceding diagram, traditional engagement is top-down, with almost no feedback mechanisms, whereas in Agile software development, teams interact closely without hierarchies. A fully engaged evolutionary collaboration with higher degrees of people interacting and engaging determines an Agile software development's success. An Agile architects' operating principles are dramatically different, such as continuous engagement, influence without authority, being part of the team, and relentlessly nurturing and growing individuals in the team. To succeed, Agile architects must acquire additional personal and interpersonal skills besides the architect's legacy competencies, such as influence and negotiation, business thinking, analytics and problem solving, and stakeholder management.

The following diagram shows the cultural and behavioral ladder for Agile architects:

Figure 13.3 – Culture and behavior ladder

In Agile software development, architects are part of an autonomous, empowered team operating with power equality. *The Global Study of Technical Engagement Report* observes that *being on a team increases engagement level by 2.3 times.*

The culture shift needs to disseminate from the structure and behavior of the organization to its teams and employees. High-performing organizations and teams are the foundation of organizational agility. We will examine this in the next section.

Examining culture in high-performing organizations

The metaphor of a large cruise ship is often associated with organizational agility. However, high-performing organizations striving to achieve business agility require much more than adopting the Agile software delivery model to successfully sail through the storm of uncertainties. A new engine in an old ship will not give the flexibility to turn the ship swiftly.

Larman's Law of Organizational Behavior observes that, in large enterprises, culture, behavior, and mindset are highly influenced by the organization's structure. To be nimble, organizations must undergo a surgical correction on their structure, leadership, and culture. In a high-performing organization, the behavior and culture of its leadership determine its transformation success. Leaders have a duty of care for their people, their health, and their wellbeing. When people have been taken care of, the mindset shifts, and the organization's agility improves dramatically.

However, the culture shift – a combination of mind shift, skills shift, and behavior shift – is one of the most challenging and beneficial parts of an organization's transformation. The *14th State of Agile Report* shows that organizational culture is one of the most significant challenge when adopting Agile software development. The same survey also observes that improved team morale is one of the benefits realized by organizations when adopting the Agile delivery model. John Kotter pushes culture shift as the last step of Agile transformation in his eight-step *Leading Change* model, observing that *people will be reluctant to change unless they see the value of the change.*

Based on the work done by Ed Schein, Frederick Laloux, Vlatka Hlupic, Peter Senge, and Alan Furlong, the *Agile consortium* defined Agile culture as *creating an environment that is underpinned by values, behaviors, and practices that enable organizations, teams, and individuals to be more adaptive, flexible, innovative, and resilient when dealing with complexity, uncertainty, and change.* High-performing learning organizations take care of individuals by providing an exquisite, safe, and enjoyable space for individuals to grow through experimentation and lifelong learning.

The following diagram compares a traditional organization's culture with a modern organization's:

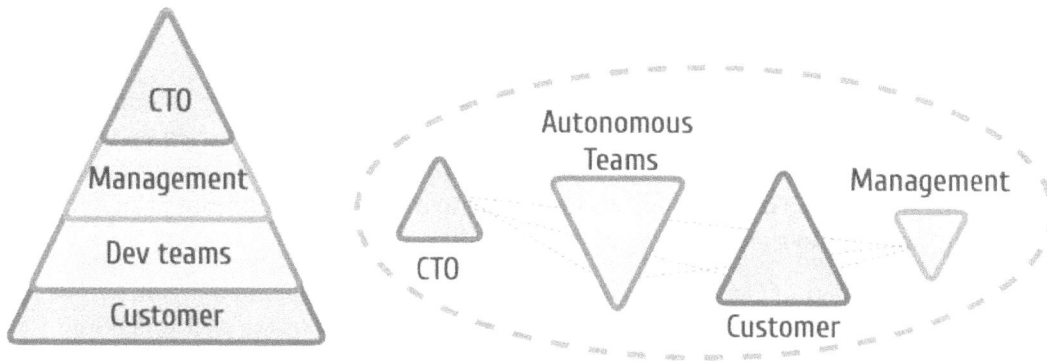

Figure 13.4 – Flattening an organization's hierarchy

Traditional organizations are rooted in hierarchies of egos, command and control, and bureaucratic decision-making, as illustrated on the left-hand side of the preceding diagram. Information flows top-down through many layers.

Modern organizations invert the triangle and develop flattened network models with a lean management layer that's interacted with through highly interactive social connections, as depicted on the right-hand side of the preceding diagram.

The culture and leadership behaviors of modern organizations are a class apart from traditional organizations. The following diagram compares different organizations based on **Agility** and **People orientation**:

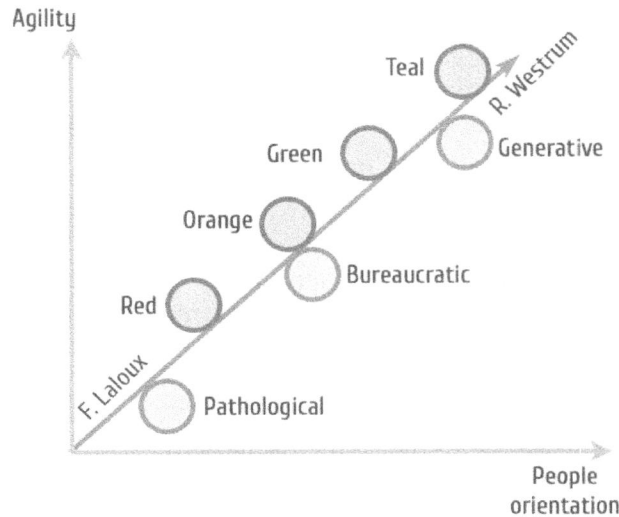

Figure 13.5 – Organizations with different types of cultures

The **Teal** organization in the preceding diagram was introduced by Frederic Laloux in his book *Reinventing Organizations* and observes that **Teal** organizations are based on values and principles, are purpose-driven, evolve evolutionarily, have shared responsibilities, and are self-managed. While **Green** organizations adhere to a few Agile manifesto principles, **Orange** and **Red** are too rigid and are not people-centric.

In the paper *A Topology of Organizational Culture*, Ron Westrum classified organizations based on culture and values and stated that **generative** organizations are well suited for agility over **pathological** and **bureaucratic** ones. A generative organization is highly focused on the mission, proactive information flow, alignment, awareness, empowerment, and learning from root causes. Netflix, a generative organization, defines their five organizational values as high performance, freedom and responsibility, context over control, and being highly aligned but loosely coupled.

Schneider's Cultural Model is a useful tool kit for understanding the organization's culture by mapping four quadrants – *collaborative*, *control*, *competency*, and *cultivation* – across two axes: people versus organization oriented and reality versus possibility oriented. The model does not weigh any parameter better than others but gives an excellent representation for reflection and adjustment. The same model can also be applied at the team level to reflect the team's culture.

High-performing teams inherit the culture, behavior, and mindsets of organizations. We will examine some of the behavioral characteristics of high-performing teams in the next section.

Understanding the behavior of high-performing teams

High-performing teams with motivated individuals are the foundation for operational agility. The behavior and characteristics of a high-performing team are summarized in the following diagram:

Characteristics of a high-performing team	
Happy workforce with high energy and enthusiasm	Accountability
High degree of team morale	Shared vision and knowledge
Trust and transparency	Clarity on roles and responsibilities
Open and honest communication	Psychological safety
Constructively managing conflicts	Eliminate wastage to focus on meaningful work
Operating with clarity	Focused on present
Driven by value and purpose	No blame culture
Demonstrate innovation and creativity	Continuous improvement
High productivity	Self organize

Figure 13.6 – Behaviors and characteristics of high-performing teams

In the book *The Five Dysfunctions of a Team*, Patrick Lencioni observes that commonly observed dysfunctions of teams are the absence of trust, fear of conflict, lack of commitment, avoidance of accountability, and inattention to results. According to Zuzana Sochova, author of *The Agile Leader*, defensiveness, blame culture, stonewalling, and contempt are toxic behaviors that teams must avoid.

Understand Team Effectiveness, published at Google Re-work, suggests that effective teams are not attributed to individuals' performance, tenure, and seniority in the team but are attributed to the team working together as a unit. In order of importance, the authors document the following variables as defining an effective, high-performing team:

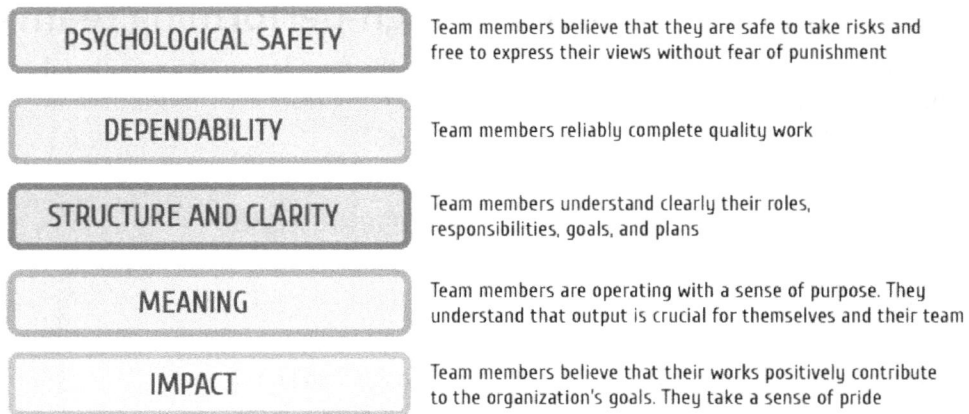

PSYCHOLOGICAL SAFETY	Team members believe that they are safe to take risks and free to express their views without fear of punishment
DEPENDABILITY	Team members reliably complete quality work
STRUCTURE AND CLARITY	Team members understand clearly their roles, responsibilities, goals, and plans
MEANING	Team members are operating with a sense of purpose. They understand that output is crucial for themselves and their team
IMPACT	Team members believe that their works positively contribute to the organization's goals. They take a sense of pride

Figure 13.7 – Team effectiveness by Google Re-work

Building high-performing teams takes time and effort and needs progressive recalibrations. Many organizations expect architects to play an advisory and leadership role; that is, facilitating and contributing to the strategic investments thus leading to a high-performing team that demonstrates sustainable technical excellence.

The rest of this chapter will discuss the personal and interpersonal qualities required for an architect who works within a high-performing team. It would be very difficult for an architect to learn and demonstrate these qualities if they operated in a non-high-performing organization and team.

Choosing the right leadership

Unlike roles such as Scrum Master and Product Owner, architects are better placed to shepherd a team of intellectual talents to achieve craftsmanship by setting high standards and growing the team toward achieving those goals. In this context, architects play a critical leadership role by assuming the following responsibilities:

- Develop high-performing individuals through a continuous, reflective learning culture.

- Collaborate, inspire, and influence individuals to build and align solutions.

- Focus on continuously improving systems and concerned processes (**Kaizen**).

- Lead prioritization to balance sustainability with lead time.

- Facilitate collective architecture decisions and resolve conflicts.

- Minimize technical constraints and help to remove technical impediments.

Good architects are like an *A player*. They set stretched goals for themselves, show a willingness to act, display personal resiliency, and embrace change quickly. A few characteristics of an A player are as follows:

- Guide the team by respectfully challenging them by using knowledge and insights.

- Understand, inspire, and motivate individuals with expertise and experience.

- Achieve satisfaction by unleashing the potential of individuals and setting them up for success.

- Anticipate changes with experience, expertise, and intellectual abilities to protect the team from losing focus.

- Backing a team's decisions and taking personal accountability for the design and architecture outcomes of the group.

Today's challenge for many enterprises is accurately reflected in the paper *Nimble Leadership*, published by Harvard Business Review. It observes that *nobody recommends command-and-control leadership, but no fully formed alternative has yet emerged*. Hence, there is no single leadership approach recommended for Agile software development as balancing innovation and discipline is still an unresolved challenge. A curated set of six leadership approaches fit for Agile software development can be seen in the following diagram:

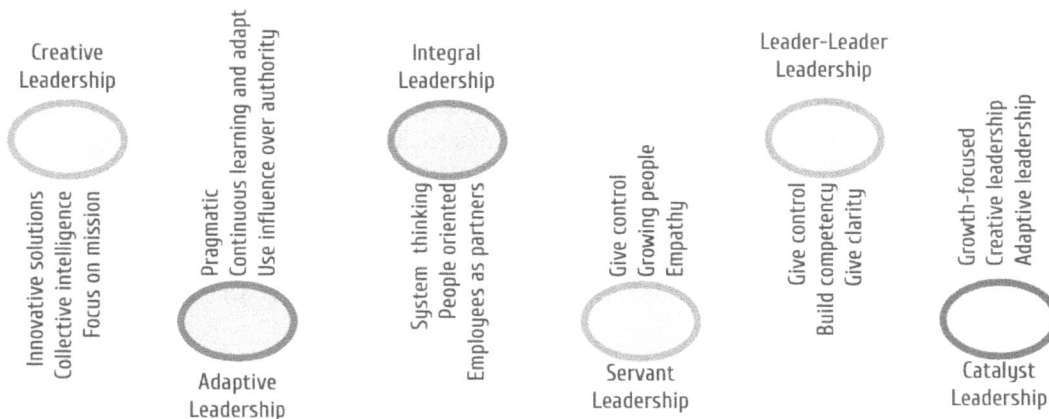

Figure 13.8 – Leadership approaches fit for Agile architects

Besides the leadership approaches shown in the preceding diagram, there are many leadership styles available. Among these, the *Six Leadership Styles* proposed by Daniel Goleman are among the most popular ones, classifying leadership behaviors into **Visionary**, **Coaching**, **Affiliate**, **Democratic**, **Pacesetting**, and **Commanding**. While the coaching style is better suited for Agile software delivery, individuals' natural leadership styles are hard to correct. Therefore, the focus must be to embrace practicing different elements from the Agile-friendly leadership approaches shown in the preceding diagram. Architects may need to switch leadership styles based on scenarios. Situational leadership, as shown in the following diagram, is useful for switching leadership styles:

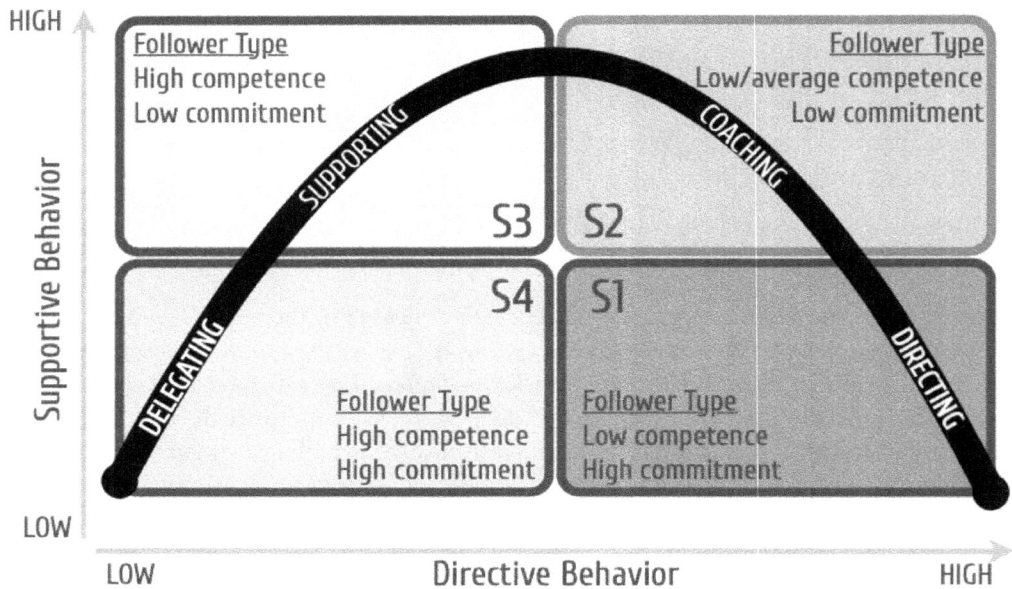

Figure 13.9 – Situational leadership, by Paul Hersey and Ken Blanchard

The model shown in the preceding diagram has been adapted from *Situational Leadership Theory*, as introduced by Paul Hersey and Ken Blanchard. They recommend adopting leadership styles based on the competency of the follower type.

Naturally, architects are not the bosses of developers. Furthermore, in Agile development, architects collaborate without authority. Applying power may demotivate individuals. The following diagram shows the link between authority and autonomy in the context of leadership:

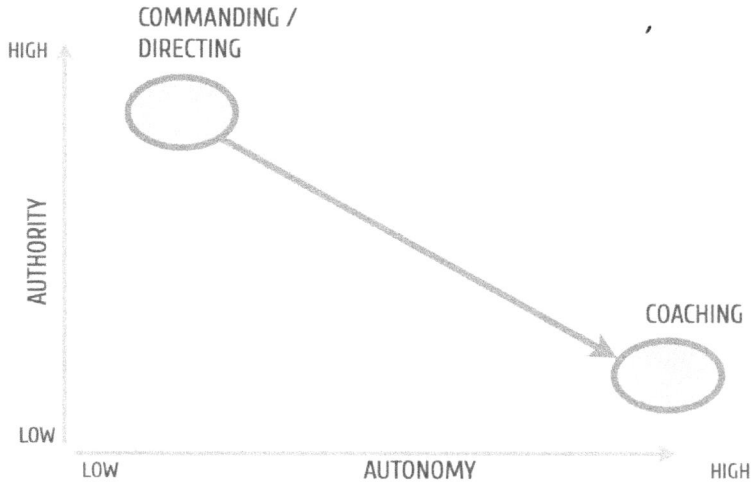

Figure 13.10 – Authority versus alignment in leadership styles

Among the six leadership styles, while commanding or directing is the most authoritative leadership, the coaching style offers autonomy and is better suited for Agile development. Collaborative leadership without authority can only be achieved if architects, as leaders, possess certain personal and interpersonal qualities. In the next section, we will go through the personal qualities required for architects.

Personality traits of an architect

Agile architects must exhibit an Agile mindset – an open and positive mindset to collaborating with the team with equality, being able to adapt to work with changing data, tirelessly focusing on customer value, and having an intense obsession with continuous learning and growth. The desired profile of Agile architects can be closely compared to E-shaped personalities.

The following diagram captures the different personality profiles:

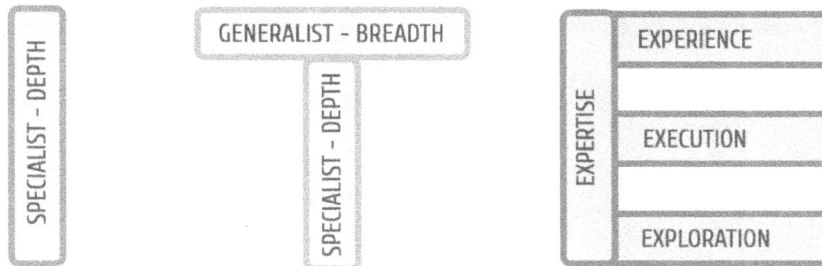

Figure 13.11 – I-, T-, and E-shaped profiles

The characteristics of the I-, T-, and E-shaped profiles are as follows:

- I-shaped individuals are specialists with deep expertise in one particular area.

- T-shaped individuals are generalized specialists who have a specialty that is the vertical stroke and breadth of their experience in other areas, as represented by the horizontal stroke.

- E-shaped individuals are next-generation specialists who combine experience, expertise, exploration, and execution.

E-shaped individuals are self-directed and creative and not only throw ideas but also translate ideas into reality. Agile architects must have an exploration mindset, deep experience and expertise on the concepts, and the ability to successfully execute ideas.

It is essential to understand and adjust behaviors based on organizational constraints and challenges as no two organizations are the same. In the subsequent sections, we will go through the top 10 personal qualities required for Agile architects to operate successfully in Agile projects.

Selflessness for relentless support

Selflessness is a catalyst for building trust – the foundation of a high-performing team. Agile architects must have the mindset to operate egolessly with their team, peers, and senior leadership. Architects should proactively understand an individual's needs and wants, and then offer unconstrained support by sharing information and knowledge to build solutions, acquire new experience, or even recover individuals from professional challenges.

Architects must develop a mindset that enables a sense of satisfaction through selfless support for individuals' betterment. The ultimate motivation for selflessness is that an architect's success is when individuals they support excel in their roles.

Selfless support can build long-lasting, deep relationships, which helps in connecting and engaging with other people better. Selflessness is a behavior you can practice by setting aside time. However, be mindful that persistence and patience are required as changes take time.

Respect, equality, and humility

Treating colleagues professionally and with respect is crucial in a cohesive team seeking collective intelligence. Every individual in the team has their own identity and status quo, and they build their professional life for a purpose. Treat everyone in the team equally and respectfully without excluding them based on their experience, competency, behavior, or other human considerations. Respect, equality, and humility are especially important for Agile architects when collaborating with the team for collective decision-making.

Encourage everyone to participate and raise questions, no matter how relevant those questions are. In the book *The Fifth Discipline*, Peter Senge observed that *when possible, indeed encouraging team members to raise the most difficult, subtle, and conflictual issues is essential to the team's work*. On the other hand, challenging individuals' viewpoints helps produce better outcomes. However, it is critical to challenge your thoughts with respect by treating people professionally with the base assumption that everyone is intelligent and has the right to express their points.

Mindfulness for managing wellbeing

One of the benefits of successful Agile software delivery is employee happiness. A happier workforce connects better with lower degrees of tension. Personal happiness can bring a sense of harmony.

Usually, architects often undergo stressful situations as their job involves frequent context switching between development teams, senior management, and business in discussions. Often, many of these meetings and discussions are politically charged. It is critical to renew an individual's energy regularly to avoid burnout.

Mindfulness is a technique beyond meditation and yoga for restoring and rebuilding minds with deliberate attention. Roman Pichler, in his article *Mindfulness Tips for Product Managers and Product Owners*, observes that *avoiding rush and panic, doing one thing at a time, taking proper breaks, keeping an open mind, being positive, and meditating* helps to keep the mind and soul calm, even in challenging and stressful situations.

Mindfulness boosts energy, empathy, kindness, and helps control a person's state of mind and emotions. It is a human aspect of preparing our minds to be calm and react appropriately to situations. Mindfulness is a behavior that's induced through daily practice that adds space and thinking time between observation and reaction. **Mindfulness-Based Stress Reduction** (**MBSR**) helps reduce anxiety and stress levels by helping us live in the present.

Curious to acquire new knowledge

In a rapidly changing technological world, architects as technical leaders need to look for ideas beyond their own team. They need to look for new avenues and sources to gain new knowledge continuously.

As eager and curious learners, architects look for new ideas, regardless of what methods they can come across. Engaging in regular, highly cognitive, intellectual conversations with developers helps in exchanging knowledge. They also create internal and external professional networks, attend conferences, participate in seminars, and engage in social networking to continuously improve their brand value. Architects are conscious of their identity and brand and continuously strive for improvement and rediscover themselves. They hate stagnancy and sense opportunities internally and externally and act toward them. It is not that they just acquire new knowledge but also that they find time to share their knowledge with the broader internal and external communities.

A good architect goes beyond symptoms and discovers the root cause of problems. They are passionate and curious to learn from setbacks. Curiosity needs passion, and it can be grown over a period of time with the right focus.

Growth mindset for positive thinking

In a fast-paced, volatile technical world, continuous learning, improvement, and innovation demands a growth mindset. The growth mindset is a positive state of mind with the belief that intelligence can be cultivated. Develop a growth mindset that enables sharing positive stories about experiences, learnings, and reactions. Architects need to develop a growth mindset to be successful as it helps an individual's cognitive growth. It also helps them earn respect.

A person with a growth mindset always seeks new challenges, shows a willingness to take the risk to activate success, and treats failure as new learning. Carol Dweck, in her book *Mindset: The New Psychology of Success*, observed *a growth mindset as a personality that embraces challenges, persists in facing setbacks, sees efforts as a path to mastery, learns from criticisms, and finds inspiration and lessons in the success of others.*

On the contrary, an individual with a fixed mindset stays with what they know, is low risk-oriented, avoids challenges, gives up easily, dismisses any criticisms, and feels insecure about others' success. A fixed mindset leads to cognitive distortion.

Intrinsic motivation for commitment

In a non-hierarchical and flat organization where growth in terms of grades, roles, and authority is thin, intrinsically motivated individuals have a better chance of success. Extrinsically motivated individuals unsettle themselves continuously by searching for money, power, and fame, and then eventually and often become frustrated and demotivated.

Intrinsically motivated people, called **autotelic** in *flow psychology*, are determined, curious, and purpose-driven. They often exhibit a *flow state*, a state of mind in which they have immersive focus on the task at hand. They focus on aspects such as work enjoyment, family life, relationships, and social connections. The work, respect, value, and reward they gain in the workplace is treated as a means to gain continuous motivation.

Intrinsic motivation can be developed cautiously through deliberately diverting attention to interest, enjoyment, positive thinking, and building cognitive intelligence.

Creativity to explore new opportunities

Delivering value through shorter innovation cycles and striving for continuous improvement in innovative ways is critical in Agile software development. When working with uncertainties and unpredictability, Agile architects need to discover creative ways to keep up a positive spirit for themselves and their team. Continuing the same job in the same style generates repetitions that lead to the demoralization of intellectual individuals.

As leaders, Agile architects must continuously reinvent approaches for improving systems, introduce solutions by experimenting with new technologies, and find inventive ways to make the continuous delivery pipeline efficient. Having the courage to take risks is one of the most important aspects for Agile architects.

Thinking outside the box and coming up with creative ideas is a characteristic that architects need to develop and demonstrate. It can be developed by exploring creative alternatives for everything.

Self-esteem to protect personal value

Self-esteem is a psychological term used for describing an individual's intrinsic personal value. It is a combination of ideas, passion, behavior, emotions, beliefs, skills, and competencies. Self-esteem is directly linked to self-motivation.

An individual with high self-esteem exhibits high motivation levels, and high confidence and spreads positive energy. A person with high self-esteem can say NO assertively when needed. As an architect, it is an important quality for negotiations and decision making.

Low-esteem individuals think others are better than them, do not express themselves, keep a low profile, have a negative outlook, have trouble accepting criticism, have low confidence, and have no clarity regarding the future. Arrogance is a negative behavior and is an outcome when people with low self-esteem try to show they are the best.

Empathy for emotional connection

Empathy is the new management mantra in Agile software development environments. Empathy is responding to another person by controlling your own emotions by connecting and understanding others' emotions. Empathy requires active listening and emotional intelligence. In an active listening scenario, the listener pays full attention to the speaker so that they can understand them and provide a better response, not just doing so for the sake of responding. Emotional intelligence helps connect personal experience with your mental state and helps you adjust your responses accordingly.

The ability to show empathy is an essential quality for any Agile architect. When developers propose solutions, architects need to approach them empathetically by enabling them and helping them instead of making harsh comments.

In a high-performance culture, Simon Sinek observed that *if someone is not performing well, it means they have a problem, not that they are a problem.* It is imperative to rally around and help such individuals, understand how they feel emotionally and physically, and guide them in an empathetic way to resolve challenges.

Executive presence for gaining attention

Agile architects actively engage in meetings with senior IT leadership and business executives in addition to other IT stakeholders. Senior executive meetings are high-profile and need to be managed differently as their time and attention span are often significantly short.

Executive presence is a critical leadership quality Agile architects should develop. The different traits of executive presence are composure, self-awareness, and understanding others. As we mentioned previously, controlling emotions by recognizing others' emotions and readjusting our responses is critical for demonstrating executive presence.

Individuals who possess executive presence connect well with others, make others comfortable, show charisma through active listening, pay attention to current events, respond at the right time with the right observations, and show confidence in communication. They capture people's attention with the right posture, eye contact, body language, and by modifying their speech. They build credibility by demonstrating their in-depth knowledge of the subject and communicate with confidence, clarity, and credibility. They avoid long speeches and articulate very clearly and concisely. Executive presence is a quality you can progressively build through practice.

Interpersonal qualities are long-term goals that need grit – the ability to chase ambition with persistence, perseverance, resilience, and tenacity. Continuously practicing these helps in achieving great success for individuals and organizations. In addition to these core skills, personal qualities add tremendous value. However, other interpersonal skills are also essential for complete success, as we will discuss in the next section.

Interpersonal traits of Agile architects

Interpersonal qualities are the mindset and behaviors required for Agile architects to interact and engage continuously with others to succeed. This section will explore 10 interpersonal skills essential for architects to work successfully in Agile teams.

Collaborating for collective ownership

In Agile software development, autonomy, empowerment, and decentralized decision-making demand architects to actively pursue evolutionary collaboration. We discussed collaboration for collective ownership in *Chapter 3*, *Agile Architects – Linchpin to Success*, and *Chapter 10*, *Lean Documentation through Collaboration*. Collaboration brings out collective intelligence, collective ownership, and collective knowledge, and also reduces excessive documentation.

Collaboration is a means of collecting diverse perspectives, and then synthesizing and agreeing on an inclusive solution. Evolutionary collaboration requires extensive and open communication to develop a shared vision, context, and purpose across all team members. Sharing the architecture vision, intentional architecture, and solution context are prerequisites for architects to collaborate with the team.

Collaboration needs excellent facilitation, conflict resolution, influencing, and negotiation skills. **Disciplined Agile (DA)** describes the role of the Agile architecture owner as a facilitator. Architects are responsible for breakdown deadlocks. A good facilitator values everyone's opinions equally and avoids dominance, such as in the case of the **Highest-Paid Person's Opinion (HIPPO)** wins. The facilitator manages the agenda for discussions and effectively manages time. Every architect may not have the skills of an effective facilitator. In larger architecture organizations, you must identify individuals with intrinsic facilitation skills for facilitating architecture discussions. Alternatively, get an external facilitator such as a Scrum Master or Agile coach for high-profile architecture meetings.

Sam Kaner's book *Facilitator's Guide to Participatory Decision-Making* suggests three principles for effective **participatory decision-making** – take advantage of the group's diversity, commit to effective listening, and manage the group's energy by providing a mixture of structured activities. In the book *Getting to Yes: Negotiating Agreement Without Giving In*, Roger Fisher, William Ury, and Bruce Patton promote **principled negotiation** for producing wise outcomes efficiently and amicably. The four elements of principled negotiation are to separate the people from the problem, focus on interests not positions, invent options for mutual gain, and insist on using objective criteria.

Customer-centric approaches such as **design thinking** add tremendous value for collaboratively developing solutions. Design thinking is a lightweight, iterative approach with the customer at the center. The process uses a double diamond approach promoted by Design Council, where the first phase identifies the problem by empathizing customers and the second phase solutions by modeling and prototyping various options. Design thinking uses a divergent-convergent approach.

Agile architects must foster conflict resolution skills to collaborate with the team to balance autonomy with alignment. We will explore this in the next section.

Resolving conflict for superior quality outcome

Conflicts, difficult conversations, and creative tension are natural when competent individuals are working hard to raise the bar with innovative intellectual thinking. Conflicts can help drive innovative and superior solutions. If there is no conflict, **groupthink** happens, where team members easily agree on solutions, but they are often of suboptimal quality.

Inquiring to augment knowledge and clarifying situations by asking **powerful questions** is the right way to lead solution discussions. **Powerful questioning** is a technique where you ask open-ended, thought-provoking questions that trigger curiosity and stimulate eagerness. Equality is one of the most critical behavioral aspects when dealing with conflicts. Avoid unprofessional toxic conversations, focus on outcomes, scope, and priorities, be respectful in confrontations, show positive intent to solve the issue, and approach conversations empathetically.

Non-Violent Communication (**NVC**), developed by Marshall Rosenberg, is an excellent approach for practicing conflict resolution when things go wrong. NVC focuses on three aspects of awareness and communication – self-empathy, empathy for others, and honest self-expression using four essential components: observations, feelings, needs and values, and requests. This only works when all the team members move to having a self-accountability and no-blame culture.

Healthy arguments are not unprofessional confrontation. Timely and constructive interjection and feedback help in resolving conflicts through conversation. If all attendees in a meeting are not on the same wavelength, it is harder to have intellectually high conversations. Level setting sessions such as coaching, 1-2-1 conversations, pairing, knowledge dissemination sessions, and adding co-pilots are useful for level setting. Usually, low-value questions also get raised in meetings. Shutting down those conversations is not healthy. Diligently conducting educational sessions and private conversations post-meeting helps keep them interested in the topic.

Understanding people based on the **DISC theory** proposed by psychologist William Moulton Marston helps engage people differently. In the case of conflicting scenarios, you must depersonalize conflicts and resolve them within 24 hours. Ginka Toegel advocates avoiding prejudging people and observes that *individuals may use different approaches and mindsets for solving issues. Their intensity of emotional reactions may be different.* Understanding team members and their styles helps with reacting to situations better.

Use one of the four approaches of Michael Tardiff's decision model specified in the previous chapter for decision-making; that is, consent, consensus, majority, and aristocratic. The outcomes of any conflict scenario can be one of four of **Thomas-Kilmann's Conflict Mode Instrument**, as shown in the following diagram:

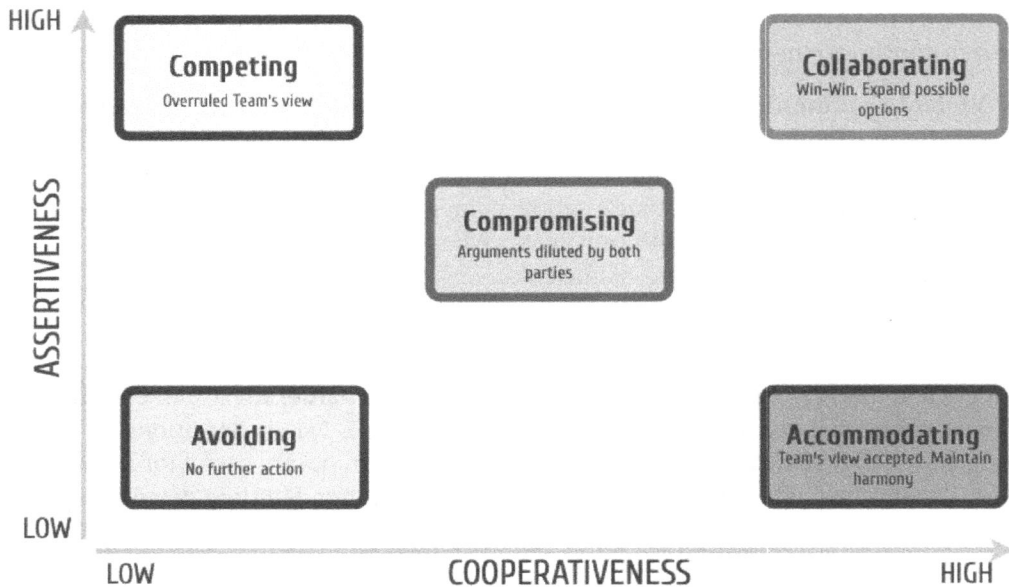

Figure 13.12 – Thomas-Kilmann's Conflict Mode Instrument for conflict resolution

While the desired model is collaborating, competing is useful when an individual is making a final executive call when there is no agreement within the team.

In summary, building a healthy environment for openly expressing conflict is vital to discovering collective intelligence. Architects, as the facilitators, need to manage such conflicts effectively. Unresolved conflicts lead to personal and professional egos, frustration, demotivation, and anger.

Communicating well using storytelling

Autonomy needs context to be continuously shared. To build sustainable solutions at pace, architects must transparently and continuously communicate vision, context, and goals with the team. Once the team thoroughly understands the customer's needs, they discover a sense of purpose in what they do, which leads to firm commitments. The **Golden Circle**, by Simon Sinek, observes that *good communication always starts with "Why" – the purpose.*

Direct, honest, and authentic communication helps team members connect better personally and emotionally, which enables trust and cohesion. Agile architects should have the mindset to work with abstract information, uncertainties, and chaos yet be confident, comfortable, and remain focused on their communication with teams.

Communication is the foundation for good, healthy teams. Breakdowns in communication cause confusion, anxiety, and frustration. When information is not communicated appropriately and promptly, individuals get into an anxious state of mind without knowing what is expected of them and what will happen next. This uncertain state can dampen morale and enthusiasm and cause stress and tension, which may even impact people's wellbeing and personal lives.

Storytelling is an excellent mechanism for effective communication. Neural coupling explains that a story activates points in the brain that allow listeners to interpret the story with their own experience. Thus, stories create long-lasting memories in the listeners' minds. Storytelling also helps human bodies release a neurochemical called oxytocin, which is linked to trust and enhances trust and empathy. In business storytelling, stories are constructed based on different personas such as customers, business departments, and so on and explain concerns by portraying them as characters in the story.

Nurturing and growing talent for a competent workforce

Decentralized decision-making needs a lot of education and coaching to ensure that teams are competent in making the right decisions. Incompetent teams making decisions can severely impact customer expectations.

Growing a community of developers for technical excellence is a vital role for any Agile architect. As leaders, architects must push individuals' boundaries beyond their comfort zones to get the best out of them. The scope of growing an individual covers behavioral, cultural, ethical, technical, and other skills and competencies, including any mental blocks.

Traditional evaluation-based and prescriptive learning approaches are not appropriate for Agile software development. These approaches create undue pressure and stress on individuals. Instead, you should create a highly motivating learning environment for individuals to freely participate in upskilling and cross-skilling themselves based on their needs and interests. Accelerated learning through empowered, trust-based and **self-directed learning** helps in nurturing curious learners. Offering platforms, learning paths, adopting playfulness, self-evaluation, and so on encourages individuals.

An individual's learning experience can be further enhanced with local meetups, internal competitions such as architecture katas, learning groups, hackathons, communities of practices, innovation sprints, open space marketplace concept, chapters, guilds, and so on.

Architects act as teachers, coaches, and mentors and help support individuals in their learning. Architects themselves must be curious and lifelong learners. To unlock and inspire individuals' intrinsic motivation, architects may have to lead from the front by playing a role model.

Often, organizations only think about the current needs of the enterprise when setting learning paths. This is not adequate in an unpredictable Agile environment. Learning paths needs to be benchmarked against an industry expert's knowledge and those with same the skill levels that are performing the same role. This approach also brings confidence and motivation for individuals.

While teams are competent enough to make decisions, Agile architects must pay extra attention to decision making. The **Tuckman model** for group development defines the four stages of leadership strategies that facilitate team development. It is an excellent model for determining when to give more freedom to teams. The following diagram captures various stages of the Tuckman model, adapted in the context of the team's competency in decision-making:

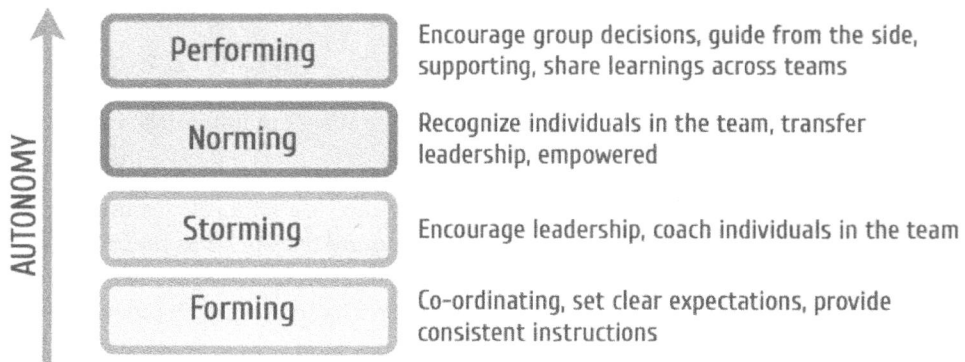

AUTONOMY		
Performing	Encourage group decisions, guide from the side, supporting, share learnings across teams
Norming	Recognize individuals in the team, transfer leadership, empowered
Storming	Encourage leadership, coach individuals in the team
Forming	Co-ordinating, set clear expectations, provide consistent instructions

Figure 13.13 – Competency maturity for decision-making using the Tuckman model

Shu-Ha-Ri is also used for learning and building Agile capabilities. *Shu* is where the team is adopting a change, *Ha* is the team practicing and making continuous improvements, and *Ri* is where they exhibit mastery.

Individuals playing architect roles may not be rounded figures who possess all the personal and interpersonal skills mentioned in this chapter. **Skill mapping** is a useful technique for identifying individuals' skills in the team, which helps determine who can complement what you're doing when the need arises.

Motivating for a better outcome

As thought leaders, architects continuously inspire and motivate people around them by transferring knowledge, enthusiasm, and energy. At times, individuals may feel demotivated, distressed, and frustrated. This causes significant stress levels, which leads to lack of interest, productivity, and quality. This often radiates to other team members. Architects, being part of the team, are well placed to sense such situations and offer counter-support.

As a technical leader, architects should motivate individuals through constant engagement. Such deep engagement, when using emotional intelligence, helps in identifying the root cause of demotivation. Recommending a break, diverting their attention to something more interesting, highlighting this to other leaders who can actively solve such issues, and so on can significantly help demotivated individuals come out of the situation.

Celebrating success together often helps motivate intrinsically motivated individuals. Inclusion through social activities brings a sense of belonging. This includes celebrating learning, early detection of failures, incremental success, recognizing good behavior of individuals, getting appreciation from others, and so on. Use celebrations as an information radiator to build visibility. Non-monetary-based kudos systems can increase employee satisfaction and retention. The Harvard Business Review report, *The Benefits of Peer-to-Peer Praise at Work*, observes that at *JetBlue, every 10% in recognition resulted in 3% increase in retention, and 2% increase in engagement.*

Reward, recognition, and appreciation can ignite individuals' intrinsic motivation. It brings happiness and positivity to the team environment. Small appreciation and thank you notes can help when individuals do their job well. Proactively sending thank you notes gives your team a positive surprise and makes them feel excited to be an individual on the team. Be genuine when you're appreciating them; don't just do it for the sake of appreciating. Understand the work they've done, connect with the topic, and use as much context as possible in appreciation notes. Saying well done and also adding context makes a huge difference.

Architects are responsible for continuously engaging in and understanding quality in an inversive way, such as looking at the code, quality dashboards, and advising team members on what needs to be corrected then and there. On-the-job mentoring and guidance can go a long way in building the right talents. Rewarding by using metrics such as code quality, architecture alignment, production faults, the effort for refactoring, and so on helps teams remain engaged and makes sure that they are encouraged to do the right thing. Use leader dashboards and game models to ensure that people who work with a craftsmanship mindset are rewarded and recognized.

In *Primed to Perform: How to Build the Highest Performing Cultures Through the Science of Total Motivation*, a model was introduced that explains motivational factors. There are six factors – three direct and three indirect – as shown in the following diagram:

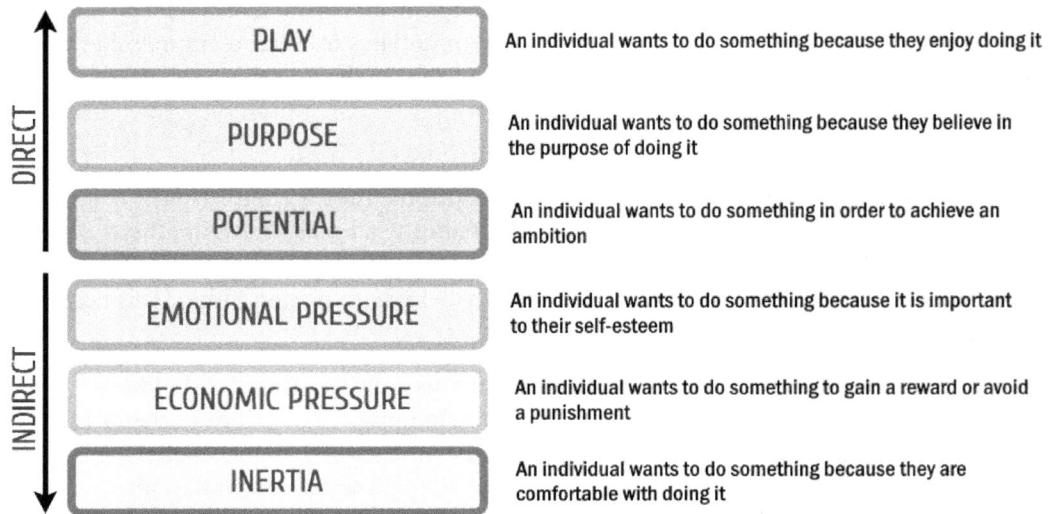

DIRECT	PLAY	An individual wants to do something because they enjoy doing it
	PURPOSE	An individual wants to do something because they believe in the purpose of doing it
	POTENTIAL	An individual wants to do something in order to achieve an ambition
INDIRECT	EMOTIONAL PRESSURE	An individual wants to do something because it is important to their self-esteem
	ECONOMIC PRESSURE	An individual wants to do something to gain a reward or avoid a punishment
	INERTIA	An individual wants to do something because they are comfortable with doing it

Figure 13.14 – Motivational factors developed by Lindsay McGregor and Neel Doshi

Busy developers may not spend time reading a lot of articles and knowledge repositories. They may find it difficult to distill information from large and extensive internet content. Architects are thought leaders who pay attention to and watch such content and curate small nuggets to inspire the development team.

Architects should also act as role models and lead by example when inspiring others by demonstrating qualities such as perseverance and tenacity, continuous striving for excellence, and so on.

Leading by example and demonstrating servant leadership

As leaders, architects consistently demonstrate high performance so that others can look at them as role models. Architects must focus on value and outcome, not the processes, and always urge action instead of analysis paralysis. As thought leaders, they produce ideas and help teams apply smart and innovative solutions to complex problems, which is highly valuable for the enterprise and its customers. They also influence people to deliver the ideas they share. Architects consistently use servant leadership principles when engaging with teams and practice what they preach.

Feedback for continuous improvement

Architects, as leaders, consistently seek feedback and fine-tune themselves. They also deliver constant feedback to other team members for their improvement. They adopt new mechanisms for feedback instead of providing frequent, across-the-table feedback interviews. Architects create deep relationships with developers to make these feedback sessions incredibly valuable. It is always good to get a sense of how people feel professionally. Informal meetings and coffee chats are better open settings for good valuable feedback than standard formal meetings.

Supporting and enabling does not mean feedback has to always be positive. Critical feedback is essential for improvement. Everyone makes mistakes at times in their work. It is the leader's responsibility to highlight and guide them to avoid such mistakes in the future. Designs and architectures can fail. Architects have a responsibility to avoid such failures. Therefore, when a design succeeds or fails, it is a shared responsibility. Instead of blaming and criticizing with an enabler mindset and positive intent, providing a supporting and helping hand is better for helping individuals improve.

The performance management of individuals is often not directly linked to architects working in the team. However, architects have the responsibility of contributing to an individual's performance reviews. It is essential to perform self-assessments as a form of measurement. Jurgen Appelo in *Management 3.0* introduced the **individual metrics rules** in which he observes that *all workers like to set their own targets and take control of achieving those targets. Empowering workers to create their own measurements improve their work and outcome.* In such cases, people will be mindful of how they perform as this information is transparent and visible. Because it is transparent, every other person can observe each other's results and respond to them if they feel they are incorrect. It also exposes the honesty and openness of individuals.

Psychological safety for trust and transparency

Psychological safety creates a trusted environment that allows individuals to feel safe to speak out about their ideas and opinions freely and fairly, without being stereotyped as incompetent, ignorant, negative, or disruptive. In such environments, everyone's voice is heard without them being punished, even if they made a mistake. Psychological safety brings openness, trust, transparency, and a fail-fast culture. It lessens individuals' stress and anxiety and promotes wellbeing.

Researches show that individuals with higher degrees of psychological safety are likely to stay longer in an organization and deliver superior quality solutions collectively. The *Global Study of Engagement Technical Report* observes that *trust in team leaders is the foundation of engagement. A worker is 12 times more likely to be fully engaged if the individual trusts the team leader.*

Trust and radical transparency are foundations for building psychologically safe teams. By observing consistent positive actions, individuals in a high-performing team organically gain trust in the system, colleagues, and their leaders. Establishing trust, transparency, and psychological safety takes time. Leaders need to carefully plant the seed of trust and must patiently wait for it to grow.

Psychological safety was discussed in *Chapter 11, Architect as an Enabler in Lean-Agile Governance*. Architects need to play the heat shield role in protecting the team from external technical and architectural challenges that could cause unnecessary pressure. Trust increases collaboration and cohesion. A well-trusted team collaborates well and therefore eliminates architecture erosion.

A team with higher psychological safety can pull the **Andon cord** when things go wrong without worrying about the impact of stopping work. The concept of the Andon cord is from the **Toyota Production System** (**TPS**). It is where you pull a cord or press a button to stop production and warn management in the case of a significant issue, since it can lead to more problems if defective parts reach customers.

Leading the change by creating multipliers

Organizational transformation involves a meticulous change management process, such as John Kotter's *8-step process for leading change*. Architects, as change leaders, have the responsibility to transition the team's mindset, behavior, and culture to embrace Agile architecture. It takes time for the organization to understand and conceive the concepts of the Agile architecture's ways of working. It starts with an individual transformation and then transforming others. The four-step model, as shown in the following diagram, captures the approach to leading this change:

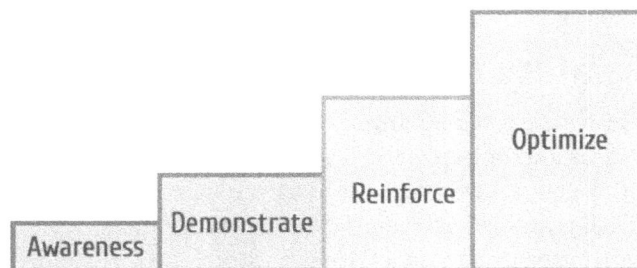

Figure 13.15 – Stages of leading change

As shown in the preceding diagram, the first step is building **Awareness** through education. In **Demonstrate step**, you must encourage everyone to follow by demonstrating what is being preached. The third step will continuously **Reinforce** the message, while the fourth step is **Optimize**, which is transparently measuring and sharing feedback for continuous improvement.

The architects need to change and transform themselves as well. The **ADKAR** model, defined by Jeffrey Hiatt, helps with self-reflecting on the progress of change. The model uses five parameters to measure individuals: *awareness of the need to change, desire to support the change, knowledge of how to change, ability to demonstrate skills and behaviors, and reinforcement to make the change stick.*

It is important to recognize that there may be a significant disruption in overall architecture capabilities during the first phase of transformation due to confusion and resurfacing legacy behaviors and cultures. David Viney explained this with a **J Curve**, observing that capability building follows the shape of a *J*; it dips first before rising. The depth of the curve depends on the effectiveness of change management and people's behavior and culture.

Architects act as champions of change for Agile architecture. However, this is not enough. They need to multiply this by identifying more and more champions. Change agents are effective in leading change. Change agents are respected individuals in the community. They think differently and are not satisfied with the status quo. Change agents form a guiding coalition for accelerating change. They iteratively identify team members who can play the role of change agents, then slowly expand the circle of influence. The following diagram captures the process of extending the circle of influence:

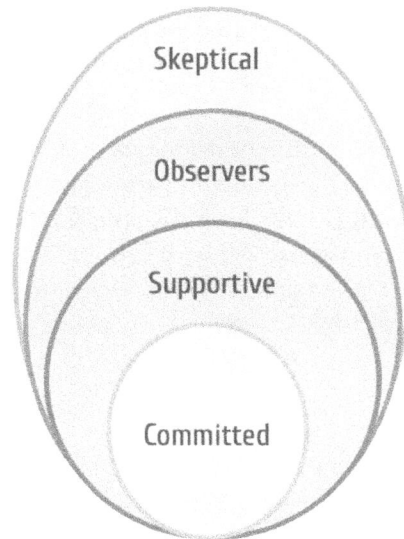

Figure 13.16 – Circle of influence for leading change

In the change agent approach, team members are profiled across four types: **Skeptical**, **Observers**, **Supportive**, and **Committed**. Identify committed individuals as initial change agents, then slowly expand the circle of influence to supportive, observers, and skeptical individuals.

Caring for building a sense of belonging

Caring helps in building deep relationships with individuals. The manifestation of care is when individuals in the team members feel like there is someone to take care of them, which brings a sense of belonging.

Connecting personally and emotionally to individuals, spending time with them, and genuinely, truthfully, and reliably caring for them helps build trust. Once architects establish a deep emotional connection with an individual, the individual will show more commitment and openness when collaborating for solution design and decision making.

It's human nature that people need to be cared for in their home and office environments. At the office, they expect leaders to care of them. In a non-hierarchical environment with no line management, there is a duty of care for leaders to care about individuals in the team. Caring goes beyond grade, role, and job status. Simon Sinek observed that *a soldier in trouble would look for help from the people on the left and the people on the right when in a war field not always look for their superiors.* In an Agile team, Scrum Masters, Product Owners, architects, and other individuals have that duty of care. When individuals realize that there are people to care for, they open up and ask for professional help when they are in trouble.

Architects can care for individuals in a team by supporting them at work. At times, extending a helping hand to a team member in debugging or helping in coding or throwing an idea while an employee is struggling, or even educating them on specific areas, are all part of caring.

Realizing that an individual's personal situations impact their work, and then connecting with them and offering them support, helps demonstrate care that can win hearts. Gavin Larkin introduced the concept of **R U OK?** Australians celebrate R U OK? day to remind everyone that every day is the day to ask R U OK? It is imperative to ask team members and peers such questions to discover their struggles. It helps break down the walls of professional egos and nurtures healthy relationships.

Summary

In this chapter, we looked at how architects change their interaction styles when dealing with Agile projects and, therefore, the need to acquire new skills and competencies. Agile architects cannot work with the new culture, behavior, and mindset if the organization and teams are not exhibiting a high-performing culture. Leaders in a learning organization must embrace a people-centric, purpose-driven approach that's often seen in generative and Teal organizations. We also examined an effective team's critical strengths based on Google research – psychological safety, dependability, structure and clarity, meaning, and impact.

Agile architects need to adopt appropriate leadership approaches such as adaptive, creative, servant, server-server, catalyst, or integral. We also learned that Agile architects may need to switch leadership styles based on situations, as explained in the situational leadership model. As E-shaped people, architects need to possess personal qualities such as empathy, a growth mindset, respectfulness, mindfulness, selflessness, curiosity, creativity, self-esteem, executive presence, and so on. We ended this chapter by looking at a set of essential interpersonal qualities, including collaboration, conflict resolution, communication, nurturing talent, motivation, caring, leading by example, and leading change. These personal and interpersonal qualities are essential to being successful when working in modern enterprises and striving for agility.

In summary, this book explored the importance of Agile architecture and Agile architect's role in Agile software development. We have also learned how to tune an architect's ways of working to focus on value delivery, enabling agility with technical excellence, evolutionary collaboration over documentation, alignment over governance, an architect's role in re-architecting organizations and finally culture and leadership traits of Agile architects.

Further reading

- **The Global Study of Engagement Report**: `https://www.adp.com/-/media/adp/ResourceHub/pdf/ADPRI/ADPRI0102_2018_Engagement_Study_Technical_Report_RELEASE%20READY.ashx`

- **Larman's Law of Organizational Behaviors**: `https://www.craiglarman.com/wiki/index.php?title=Larman%27s_Laws_of_Organizational_Behavior`

- **Towards an Agile Culture**: `https://cdn.ymaws.com/www.Agilebusiness.org/resource/resmgr/documents/whitepaper/towards_an_Agile_culture.pdf`

- **Understand Team Effectiveness**: `https://rework.withgoogle.com/guides/understanding-team-effectiveness/steps/introduction/`

- **Nimble Leadership**: `https://hbr.org/2019/07/nimble-leadership`

- **How to preempt team conflicts**: `https://hbr.org/2016/06/how-to-preempt-team-conflict`

- **Conflict and challenge, The Thomas Kilmann Model**: `https://challengingcoaching.co.uk/conflict-and-challenge/`

- **Accelerate, State of DevOps 2019**: https://services.google.com/fh/files/misc/state-of-devops-2019.pdf
- **Design Thinking using Double Diamond**: https://www.designcouncil.org.uk/news-opinion/what-framework-innovation-design-councils-evolved-double-diamond
- **Busines Storytelling**: https://medium.com/@marketingmanegoa/the-ultimate-guide-to-business-storytelling-in-2019-40793c9a7836
- **ADKAR framework**: https://www.prosci.com/adkar
- **Servant Leadership and Inverted Pyramid**: https://www.trig.com/tangents/leadership-and-the-inverted-pyramid
- **A Topology of Organizational Cultures**: https://www.researchgate.net/publication/8150380_A_Typology_of_Organisational_Cultures
- **Participatory Decision Making**: https://www.researchgate.net/publication/46472555_Participatory_decision-making_The_core_of_multi-stakeholder_collaboration
- **14th State of Agile Report**: https://stateofAgile.com
- **Mindfulness Tips for Product Managers and Product Owners**: https://www.romanpichler.com/blog/mindfulness-tips-for-product-managers-and-product-owners
- **Non-Violent Communication**: https://www.nonviolentcommunication.com/learn-nonviolent-communication/4-part-nvc/
- **DISC Theory**: https://discinsights.com/disc-theory
- **The Science Behind Powerful Questioning**: https://www.researchgate.net/publication/320826477_The_Science_Behind_Powerful_Questioning_A_Systemic_Questioning_Framework_for_Coach_Educators_and_Practitioners
- **The Tuckman Model**: http://www.mspguide.org/tool/tuckman-forming-norming-storming-performing
- **Shi-Ha-Ri Model**: https://www.pmi.org/learning/library/becoming-Agile-with-shuhari-9649

- **The Benefits of Peer-to-Peer Praise at Work**: https://hbr.org/2016/02/
  the-benefits-of-peer-to-peer-praise-at-work

- **The J Curve**: https://journal.jabian.com/wp-content/
  uploads/2017/10/Consider-the-Human-Element-Before-Racing-
  Your-Next-Initiative-to-the-Finish-Line-Jabian-Journal.pdf

- **Powerful questions**: https://www.lwv.org/sites/default/
  files/2018-07/Powerful%20Questions.LWV%20DEI%20Training%20
  Resource.pdf

Packt>

Packt.com

Subscribe to our online digital library for full access to over 7,000 books and videos, as well as industry leading tools to help you plan your personal development and advance your career. For more information, please visit our website.

Why subscribe?

- Spend less time learning and more time coding with practical eBooks and Videos from over 4,000 industry professionals

- Improve your learning with Skill Plans built especially for you

- Get a free eBook or video every month

- Fully searchable for easy access to vital information

- Copy and paste, print, and bookmark content

Did you know that Packt offers eBook versions of every book published, with PDF and ePub files available? You can upgrade to the eBook version at packt.com and as a print book customer, you are entitled to a discount on the eBook copy. Get in touch with us at customercare@packtpub.com for more details.

At www.packt.com, you can also read a collection of free technical articles, sign up for a range of free newsletters, and receive exclusive discounts and offers on Packt books and eBooks.

Other Books You May Enjoy

If you enjoyed this book, you may be interested in these other books by Packt:

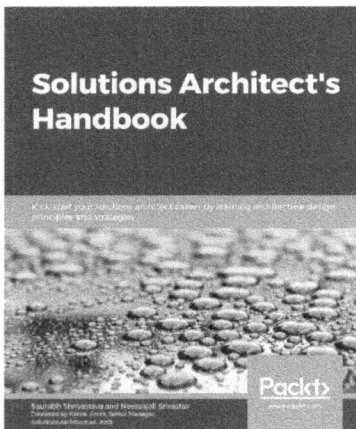

Solutions Architect's Handbook

Saurabh Shrivastava, Neelanjali Srivastav

ISBN: 978-1-83864-564-9

- Explore the various roles of a solutions architect and their involvement in the enterprise landscape

- Approach big data processing, machine learning, and IoT from an architect's perspective and understand how they fit into modern architecture

- Discover different solution architecture patterns such as event-driven and microservice patterns

- Find ways to keep yourself updated with new technologies and enhance your skills

- Modernize legacy applications with the help of cloud integration
- Get to grips with choosing an appropriate strategy to reduce cost

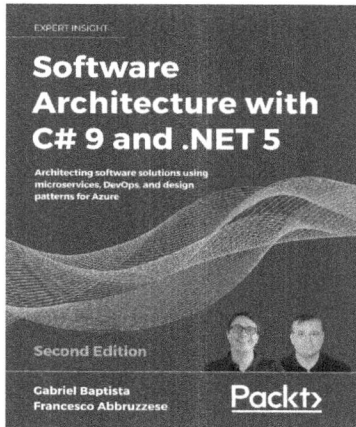

Software Architecture with C# 9 and .NET 5 – Second Edition

Gabriel Baptista, Francesco Abbruzzese

ISBN: 978-1-80056-604-0

- Use different techniques to overcome real-world architectural challenges and solve design consideration issues
- Apply architectural approaches such as layered architecture, service-oriented architecture (SOA), and microservices
- Leverage tools such as containers, Docker, Kubernetes, and Blazor to manage microservices effectively
- Get up to speed with Azure tools and features for delivering global solutions
- Program and maintain Azure Functions using C# 9 and its latest features
- Understand when it is best to use test-driven development (TDD) as an approach for software development
- Write automated functional test cases
- Get the best of DevOps principles to enable CI/CD environments

Packt is searching for authors like you

If you're interested in becoming an author for Packt, please visit `authors. packtpub.com` and apply today. We have worked with thousands of developers and tech professionals, just like you, to help them share their insight with the global tech community. You can make a general application, apply for a specific hot topic that we are recruiting an author for, or submit your own idea.

Leave a review - let other readers know what you think

Please share your thoughts on this book with others by leaving a review on the site that you bought it from. If you purchased the book from Amazon, please leave us an honest review on this book's Amazon page. This is vital so that other potential readers can see and use your unbiased opinion to make purchasing decisions, we can understand what our customers think about our products, and our authors can see your feedback on the title that they have worked with Packt to create. It will only take a few minutes of your time, but is valuable to other potential customers, our authors, and Packt. Thank you!

Index

www.ingramcontent.com/pod-product-compliance
Lightning Source LLC
Chambersburg PA
CBHW080714220326
41598CB00033B/5417